Sustainable Landscape Planning and Natural Resources Management

IEREK Interdisciplinary Series for Sustainable Development

Alexandria, Egypt

Sustainable Landscape Planning and Natural Resources Management (SLNR) is a peer-reviewed book series that presents groundbreaking research through the intersection of two main areas that of which are 'Sustainable Landscape Planning' and 'Natural Resources Management'. The series initially covers a wide range of academic disciplines in sustainability, including landscape ecology, green urbanism, urban regeneration, land restoration and renaturation, energy management, and nature conservation. Moreover, it also incorporates the latest developments and research in engineering topics that contribute to more sustainable cities, such as Geosciences, infrastructure, civil engineering, geology, mining, water, and natural resources conservation and management.

By showcasing methods for discovering, managing, and employing natural resources, the series aims to provide approaches that enable the responsible use of these resources. Ultimately, the SLNR series is committed to advancing the United Nations' Sustainable Development Goals for 2030.

The wide range of topics covered in this series are primarily related to sustainable development, involving durable infrastructure projects, civil engineering, urban regeneration, the built, sustainability, climate change mitigation, energy management, geosciences, geology, and the future of cities, all of which depend on the availability of natural resources and effective wealth management. The series also features conceptual and empirical contributions from various interrelated fields of earth science, with an emphasis on the era of new discovery. It focuses on providing practical solutions to ensure sustainable development and provides new case studies that offer concrete examples of how to manage natural resources effectively.

The series encompasses the best research papers from various international conferences organized by IEREK conferences, also accepts proposals from other international conferences, including interdisciplinary contributions. It exhibits innovative approaches and highlights how they can best support both economic and sustainable development through the responsible use of natural resources and global, local, and individual activities for the welfare of all societies on Earth.

The SLNR series is a valuable source for experts in government, international organizations, consultancies, industry, business sectors, and academics. Published in cooperation with IEREK, the Springer SLNR series displays important new research in various sectors, delivering valuable insights and practical solutions to the complex challenges of sustainable development for the readers.

Javid Ahmad Parray • Nowsheen Shameem
A. K. Haghi
Editors

Environmental Landscape and Sustainable Biodiversity for Healthy Green Growth

Springer

Editors
Javid Ahmad Parray
Environmental Science
Govt. Degree College Eidgah Srinagar
Srinagar, Jammu and Kashmir, India

Nowsheen Shameem
Department of Environmental Science
Sri Pratap College
Srinagar, Jammu and Kashmir, India

A. K. Haghi
Research Chemistry Centre
Edinburgh, UK

ISSN 2948-1910 ISSN 2948-1929 (electronic)
Sustainable Landscape Planning and Natural Resources Management
ISBN 978-3-031-75407-4 ISBN 978-3-031-75405-0 (eBook)
https://doi.org/10.1007/978-3-031-75405-0

This Springer imprint is published by the registered company Springer Nature Switzerland AG
The registered company address is: Gewerbestrasse 11, 6330 Cham, Switzerland

Contents

Sustainability Challenges, Developments, and Solutions Related to Organic Products for Good Health and Well-Being

Mukta Sharma, Garima Bartariya, and Anandveer Singh Sindhu

Abstract

This chapter explores how the ecological impact of organic food contributes to sustainable development and human health. The consumption of more organic food promotes mental and physical health. Compared to conventional foods, organic food has a better nutritional value. Because organic farming does not utilize harmful chemical pesticides and maintains the integrity of the soil, it is also beneficial to the environment. Therefore, from the production and consumption standpoint, organic food enhances sustainable growth. Organic and traditional conventional foods possess distinct nutritional compositions. For example, organic foods have more omega-3 fatty acids and more potent antioxidants in their agricultural products, reducing cadmium and pesticides. Given that it contributes significantly to the emergence of antibiotic resistance in society, there is an essential cause of worry over the widespread use of antibiotics in traditional livestock farming. This chapter discusses the many recognized and potential health benefits of producing and cultivating various sustainable organic food products. Utilization of these production techniques is anticipated to yield positive outcomes in traditional agriculture and sustainable development.

Keywords
Organic food products · Organic farming · Sustainable organic food · Good health

M. Sharma (✉) · G. Bartariya
School of Life Science and Technology, IIMT University, Meerut, Uttar Pradesh, India

A. S. Sindhu
Department of Chemistry, Meerut College, Meerut, Uttar Pradesh, India

1 Introduction

Numerous foods are available to meet worldwide dietary requirements, as evidenced by the growth of novel foods, agricultural products, and facilities. However, imbalance and inconsistency make this genuinely unstable. In the terms FAO uses, sustainable food impacts and assists in promoting sustainable development.

Numerous intergovernmental bodies prioritize creating sustainable food systems (WHO 2012; Burlingame and Dernini 2012; Sustainable Food Systems Programme 2016). Many agricultural farming practices may impact the sustainability of food systems, as these practices may harm the well-being of animals, food security, health of human beings, and environmental sustainability. This chapter reviews the current scientific evidence about the sustainability challenges, developments, and solutions related to organic products and human health.

When questioned about sustainable or organic farming and sustainable foods, most potential consumers will refer to it regarding what organic producers do not use. Responses will likely include words like "natural," "no fertilizers," and "no pesticides" (Zanoli 2004). Although it is OK to characterize organic methods in this manner, it minimizes organic producers' basic concepts and techniques (Lampkin 2003).

The cultivation of organic foods as an integrated notion has its roots in concepts and values that date back nearly a century (Lockeretz 2007). The many founders examined and understood the primary issues facing conventional agriculture, such as the necessity of recycling nutrients and viewing the farm as an integrated whole, while also sharing an appreciation of farming and gardening. In this context, the word "organic" was first used in the 1940s to describe the idea of the farm as a living thing, whereby each of its components, like minerals of the soil, organic elements, microbes, plants, animals, insects, and humans, combines their efforts to create an enduring and harmonious intact, not just the kind of inputs that are used.

J. A. Parray et al. (eds.), *Environmental Landscape and Sustainable Biodiversity for Healthy Green Growth*, Sustainable Landscape Planning and Natural Resources Management, https://doi.org/10.1007/978-3-031-75405-0_1

The completion of cycles of nutrients and a need for local resources are essential to the concept itself. In the meantime, various issues have gained prominence at multiple points. These include the 1930s dust bowls and soil conservation (Howard 1940), pesticides and the Silent Spring phenomenon (Carson 1962), energy after the 1973 oil crisis (Lockeretz 1977), and current concerns regarding the welfare of animals, loss of biodiversity, global warming, oil peaking, peak phosphate, and security of food.

The labels "biological" or "ecological" can identify these, which describe organic farming in several European countries. This represents a focus on using "biological" and "ecological" techniques instead of external inputs to stimulate and enhance self-regulatory mechanisms and intensify agricultural systems (Vogt 2000).

The four core tenets of sustainable farming—well-being and health, ecosystems, equity, and care—embody these concepts (IFOAM 2005). IFOAM established the previously stated principles while working with stakeholders and embracing democratic membership acceptance.

The effort to build solid customer relationships and partnerships, initially mainly through face-to-face farm sales, has affected the growth and development of sustainable farming and the discussions around it. Growers are paying attention to customers' willingness to pay more significant amounts for the alleged advantages of organic food to preserve the financial viability of organic farming practices. This reflected more "self-interested" concerns about food safety and quality, particularly pesticide residues and individual health. Still, it also sometimes reflected increased generous worries about environmental issues, safeguarding animals, and societal matters (Aertsens et al. 2009).

Classifying food production processes can be challenging. This complexity results from the overlap between ordinary and organic agriculture systems and their variety and diversity. The European Union (EU) designates general agricultural practices as the primary farming technique. This is indicated by a significant amount of conventionally manufactured concentrates for feed used in livestock breeding and a considerable quantity of chemical-based insecticides. In contrast, organic agriculture complies with EU regulations or similar guidelines regarding organic farming; these consist of employing organic fertilizers such as a farmyard and green and natural compost, heavily depending on ecosystem services, using nonchemical and biological pest management methods, and allowing livestock accessibility outdoors.

Regardless of certification, any farmer can implement organic farming practices. The combined application of multiple factors and practices to achieve significant objectives such as safety, security, well-being, the health of the public, and the preservation of the environment characterizes the natural systems methodology. This approach has numerous advantages for the community's well-being (Lampkin 2010).

2 Organic Food Systems

Food items derived from forest products, fishery, farming, and components of the overall social, cultural, and environmental contexts in which these foods are located are produced, aggregated, processed, distributed, consumed, and disposed of. The entire range of participants and their interrelated activities that add value constitute organic food systems (FAO 2018a).

Smaller structures (which include agricultural activities, input supply, waste management, and many more) that collaborate with other more enormous structures (including trade, energy, health, and energy systems) constitute the organic food system. Therefore, a change in a single system can cause an alteration to another. As an illustration, the food and beverage industry would significantly promote increasing biofuel use in the energy system (Nicolopoulou-Stamati et al. 2016).

A sustainable food system (SFS) fulfills consumers' nutrition, safety, and food security demands while sustaining the economic, social, and environmental bases required to meet the requirements of future generations. This suggests that it has no adverse effects on the environment or only positive effects on it (ecological sustainability), is helpful to society as a whole (social sustainability), and is beneficial overall (economic sustainability) (FAO 2018a).

The Sustainable Development Goals set by the United Nations are primarily focused on establishing a sustainable food system.

Adopted in 2015, Sustainable Development Goals mandate modifications in farming and agriculture systems to accomplish appropriate food supply, reduce hunger, and enhance overall health, well-being, and nutrition by 2030.

The world's food system must be changed to achieve the SDGs. It must become more robust and environmentally sustainable, more inclusive of the impoverished and excluded, and capable of providing nutritious food to everyone. These are complicated, systemic problems that need coordinated local, state, federal, and international efforts (Niggi 2014).

Increasing population, wealth, urbanization, changing buying patterns, climate change, the rise of globalization, and subsequent depletion of natural resources are just a few issues that need to be considered when examining a food system. Food system improvements have generated various beneficial outcomes over the last three decades, especially in developing countries (Miller et al. 2021). Among these outcomes are the growth of nonfarm jobs as the food industry has progressed and various food options beyond regional specialties, satisfying consumers' gastronomic, form, and quality preferences.

However, these rapid advancements have also resulted in increasing grave issues, which may eventually significantly affect nutrition, health, and food security.

Numerous heavily refined foods that are inadequate in nutrients and heaviest in calories, which are currently frequently eaten, are among them referred to as the challenges small-scale farmers and agribusinesses have in locating advantageous markets, the increased volume of waste, an increase in food safety incidences and issues impacting human health, and growing demand for energy and environmental impact caused by the growth and food supply systems are becoming more industrialized (Miller et al. 2021). Consequently, an improved comprehension of the functioning of various food systems is crucial to ensure that these systems grow in a manner that limits their detrimental effects and optimizes their advantageous effects.

The complicated nature of food systems demands a systematic and perfectly aligned approach. Nutrition and food safety are complex issues with multifaceted approaches that cross disciplinarians, divisional ones, and organizational boundaries.

These challenges emerge in food systems that are becoming more integrated on many levels and in several aspects.

Collaboration among stakeholders from both the public and private sectors at all levels local, national, regional, and global—as well as on several fronts—trade, regulations, wellness, environment, gender stereotypes, education, transportation, infrastructure, and agriculture are required to address them. There must be a favorable confluence of ideas rather than a negative collision between these diverse points of view (Niggi 2014).

Conventional food security initiatives often have a production-focused stance, aiming to directly impact food security by augmenting the food supply. Inadequate food production remains the primary source of hunger and malnutrition in some areas, particularly in Sub-Saharan Africa. However, during the past few decades, the food system has undergone tremendous transformation, creating complex linkages and feedback systems resulting in repercussions on food security and nutrition (Ingram 2011). The attention to food cultivation sometimes results in neglecting other sectors, many of which have the leverage points to effect the most significant changes and the underlying causes of the food system's underperformance. Additionally, because of the complex connections and feedback within the food system, other systems-thinking-based methodologies are gaining popularity. For instance, a value-chain (VC) development plan extensively investigates how producers and other stakeholders—such as clients, governments, and labor unions—create and retain ability. To improve the chain's performance, systemic assessments and integrated actions constitute the focus of VC development. However, because the VC development approach concentrates on a single commodity, it frequently ignores relationships among other VCs. Farmers, mainly smallholders, often farm a range of crops in addition to raising livestock, engaging in fishing, and forestry, and the consumers' nutritional balance is dependent on diets that include a diversity of commodities. Therefore, obtaining wide-ranging developmental effects necessitates examining more closely how all food VCs interact (Iacovidou et al. 2021). To deal with systemic limits to market links that can affect many value chains, a further approach is becoming increasingly popular; the market systems approach recognizes markets as complex adaptive systems boosting ties to financial services for farmers. The market systems approach is susceptible to a similar restricted perspective difficulty as the VC technique despite overcoming the "one chain at a time" barrier of the VC approach (FAO 2018a). The primary objectives of the Food and Agriculture Organization, which include nutrition, food security, and poverty reduction, are at the core of the food system framework. These refer to sustainability's economic, social, and environmental aspects and are ingrained in the system's overall performance. The activities of different food system competitors or consumers determine this type of efficiency. The system's framework, which consists of societal, natural, and fundamental elements, reflects this behavior. The primary system consists of various resources (supporting the flow) and a series of events like food production, aggregation and processing, distribution and consumption of food, and then waste disposal. The natural world and the social milieu surround these activities (Gamage et al. 2023). The initial category encompasses all relevant laws, rules, infrastructures, organizations, and sociocultural norms. The latter include ecosystems, genetics, climate, water, soils, and air (Fig. 1).

Producing food organically means not using artificial chemicals or genetically engineered ingredients. The cultivation and consumption of food are reaching the limits on the earth, even with the claims of sustainability promoted by the majority of the food industries and many farming practices. Many scientists claim that agricultural practices should be enhanced by the expected 9.6 billion individuals by 2050 (Godfray et al. 2014). The organic food system might serve as an alternative approach to attain sustainability. Research on organic farming shows that using organic techniques may minimize environmental damage, promote societal unity, and increase the economy's resilience. Some scientists suggest that the average yield of organic systems is around 20% lower than comparable conventional systems (Seufert et al. 2017). This could have a more significant negative influence on the environment, the use of land, and natural ecosystems, risking the food supply worldwide (Tuomisto et al. 2012).

Three distinct strategies can be determined to resolve these concerns-efficiency, consistency, and sufficiency

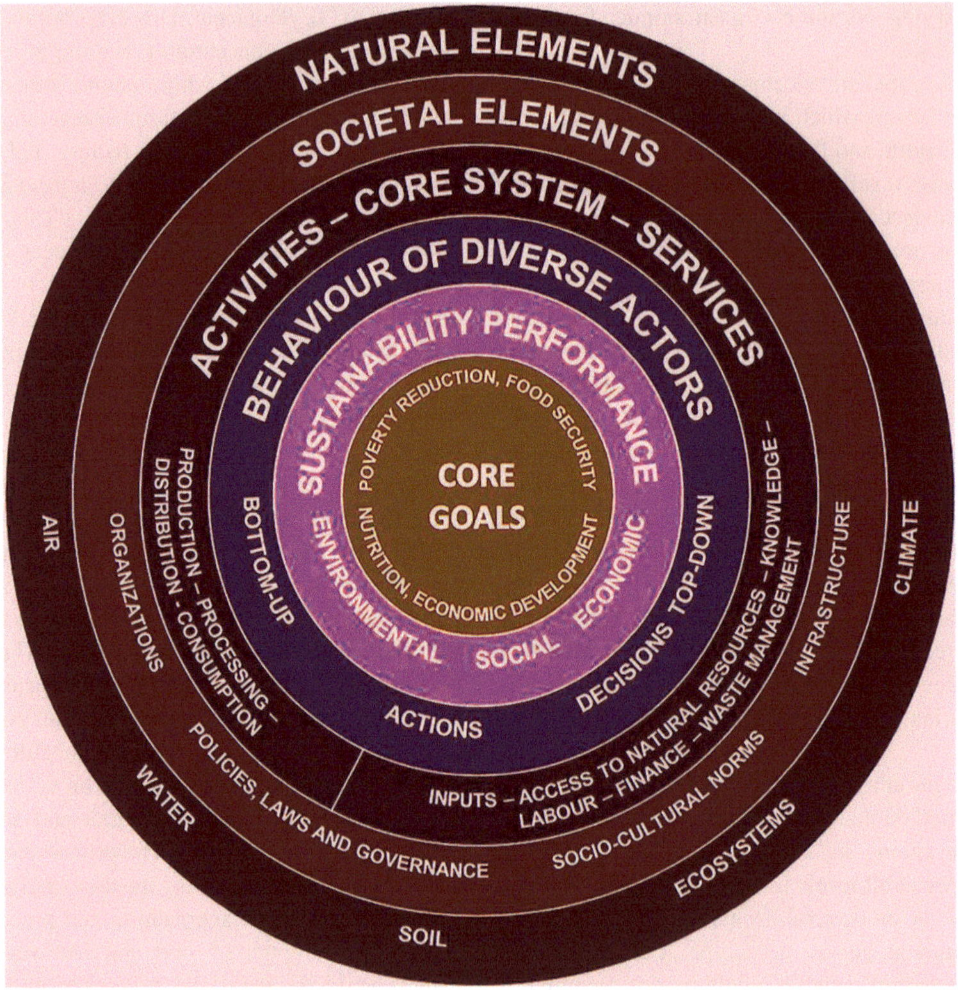

Fig. 1 The food system wheel

(Seufert et al. 2017). The efficiency approach aims to determine the most significant correlation between a system's adverse effects and positive outcomes. The consistency approach aims to move industrial systems toward the principles of sustainability or natural systems. Subsequently, the sufficiency approach tackles consumption by lessening adverse effects on resources. Systems of organic food offer an alternative approach to attaining sustainability. Sustainable agricultural techniques can lessen environmental harm, enhance social well-being, and boost economic resilience (Schader et al. 2012).

3 The Performance of Organic Products in Terms of Sustainability at Various Levels

Multiple views of sustainability have led to confusion among food producers, consumers, and even scientists as the concept becomes more and more critical in discussing future food systems (FAO 2013; Godfray and Garnett 2014; McDonald and Oates 2006). Variability in the depth and viewpoint of the assessments may be a significant contributing factor to this uncertainty (Schader et al. 2014). Analytical variations in the evaluation produce disparate and opposing outcomes, particularly when assessing complex systems like organic food systems, which affect production and consumption (Meier et al. 2015).

To reduce uncertainty and clarify the differences between the most popular methods and instruments for evaluating sustainability, efforts have been made to classify sustainability assessment techniques (Schader et al. 2014) and define sustainability more broadly (FAO 2013). This chapter highlights distinctions between the operator, product, and spatial/policy level.

At the operator level, indicator sets assess an organization's or farm's sustainability performance. FAO's 2013 SAFA recommendations, which address four categories (good administration, integrity of the environment, financial stability, and social well-being) and 58 (sub)themes, offer an internationally adaptable structure that provides a comprehensive view of sustainability. The SAFA guidelines state that sustainability evaluations must consider the operator's region and the other farms or businesses impacted by its decisions.

For example, a dairy corporation can have significant control over the production methods of its suppliers. Because a substantial portion of the environmental and social impacts may be linked to agricultural farming processes, dairy farms can be included in an operator-level sustainability evaluation. (Bystricky et al. 2014).

Life cycle assessment (LCA), as proposed by Finkbeiner et al. (2010), is the predominant method for performing assessments at the product level. The term "functional unit" refers to the supply of a set quantity of a particular type of food, and LCAs link the environmental effects of a production system to this unit (ISO (International Organization for Standardization) 2006a, b). Environmental impact categories typically covered by life cycle assessments (LCAs) include global warming potential, eutrophication, and acidification; however, impacts on biodiversity and soil fertility, for example, are more challenging to link to a functional unit related to a product (De Baan et al. 2013). These features are crucial for distinguishing between conventional and organic food systems. Additionally, theories for assessing the social and economic effects of an LCA framework are still in their early stages of development (Finkbeiner et al. 2010).

Efficiency is the primary focus of a product-level assessment regarding the three key sustainability strategies (consistency, sufficiency, and efficiency). One can only consider a limited number of consistency-related factors. The sufficiency technique would entail establishing a more generic functional unit or examining other functional units instead of rejecting the idea of a standard functional unit.

When evaluating at the spatial or policy level, all operators in a given geographic area or the effects of a policy on all impacted operators are assessed rather than just one operator or product. For example, analyzing a conversion to organic farming at the spatial or policy level requires more than examining individual goods or farms. However, in addition to providing a broad overview of a region, this degree of evaluation also enables a closer examination of certain goods or farm kinds. All environmental and economic challenges can be handled at the spatial/policy level. Social difficulties, except specific general considerations arising from shifts in product prices and input (labor, for example) demand, are frequently context-specific and thus rarely generalizable. In contrast to the other two levels, the spatial/policy level permits the examination of sufficiency factors in addition to problems of efficiency and consistency. This is because it allows the assessment of nutrition patterns and the integration of economic factors related to shifting supply and demand patterns into the investigation (Schader et al. 2015).

4 Advantages and Disadvantages of the Organic and Conventional Food Products

4.1 Disparities in Nutrition

Despite conflicting evidence, consumers generally believe that organic food is healthier than conventionally grown crops. There have been no discernible variations in the amounts of vitamins, minerals, or carbohydrates. Because nitrates have been linked to both gastrointestinal cancer and methemoglobinemia in infants, eating more organic food may contain fewer nitrates than eating conventional food. Organic green vegetables, such as spinach, lettuce, and cabbage, exhibited higher vitamin C concentrations than conventionally cultivated vegetables in 21 out of 36 tests (Williams 2002). According to other studies, organic veggies had higher total phenol content than those grown conventionally, which may have antioxidant benefits (Asami et al. 2003). Research contrasting conventional with organic food has produced various results (Forman et al. 2012). A comprehensive study conducted in 2009 revealed that several factors, including farm location, soil characteristics, season climate, harvest maturity, storage, and testing time, affected the nutritional value of food. Since so many nutrients are listed in many articles, the authors have divided them into 11 groups. Surprisingly, research comparing conventional and organic food revealed differences in only three nutrient categories: phosphorus, titratable acidity, and nitrogen content (Dangour et al. 2009). In 2008, scientists compared the nutritional composition of 236 matched pairs of organic and conventional food samples. They revealed that organic foods' total phenolics, vitamin E, vitamin C, quercetin, and total antioxidant capacity were 80% higher than traditional foods (Duval et al. 2021). Conventional products have reduced total protein content but have higher potassium, phosphorus, and total protein levels—all essential components of conventional fertilizers (Benbrook et al. 2008). Moreover, Lombardi-Boccia et al. (2004) found that, in comparison to organic plums, conventional plums were the largest providers of total polyphenols. Quercetin (54.1%) was more prevalent in traditional plums, although myricetin (22.2%) and kaempferol (183.3%) were more prevalent in organic plums; the most abundant constituents in both organic and conventional fruits were caffeic acid, chlorogenic acid, and quercetin (Lombardi-Boccia et al. 2004); Gąstoł 2013. In a split-plot factorial system experiment, Valverde et al. (2015) reported no variations in the total phenols and flavonoid levels in broccoli grown over two years (Valverde et al. 2015). According to Granato et al. (2015), there is no difference between conventional and organic purple grape juices' total phenolic content (Granato et al. 2015).

4.2 Natural Toxins

Because organic farming doesn't apply fungicides, which could encourage fungal infections, and because stressed plants launch defensive responses against pests in the absence of pesticide protection, organic agriculture may be more vulnerable to natural toxins like phytotoxins and mycotoxins (Mithöfer and Boland 2012). Furthermore, compared with conventional farming, organic farming is thought to increase species richness by about 30%, which benefits biodiversity. Although this is good for environmental diversity, it might also lead to more toxins linked to natural defenses being synthesized, which would lessen plant damage (Tuck et al. 2014). Furthermore, while conventional farming concentrates on high-yielding crop strains, organic farming frequently uses resistant crop varieties to compensate for the lack of synthetic pesticides and consequently prevent plant ailments. Compared to conventional vegetation, which might use that energy to grow, organic plants most likely invest more power in producing natural toxins. According to these disputes, crops grown organically might have higher levels of naturally occurring toxins than conventional crops (Brandt and Mølgaard 2001). Fungi and molds like *Aspergillus, Penicillium*, and *Fusarium*, commonly found in organic agricultural foods and feeds, are other toxins linked to organic agriculture (Vrček et al. 2014). Mycotoxins produced by these fungi include aflatoxins (B1, B2, G1, and G2), which are cytotoxic and carcinogenic (Liu et al. 2015). Moreover, a majority of mycotoxins exhibit heat resistance and have the potential to be transmitted throughout the food chain (Vrček et al. 2014), posing severe risks to human health and resulting in substantial financial losses (Gourama 2015). Considering this, much study and documentation has been done on mycotoxin contamination, particularly by the European Food Safety Authority (EFSA). Concerns have arisen that organic farming may be more contaminated than conventional farming since it does not use fungicides (Escobar et al. 2013).

4.3 Flavor and Taste of the Food

Theuer (2006) shows no significant differences in Taste and organoleptic quality between customers' opinions of the flavor and Taste of organic foods (Theuer 2006). Reganold et al. (2010) claim that organic strawberries have a superior flavor (Reganold et al. 2010). Furthermore, many intricate cognitive processes, such as expectations, training, informative framing, and opinions (such as environmental factors), impact perception (Sörqvist et al. 2013).

4.4 Environment-Related Characteristics

One of the main points of contention in the organic controversy is whether organic agricultural practices are more economical, productive, and less damaging to the ecosystem than conventional methods. Several studies and surveys have contrasted conventional versus organic farming methods. Since organic farming does not use or release synthetic pesticides into the environment, many people believe that it is healthier for the environment. Some of these pesticides can potentially harm animals in the area, water, and soil (Oquist et al. 2007). Because organic farms use practices like crop rotation, they are also thought to be more successful than conventional farms at sustaining different ecosystems, including populations of plants, insects, and animals. According to Hansen et al. (2001), organic farming uses less energy and produces less waste per land unit or output (Hansen et al. 2001). Because organic soil is of higher quality and holds more water, it can make more in dry years. The following is a summary provided by Environmental-conscience.com (2021) of the primary concerns regarding the ecological impacts of traditional agriculture and the production of food: The overuse of chemical fertilizers and pesticides in conventional agriculture can erode the soil gradually, making it impossible for farmers to grow sufficient crops in the future. Additionally, it can allow bugs to get resistant, and after some years, traditional insecticides might stop working. Because of this, many farmers might not be able to keep pests out of their fields and might lose a lot of their yearly yield to pests. Likewise, genetic engineering of conventional food-producing plants to increase food output and productivity may cause them to become relatively weak and disease-prone. Diseases that affect plants can occasionally spread quickly and significantly lower food production for years or even decades. Furthermore, because everything in standard agriculture is designed for income and output maximizing, it is frequently criticized for diminishing the earth's assets. Furthermore, almost all traditional farmers show little concern for the harm that traditional farming causes to our ecological systems. Furthermore, numerous species could quickly grow endangered or even extinct due to conventional farming, pesticides, chemical fertilizers, and deforestation linked to intensive agriculture. The loss of biodiversity overall may eventually result in an ecological imbalance with relatively unknown effects on our planet and humanity.

4.5 Economic Aspects

It is crucial to understand that conventional farming may increase job opportunities, speed up output, lower farming expenses, lessen the likelihood of international conflict, and enhance the nutritional balance of the world. However, because of their higher costs, the fact that few consumers are willing to pay more significant amounts for organic food, and the difficulty and expense of advertising and marketing their products, organic food manufacturers may find it difficult to establish themselves with their competitors. However, the primary motivation behind buying organic products is not health-related. It is a more complex idea that is impacted by numerous factors. They are based on ecological, ethical, and political beliefs and extend beyond health to encompass a profound concern for environmental sustainability and food system resilience, perceptions of risk, and social conventions (Kahl et al. 2012; Læssøe et al. 2014). Therefore, by focusing their marketing efforts on people who feel satisfied with their products, organic food producers can cultivate a favorable mindset toward their organic products and services and ultimately win over loyal customers (da Dias et al. 2016; Naz et al. 2021).

5 Interrelationship Between Organic Production and Sustainability

Organic farming is among the most appreciated and quickly growing sustainability initiatives. Often referred to as "regenerative farming," this idea is the foundation of many sustainable food and beverage ventures. The program's objectives are to decrease or do away with chemical-based fertilizers, increase crop yields and productivity, promote ecosystems, protect water resources, and sequester carbon in the soil. Organic farming methods are frequently associated with it (Pretty et al. 2018). One advantage of organic farming to sustainable agricultural methods is integrating pest management (IPM) strategies, which lower the need for pesticides. The application of no-till or reduced-tillage methods of agriculture to decrease soil compaction, crop rotation, planting nitrogen-fixing, cultivating cover crops, and mulching can all help improve soil health. Additionally, using soil samples as a guide to figure out the quantities of organic and chemical fertilizer to be utilized, encouraging the utilization of organic fertilizers over chemical fertilizers, avoiding excessive application of chemical-based fertilizers, and promoting precision agriculture as well as the development of innovative irrigation designs that minimize water waste, erosion, and salinity while optimizing crop or pasture yield (Chaichi et al. 2018; Salim et al. 2020; Merah et al. 2021). Along with stressing the preservation of the environment, it also ensures that ground cover, mulching, and pesticide application are minimized to avoid soil erosion. Growing in popularity, organic farming offers several health advantages as well as significant benefits for the environment. Because of the possible health hazards, people are increasingly worried about organic products and, therefore, are prepared to spend more money on them. Preservation of resources and a positive effect on organic expenditures were further outcomes of environmental consciousness and a commitment to ecological sustainability.

5.1 The Role of Organic Industries in Promoting Sustainable Diets

Greenhouse gas emissions, utilization of land, impacts on the environment, and consumption of water are frequently used to characterize environmentally friendly farming practices. However, this approach must consider that all food is grown to provide nutrition. Diets and environmental sustainability have recently been found to be linked. The four categories of health for the community, the environment, affordability of food, and societal significance are the focus of most acknowledged models of "sustainable diets." The FAO states that genuinely sustainable diets should include foods that are widely accessible, high in nutrients, manufactured from resources that are grown responsibly, and that people want to eat.

Finding strategies that target all four domains at once is still tricky. Moreover, as food cost and accessibility remained the main issues, the recent global epidemic revealed the vulnerability of the food supply. Demands for rich in nutrients and environmentally friendly foods have been growing. Nevertheless, consumers' role in adopting and integrating such foods within their dietary habits has received a different recognition.

Dietary recommendations that promote sustainability often need to pay more attention to the motivations and processes underlying people's dietary decisions. Taste, price, and nutritional value are important considerations for consumers when making food choices, and the food system must consider them. Since many perceive them as expensive, sustainable foods are not commonly incorporated into diets, a prerequisite for meaningful change. Food systems and supply chains must undergo revolutionary change to resolve complicated problems relating to cost, flavor, and sustainability. Consumers may now purchase environmentally friendly foods at a reasonable price because of a growing trend known as regenerative agriculture, which is an all-encompassing, natural agricultural method. A broad alliance, including government, academia, and the food industry, can be formed to guarantee that sustainability and nutritional value are given equal priority in the food supply within the context of foodstuffs that consumers want. Additionally, the coalition can ensure that consumers generally choose sustainable diets. The food industry's and its partners' potential and obstacles to providing a total supply of nutrients and a sustainable food supply while satisfying consumer expectations shall constitute the main topics of discussion from this viewpoint (Miller et al. 2021).

5.2 Modern Trends in Organic Product Sustainability

Sustainable consumption and production (SCP) have made dietary intake a central concern due to its effects on the environment, community and personal well-being, social harmony, and economic growth (Reisch et al. 2013). Consumer preferences are a significant factor in how production is oriented; they choose items based on the company that produces them, the area of origin, or the production methods. Customers also significantly impact how food is purchased, transported, stored, prepared, and consumed. However, feeding is one of the primary ways that the globe's resources are employed, and it also results in the development of elements that are bad for the ecosystem (Baldi et al. 2012). In this context, eating organic food is becoming increasingly popular as a sustainable trend (Colucci et al. 2012). Organic farming maintains the well-being of individuals, ecosystems, and lands. Instead of utilizing hazardous substances, it depends on ecological processes, biodiversity, and cycles adjusted to the local environment. Organic cultivation is a practice that integrates science, creativity, and tradition to improve the everyday environment, foster just interactions, and improve the overall standard of life. In simpler terms, the goal of organic cultivation is to produce high-quality, healthful food without the use of artificial chemicals. Therefore, organic farming protects the environment and enhances public health, significantly impacting rural communities' economies and social bonds (IFOAM 2012; De Magistris and Gracia 2008; Falguera et al. 2012) (Fig. 2).

The agricultural sector is often acknowledged as one of the most essential elements of long-term sustainable development. With a direct contributor to the Sustainable Development Goals by 2030, small-scale agricultural producers in developing nations must be involved in changing farming and food systems (FAO 2018b). Goal 2 of the 2030 sustainable goal, which aims to end hunger, revolves around farming. However, the farming industry has a role in the other SDGs. In response to the growing number of malnourished and impoverished people, the first Green Revolution emerged, which made it possible for farming products like fertilizers and insecticides to expand to developing nations.

The Zero Hunger Challenge is supported by several member countries and additional institutions by UN Sustainable Development principles (Oluwole et al. 2023). There should be no malnourished kids below the age of 2 years, entirely throughout the year, access to sufficient food, 100% improvement in agricultural production and income, and a 100% reduction in food loss or waste. All food systems should also be sustainable. Organic farming is the only one that reflects the trend of improving sustainable development through agriculture. Although sustainable organic farming has gained global recognition, it faces obstacles and constraints.

5.3 Organic Product Sustainability: Challenges and Limitations

Reducing the compromises between production and long-term sustainability is the most significant agricultural problem. Therefore, analyzing organic cultivation, a specific set of farming techniques that emphasize ecological sustainability, is fascinating. One way to describe organic agriculture is as being more integrated into ecosystem activities and less dependent on inputs outside the farm. There is abundant research on organic agriculture's noncommodity outputs and public goods (Gamage et al. 2023). Most papers discuss the benefits of organic farming for soil fertility, biodiversity preservation, and the preservation of soil, water, and air resources. Because organic agriculture prioritizes public values, it is less productive.

According to meta-analyses, the average yields of organic farming fall within the range of 0.75–2/8 compared with traditional farming. Exemplary depictions from underprivileged locations and climates demonstrate that organic farming can be just as productive as conventional farming, or perhaps more so. Therefore, organic farming is an excellent example of sustainable and fruitful food production. The primary reason for yield disparities in crops and livestock is insufficient funding for research and development aimed at resolving certain obstacles in organic farming. Consequently, there is a great deal of promise for enhancing organic farming's performance through research on agriculture (Alvarez 2021).

However, organic farming still occupies an exclusive position in most countries; in several countries, it is on the point of becoming universal. Over the past 20 years, consumer interest and demand have increased, and it appears that this will continue the growth of organic farming in the future (Giller et al. 2021).

5.4 Relationship Between Consuming More Organic Products and Overall Health and Well-Being

Scientists are becoming increasingly concerned about how eating decisions affect one's well-being and health in response to social issues (Block et al. 2011; Bublitz et al. 2019). Many consumers highlight organic food when asked about food and well-being (Ares et al. 2015). According to researchers (2014), consumers appear to link the intake of organic food items with feelings of contentment and pleasure (Vanessa et al. 2018). Nonetheless, there is still a shortage of research on the hedonistic impacts of eating organic food. Does eating organic food improve one's overall health? There hasn't been enough convincing empirical data to support the idea that eating organic food has a causal effect on

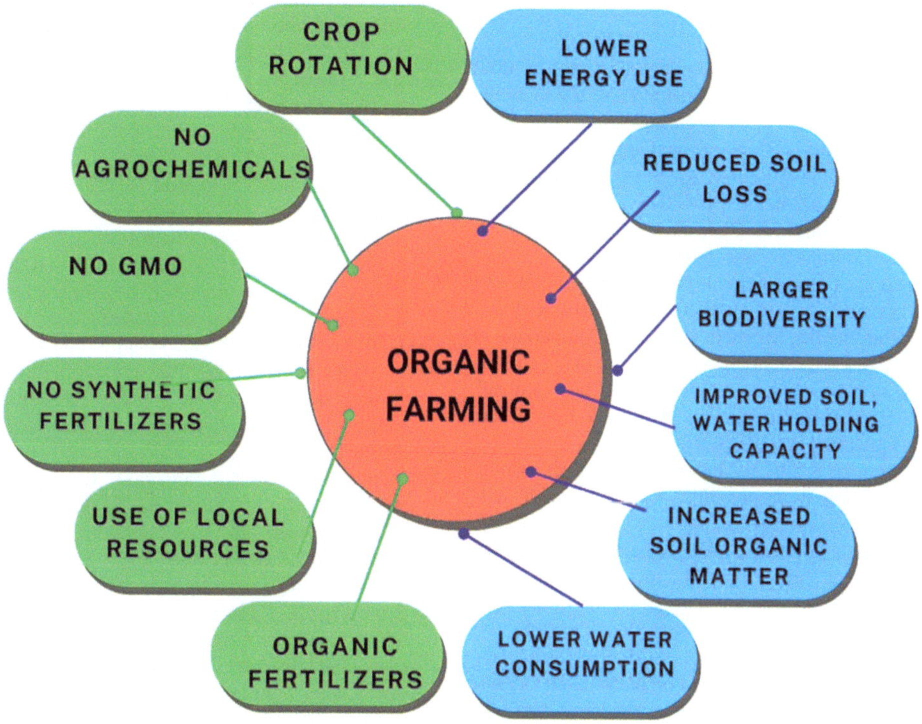

Fig. 2 The fundamental techniques and consequences of sustainable agriculture

one's later perception of well-being. Until now, research on well-being has primarily focused on organic food intake as a motivating precursor rather than a result. Indeed, consumers' acceptance of organic food is significantly influenced by their indulgence and desire for pleasure (Zanoli and Naspetti 2002; Bauer et al. 2013). Determining if eating organic food improves one's perception of well-being would have critical theoretical ramifications for understanding the factors and mechanisms that influence a consumer's well-being.

Furthermore, there would be significant real-world ramifications for marketers of organic food items and for consumer policies targeted at improving consumer wellness and encouraging the use of organic foods. Even if this kind of connection was formed, there are still a lot of questions that still need to be answered. Which mechanisms of the process are at play here? Can a branding impact centered around expectation belief impact explain the organic–wellness relationship? Do the consumer's perceptions and worries about their health affect this relationship? Examining the connections between the frequency of eating organic food, personal health concerns, health attitudes, and well-being. Specifically, we tackle the question of whether the purported impact of an organic diet on mental health is the result of intentional cognitive processing of health beliefs—that is, because organic consumers believe their diet is healthier—or if this effect is merely an automatic emotional response in the absence of substantial cognitive elaboration (Gundala and Singh 2021).

6 Conclusions

Changes in agricultural techniques are crucial for achieving the UN's Sustainable Development Goals by 2030. The appropriate strategy for achieving the SDGs is a highly contentious question. According to studies and disputes, sustainable food systems and agriculture must provide sufficient healthy food for all people while minimizing harmful environmental effects and enabling farmers to earn their livelihoods. Understandably, agriculture and food systems must adapt highly efficiently to accomplish several SDGs while staying within ecological restrictions. Nevertheless, the most effective method to achieve this is strongly contested, with two main approaches: agroecologically based, revolutionary restructuring of organic agricultural systems, and gradual steps to increase conventional agriculture's efficiency while reducing adverse environmental impacts. To mitigate the restrictions and challenges of organic farming, a combination of enhanced sludge, biochar with organic fertilizer, biofertilizers, organominerals, and digital technology is essential. Organic farming is a creative and sustainable method that improves many producers' quality of life and agricultural output while being environmentally beneficial.

References

Aertsens J, Verbeke W, Mondelaers K, Huylenbroeck GV (2009) Personal determinants of organic food consumption: a review. Br Food J 111:1140–1167

Alvarez R (2021) Comparing productivity of organic and conventional farming systems: a quantitative review. Arch Agron Soil Sci 68. https://doi.org/10.1080/03650340.2021.1946040

Ares G, de Saldamando L, Giménez A, Claret A, Cunha LM, Guerrero L, Deliza R (2015) Consumers' associations with well-being in a food related context: a cross-cultural study. Food Qual Prefer 40(PB):304–315. https://doi.org/10.1016/j.foodqual.2014.06.001

Asami DK, Hong YJ, Barrett DM, Mitchell AE (2003) Comparison of the total phenolic and ascorbic acid content of freeze-dried and air-dried Marionberry, strawberry, and corn grown using conventional, organic, and sustainable agricultural practices. J Agric Food Chem 51(5):1237–1241. https://doi.org/10.1021/jf020635c

Baldi L, Banterle A, Stranieri S (2012) Alimentazione, salute e ambiente: consumatori e imprese" sostenibili". Rivista di studi sulla sostenibilita'

Bauer HH, Daniel H, Daniela B, Schäfer (2013) The effects of organic labels on global, local, and private brands: more hype than substance? J Bus Res 66(8):1035–1043.,ISSN 0148-2963. https://doi.org/10.1016/j.jbusres.2011.12.028

Benbrook C, Zhao X, Yáñez J, Davies N, Andrews P (2008) New evidence confirms the nutritional superiority of plant-based organic foods. https://www.organic-center.org/newevidence-confirms-nutritional-superiority-plant-based-organic-foods

Block LG, Grier TL, Childers B, Davis JE, Ebert S, Kumanyika MN, van Ginkel B (2011) From nutrients to nurturance: a conceptual introduction to food well-being. J Public Policy Mark 30(1):5–13. https://doi.org/10.1509/jppm.30.1.5

Brandt K, Mølgaard JP (2001) Organic agriculture: does it enhance or reduce the nutritional value of plant foods? Nutritional value of organic plants. J Sci Food Agric 81(9):924–931. https://doi.org/10.1002/jsfa.903

Bublitz MG, Hansen J, Peracchio LA, Tussler S (2019) Hunger and food well-being: advancing research and practice. J Public Policy Mark 38(2):136–153. https://doi.org/10.1177/0743915619827012

Burlingame B, Dernini S (2012) Food and Agriculture Organization of the United Nations (FAO). In: Burlingame B, Dernini S (eds) Sustainable diets and biodiversity. Directions and solutions for policy, research, and action

Bystricky M, Nemecek T, Alig M, Gaillard G (2014) Challenges of comparing food and feed products from different countries of origin. Available at http://lcafood2014.org/papers/160.pdf)

Carson R (1962) Silent Spring. Houghton Mifflin, Boston

Chaichi W, Djazouli Z, Zebib B, Merah O (2018) Effect of vermicompost tea on faba bean growth. Compost Sci Util 26(4):279–285. https://doi.org/10.1080/1065657X.2018.1528908

Colucci F, Menegoni P, Nocenzi M, Presenti O (2012) L'evoluzione dei modelli di consumo e degli stili alimentari tra sostenibilità e benessere. Rivista Di Studi Sulla Sostenibilita'

da Dias VV, da Schuster MS, Talamini E, Révillion JP (2016) The scale of consumer loyalty for organic food. Br Food J 118(3):697–713. https://doi.org/10.1108/BFJ-09-2015-0332

Dangour AD, Dodhia SK, Hayter A, Allen E, Lock K, Uauy R (2009) Nutritional quality of organic foods: a systematic review. Am J Clin Nutr 90(3):680–685. https://doi.org/10.3945/ajcn.2009.28041

De Baan L, Alkemade R, Koellner T (2013) Land use impacts on biodiversity in LCA: a global approach. Int J Life Cycle Assess 18(6):1216–1230

De Magistris T, Gracia A (2008) The decision to buy organic food products in Southern Italy. Br Food J 110(9):929–947

Duval J, Cournut S, Hostiou N (2021) Livestock farmers' working conditions in agroecological farming systems. A review. Agron Sustain Dev 41(2):22. https://doi.org/10.1007/s13593-021-00679-y

Escobar J, Lorán S, Giménez I, Ferruz E, Herrera M, Herrera A, Ariño A (2013) Occurrence and exposure assessment of Fusarium mycotoxins in maize germ, refined corn oil and margarine. Food Chem Toxicol 62:514–520. https://doi.org/10.1016/j.fct.2013.09.020

Falguera V, Aliguer N, Falguera M (2012) An integrated approach to current trends in food consumption: moving toward functional and organic products? Food Control 26(2):274–281

FAO (2013) Sustainability Assessment of Food and Agriculture Systems (SAFA). Rome (available at http://www.fao.org/nr/sustainability/sustainability-assessments-safa/en/

FAO (2018a) Sustainable food systems: concept and framework. Rome

FAO (2018b) Transforming food and agriculture to achieve the SDGs: 20 interconnected actions to guide decision-makers, technical reference document. Available at: http://www.fao.org/3/I9900EN/i9900en.pdf

Finkbeiner M, Schau E, Lehmann A, Traverso M (2010) Towards life cycle sustainability assessment. Sustain For 2(10):3309–3322

Forman J, Silverstein J, Committee on Nutrition, Council on Environmental Health Bhatia, J. J. S, Abrams SA, Corkins MR, de Ferranti SD, Golden NH, Silverstein J, Paulson JA, Brock-Utne AC, Brumberg HL, Campbell CC, Lanphear BP, Osterhoudt KC, Sandel MT, Trasande L, Wright RO (2012) Organic foods: health and environmental advantages and disadvantages. Pediatrics 130(5):e1406–e1415. https://doi.org/10.1542/peds.2012-2579

Gamage A, Gangahagedara R, Gamage J, Jayasinghe N, Kodikara N, Suraweera P, Merah O (2023) Role of organic farming for achieving sustainability in agriculture. Farm Syst 1(1):100005, ISSN 2949-9119. https://doi.org/10.1016/j.farsys.2023.100005

Gąstoł M (2013) Comparing nutritional content of fruits, vegetables, and juices from organic and conventional crops. CAB Rev 8(060). https://doi.org/10.1079/PAVSNNR20138060

Giller KE, Delaune T, Silva JV et al (2021) The future of farming: who will produce our food? Food Sec 13:1073–1099. https://doi.org/10.1007/s12571-021-01184-6

Godfray HCJ, Garnett T (2014) Food security and sustainable intensification. Philos Trans R Soc B Biol Sci 369:20120273

Godfray H, Charles J, Tara G (2014) Food security and sustainable intensification. Phil Trans R Soc B369:20120273. https://doi.org/10.1098/rstb.2012.0273

Gourama H (2015) A preliminary mycological evaluation of organic and conventional foods. Food Prot Trends 35(5):385–391

Granato D, Margraf T, Brotzakis I, Capuano E, van Ruth SM (2015) Characterization of conventional, biodynamic, and organic purple grape juices by chemical markers, antioxidant capacity, and instrumental taste profile. J Food Sci 80(1):C55–C65. https://doi.org/10.1111/1750-3841.12722

Gundala RR, Singh A (2021) What motivates consumers to buy organic foods? Results of an empirical study in the United States. PLoS One 16(9):e0257288. https://doi.org/10.1371/journal.pone.0257288. PMID:34506582; PMCID: PMC8432837

Hansen B, Alrøe HF, Kristensen ES (2001) Approaches to assess the environmental impact of organic farming with particular regard to Denmark. Agric Ecosyst Environ 83(1–2):11–26. https://doi.org/10.1016/S0167-8809(00)00257-7

Howard A (1940) An agricultural testament. Oxford University Press, London

Iacovidou E, Hahladakis JN, Purnell PA (2021) Systems thinking approach to understanding the challenges of achieving the circular economy. Environ Sci Pollut Res 28:24785–24806. https://doi.org/10.1007/s11356-020-11725-9

IFOAM (2005) Principles of organic agriculture Bonn: International Federation of Organic Agriculture Movements

IFOAM (2012) Definition of organic agriculture. Retrieved April 05, 2012, from IFOAM, International Federation of Organic Agriculture Movements: http://www.ifoam.org/growing_organic/definitions/doa/index.html
Ingram J (2011) A food systems approach to researching food security and its interactions with global environmental change. Food Secur 3(4):417–431
ISO (International Organization for Standardization) (2006a) Environmental management – life cycle assessment – requirements and guidelines. Switzerland, Geneva
ISO (International Organization for Standardization) (2006b) Environmental management – life cycle assessment – principles and framework. Switzerland, Geneva
Kahl J, Baars T, Bügel S, Busscher N, Huber M, Kusche D, Rembiałkowska E, Schmid O, Seidel K, Taupier-Letage B, Velimirov A, Załęcka A (2012) Organic food quality: a framework for concept, definition, and evaluation from the European perspective: organic food quality. J Sci Food Agric 92(14):2760–2765. https://doi.org/10.1002/jsfa.5640
Læssøe J, Ljungdalh AK, Alrøe HF, Noe E, Christensen T, Dubgaard A, Olsen SB, Kærgård N, Kastberg P (2014) Three perspectives on motivation and multicriteria assessment of organic food systems. Ecol Soc 19(3) https://www.jstor.org/stable/26269638
Lampkin NH (2003) Organic farming. In: Soffe RJ (ed) Primrose McConnell's agricultural notebook, 20th edn. Blackwell Science, Oxford
Lampkin N (2010) Organic farming myth and reality. World Agric 1:46–53
Liu C, Shen H, Yi L, Shao P, Soulika AM, Meng X, Xing L, Yan X, Zhang X (2015) Oral administration of aflatoxin G1 induces chronic alveolar inflammation associated with lung tumorigenesis. Toxicol Lett 232(3):547–556. https://doi.org/10.1016/j.toxlet.2014.11.002
Lockeretz W (ed) (1977) Agriculture and energy. New York Academic Press
Lockeretz W (ed) (2007) Organic farming: an international history. CABI, Wallingford
Lombardi-Boccia G, Lucarini M, Lanzi S, Aguzzi A, Cappelloni M (2004) Nutrients and antioxidant molecules in yellow plums (Prunus domestica L.) from conventional and organic productions: a comparative study. J Agric Food Chem 52(1):90–94. https://doi.org/10.1021/jf0344690
McDonald S, Oates CJ (2006) Sustainability: consumer perceptions and marketing strategies. Bus Strateg Environ 15(3):157–170
Meier MS, Stoessel F, Jungbluth N, Juraske R, Schader C, Stolze M (2015) Environmental impacts of organic and conventional agricultural products – are the differences captured by life cycle assessment? J Environ Manag 149:193–208
Merah O, Djazouli Z-E, Zebib B (2021) Aqueous extract of Algerian Nettle (Urtica dioïca L.) as a possible alternative pathway to control some plant diseases. Accepte pour publication dans. Iranian J Sci Technol Trans A Sci 45:463–468. https://doi.org/10.1007/s40995-021-01061-z
Miller KB, Eckberg JO, Decker EA, Marinangeli CPF (2021) Role of food industry in promoting healthy and sustainable diets. Nutrients 13(8):2740. https://doi.org/10.3390/nu13082740. PMID: 34444900; PMCID: PMC8398875
Mithöfer A, Boland W (2012) Plant defense against herbivores: chemical aspects. Annu Rev Plant Biol 63(1):431–450. https://doi.org/10.1146/annurev-arplant042110-1038
Naz F, Alshaabani A, Rudnák I, Magda R (2021) Role of service quality in improving customer loyalty towards telecom companies in Hungary during the COVID-19 pandemic. Economies 9(4):200. https://doi.org/10.3390/economies9040200
Nicolopoulou-Stamati P, Maipas S, Kotampasi C, Stamatis P, Hens L (2016) Chemical pesticides and human health: the urgent need for a new concept in agriculture. Front Public Health 4. https://doi.org/10.3389/fpubh.2016.00148
Niggi U (2014) Sustainability of organic food production: challenges and innovations. Proc Nutr Soc 74(1):83–88. https://doi.org/10.1017/S0029665114001438
Oluwole O, Ibidapo O, Arowosola T, Raji F, Zandonadi RP, Alasqah I, Lho LH, Han H, Raposo A (2023) Sustainable transformation agenda for enhanced global food and nutrition security: a narrative review. Front Nutr 10:1226538. https://doi.org/10.3389/fnut.2023.1226538. PMID: 37599683; PMCID: PMC10433737
Oquist KA, Strock JS, Mulla DJ (2007) Influence of alternative and conventional farming practices on subsurface drainage and water quality. J Environ Qual 36(4):1194–1204. https://doi.org/10.2134/jeq2006.0274
Pretty J, Benton TG, Bharucha ZP, Dicks LV, Flora CB, Godfray HCJ, Goulson D, Hartley S, Lampkin N, Morris C, Pierzynski G, Prasad PVV, Reganold J, Rockstrom J, Smith P, Thorne P, Wratten S (2018) Global € assessment of agricultural system redesign for sustainable intensification. Nat Sustain 1:441–446. https://doi.org/10.1038/s41893-018-0114-0
Reganold JP, Andrews PK, Reeve JR, Carpenter-Boggs L, Schadt CW, Alldredge JR, Ross CF, Davies NM, Zhou J (2010) Fruit and soil quality of organic and conventional strawberry agroecosystems. PLoS One 5(9):e12346. https://doi.org/10.1371/journal.pone.0012346
Reisch L, Eberle U, Lorek S (2013) Sustainable food consumption: where do we stand today? An overview of issues and policies. Sustain Sci Pract Policy 9:1
Salim D, De Caro P, Merah O, Chbani A (2020) Control of postharvest citrus green mold using Ulva lactuca extracts as a source of active substances. Int J Biodivers Sci Manag 11(3):287–296. https://doi.org/10.23910/1.2020.2107
Schader C, Stolze M, Gattinger A (2012) Environmental performance of organic farming. In: Boye JI, Arcand Y (eds) Green technologies in food production and processing. Springer, New York, pp 183–210
Schader C, Grenz J, Meier MS, Stolze M (2014) Scope and precision of sustainability assessment approaches to food systems. Ecol Soc 19(3):42
Schader C, Stolze M, Niggli URS (2015) How the organic food system contributes to sustainability. In: Meybeck A, Redfern S, Paoletti F, Strassner C (eds) Assessing sustainable diets within the sustainability of food systems. Proceedings of an international workshop, 15–16 September 2014, CREA, Rome, Italy. Food and Agriculture Organization of the United Nations (FAO), pp 27–36
Seufert V, Ramankutty N, Mayerhofer T (2017) What is this thing called organic? – how is organic farming codified in regulations? Food Pol 68:10–20. https://doi.org/10.1016/j.foodpol.2016.12.009
Sörqvist P, Hedblom D, Holmgren M, Haga A, Langeborg L, Nöstl A, Kågström J (2013) Who needs cream and sugar when there is eco-labeling? Taste and willingness to pay for "eco-friendly" coffee. PLoS One 8(12):e80719. https://doi.org/10.1371/journal.pone.0080719
Sustainable Food Systems Programme (2016). http://www.unep.org/10yfp/programmes/sustainable-food-systems-programme
Theuer RC (2006) Do organic fruits and vegetables taste better than conventional fruits and vegetables? https://www.yumpu.com/en/document/view/25040542/executive-summary-theorganic-cente
Tuck SL, Winqvist C, Mota F, Ahnström J, Turnbull LA, Bengtsson J (2014) Land-use intensity and the effects of organic farming on biodiversity: a hierarchical meta-analysis. J Appl Ecol 51(3):746–755. https://doi.org/10.1111/1365-2664.12219
Tuomisto H, Hodge I, Riordan P, Macdonald D (2012) Does organic farming reduce environmental impacts? A meta-analysis of European research. J Environ Manag 112:309–320
Valverde J, Reilly K, Villacreces S, Gaffney M, Grant J, Brunton N (2015) Variation in bioactive content in broccoli (Brassica oleracea

var. italica) grown under conventional and organic production systems: bioactive content in broccoli under different production systems. J Sci Food Agric 95(6):1163–1171. https://doi.org/10.1002/jsfa.6804

Vanessa A, Patrick H, D'Souza C, Cristina ML (2018) Eat organic – Feel good? The relationship between organic food consumption, health concern and subjective well-being. Food Qual Prefer 63:51–62., ISSN 0950-3293. https://doi.org/10.1016/j.foodqual.2017.07.011

Vogt G (2000) Entstehung und Entwicklung des ökologischen Landbaus im deutschsprachigen Raum, Bad Duerkheim, Stiftung Ökologie und Landbau

Vrček IV, Čepo DV, Rašić D, Peraica M, Žuntar I, Bojić M, Mendaš G, MedićŠarić M (2014) A comparison of the nutritional value and food safety of organically and conventionally produced wheat flours. Food Chem 143:522–529. https://doi.org/10.1016/j.foodchem.2013.08.022

WHO (2012) Health indicators of sustainable agriculture, food and nutrition security in the Rio+20 UN Conference on Sustainable Development context. www.who.int/hia/green_economy/indicators_food.pdf. Accessed 2017–09-11

Williams CM (2002) Nutritional quality of organic food: shades of grey or shades of green? Proc Nutr Soc 61(1):19–24. https://doi.org/10.1079/PNS2001126

Zanoli R (ed) (2004) The European consumer and organic food. Aberystwyth School of Management and Business, University of Wales

Zanoli R, Naspetti S (2002) Consumer motivations in purchasing organic food: a means-end approach. Br Food J 104(8):643–653. https://doi.org/10.1108/00070700210425930

Green Process Technologies: A Necessity for a Sustainable Future and Economic Growth

Chandrarati Das, Ankan Acharjee, and Sahana Ghosh

Abstract

The increased demand for natural resources has soared in recent technological and societal growth and development. The anthropogenic process has led to the deterioration of environmental standards without considering the negative consequences. To minimize these detrimental effects that have been in action for so long, green process technologies (GPT) can be implemented to upgrade the economy alongside advancing environmental and societal well-being. In terms of sustainability, these technologies play a vital role in dealing with (a) climate change, (b) green energy efficiency and usage, and (c) access to clean and safe drinking water and food, which are essential for a healthy life. By incorporating environment-friendly practices across various industries, green process technologies aim to minimize the global carbon footprint of different industrial processes. The principle of reuse and recycling also forms a cornerstone for sustainable practices to reduce the overconsumption of raw materials, thereby extending the lifetime of the same through refurbishment, remanufacturing, and repurposing. Advanced recycling technologies, sensor technologies to analyze and remotely operate the environmental metrics, bioremediation, and precision agriculture are the various building blocks of this approach toward a sustainable future. These methods play pivotal roles in fostering economic growth. The alignment of ecological systems and financial structures in the sustainable development goals reinforces the necessity for transitioning toward a green economy. In navigating the intertwined challenges of ecology and economy, a green economy emerges as a guiding principle to achieve a harmonious equilibrium between environmental responsibility and socio-economic advancement.

Keywords

Green process technologies · Green economy · Sustainable development · Green energy

1 Introduction

Green process technologies (GPT) are a set of techniques and practices designed to minimize the environmental carbon footprint of industrial processes, maximizing resource efficiency and leading to sustainable growth and development. The wish to enhance sustainability across various fields, including science, industry, society, and policy, has been recognized as one of the century's most significant problems. The concept of sustainable development commenced with the release of the report "Our Common Future" by the Brundtland Commission in 1987. Since that moment, it has undergone consecutive refinement and had to undergo a lot of agreements, national programs, and scientific investigations. Thus, sustainability has become highly complex, with continuous definition modifications (Hessel et al. 2008; Brundtland 1987; United Nations Conference on Environment and Development (UNCED) 1992). In recent times, industries have been using an excess amount of energy that is not necessary, resulting in excessive pollution levels. Thus, a managerial system centered around green processes and products is required to reduce these pollution levels (Iravani et al. 2017).

Various techniques have been implemented in this field, and some have been discussed further. Renewable energy utilization utilizes renewable energy sources such as solar, wind, hydroelectric, and biomass in various industrial processes, thus reducing reliance on fossil fuels and lowering greenhouse gas emissions (Banerjee et al. 2014). There are specific energy efficiency measures emphasizing the implementation of technologies such as energy-efficient machinery, improved insulation, and optimized process design to minimize energy consumption and generation of waste. GPT

C. Das · A. Acharjee · S. Ghosh (✉)
Department of Microbiology, Vijaygarh Jyotish Ray College, Jadavpur, Kolkata, West Bengal, India

J. A. Parray et al. (eds.), *Environmental Landscape and Sustainable Biodiversity for Healthy Green Growth*, Sustainable Landscape Planning and Natural Resources Management, https://doi.org/10.1007/978-3-031-75405-0_2

also aims to reduce waste generation through strategies like process optimization and recycling of materials and by-products. This ultimately aids in the conservation of products, lessens landfill waste, and decreases pollution. Green chemistry principles have been incorporated, which use chemical reactions that are safe and eco-friendly and minimize the generation of toxic by-products. Scrubbers, filters, and catalytic converters are employed for pollution prevention and controlled to treat pollutants emitted during industrial processes (Anastas and Warner 1998; Anastas and Williamson 1996). GPT also includes green, imperishable building and infrastructure practices such as energy-efficient roofs and building design to reduce energy consumption, urban heat island effects, and enhance ecosystem amenities (Habibi et al. 2020). The striking advantages of GPT involve the reduction of detrimental emissions into the environment or atmosphere and the replenishable nature of the sources.

Furthermore, it requires cost-effective raw materials, reducing the need for significant financial investment to operate them. Despite the identified benefits, there are certain drawbacks to such technologies. In certain areas, the establishment of these facilities impacts the ploughland. Few green energy sources may not be viable in specific geographical locations, such as wave energy requiring waves of at least 16 feet or geothermal energy being restricted to sulfur-rich areas only geologically. High implementation costs, lack of known alternative chemical or raw material inputs or alternative process technology, and uncertainty of an appropriate income hinder the approach (Iravani et al. 2017).

2 Climate Change Mitigation

Climate change is a broad concept based on the alterations in the Earth's climate on various scales, compelled primarily by human activities in industries. This includes burning fossil fuels alongside deforestation, significantly increasing atmospheric carbon emissions (Mehmood et al. 2024). Greenhouse gases (GHGs) like methane and carbon are emitted predominantly by the transportation, industrial, and agricultural sectors, leading to heat entrapment and longer atmospheric wavelengths (Grajzl and Murrell 2021; Sovacool et al. 2021). Gradual industrialization resulted in global warming with an immense surge in greenhouse gas pollution. After the Industrial Revolution, industry expansion released large quantities of carbon dioxide and other greenhouse gases into the air. All these cumulative effects have led to rising sea levels and disturbances to ecosystems and biodiversity (Pandit 2021). To deal with global warming challenges, there is a need for collective efforts to regulate greenhouse gas emissions and eventually shift to renewable energy alternatives, enhance better energy utilization mechanisms, and adopt sustainable ways of operation in all industries. Adopting green process technologies is a crucial strategy to mitigate adverse effects. The adoption of green technologies is driven by environmental, economic, and regulatory factors (Klein et al. 2015). Adopting green technologies depends on regulatory frameworks, including carbon pricing mechanisms, emissions trading schemes, renewable energy targets, and energy efficiency standards (Pradhan et al. 2017). Lastly, financial considerations such as the decreasing cost of renewable energy technology, increasing energy efficiency, and possible cost reductions from transportation and energy and their reduced usage drive the adoption of green technologies (Sovacool et al. 2020). Sustainable technology is intensely interested in pure and permanent knowledge due to growing awareness of the need to address climate change and environmental issues. Green technology projects need to be swiftly integrated into sectors due to some factors, including customer demand for environment-friendly products and services, technological advancements, and the falling prices of alternative energy.

Over the years, green technologies have undergone various innovations that have impacted society's needs. Reducing greenhouse gas emissions (GHGs), such as methane (CH_4), carbon dioxide (CO_2), and carbon monoxide (CO), are the primary issues that still have to be addressed. They trap heat in the Earth's atmosphere and contribute to climate change. These problems must be considered, and easily accessible renewable energy sources such as hydroelectricity, solar, wind, and biomass must be implemented (Sovacool et al. 2021). Renewable energy sources like wind turbines and solar photovoltaic cells produce electricity without causing pollution and releasing greenhouse gases to the environment (Ribbing and Xydis 2021; Fischhendler et al. 2021). Energy-efficient appliances and LED lighting use less energy to perform the same tasks but with reduced electricity consumption and emissions. Proper insulation and intelligent building technologies optimize energy use by lowering specific heating, cooling, and lighting needs in buildings. Combined heat and power (CHP) systems simultaneously generate electricity and proper heat from a single fuel source, maximizing energy efficiency (Liu et al. 2014). In Industrial sectors, industrial processes can be optimized to minimize energy consumption and capture the generated waste heat for reuse in other methods, thus improving overall efficiency and reducing emissions. Recently, hybrid and electric vehicles and improved fuel efficiency standards have gained popularity and reduced GRH gas emissions from the transportation divisions (Wu and Zhang 2017).

Furthermore, advancements in energy storage technologies enable better integration of discontinuous renewable energy sources into the grid, ensuring a reliable power supply. Taking approaches to improve energy efficiency in buildings, industries, and transportation systems is another effective strategy for reducing greenhouse gas emissions. Green building designs utilize energy-efficient materials and intelligent technologies to minimize energy consumption and reduce the carbon footprint. Similarly, adopting fuel-efficient vehicles, public transportation systems, and eco-friendly manufacturing processes helps mitigate the emissions associated with conventional energy use. Agriculture contributes significantly to greenhouse gas emissions through deforestation, livestock rearing, and chemical fertilizers. Implementing sustainable agricultural practices such as precision farming, organic farming, and agroforestry can significantly reduce emissions while promoting soil health and biodiversity. Precision agriculture uses data-driven technologies where the availability of resources is optimized to minimize GHG emissions from fertilizers and water used. Organic farming methods can also be administered to avoid the misuse of synthetic fertilizers and pesticides, thus reducing emissions from agricultural production and promoting positive soil health. Another strategy, agroforestry, can be used to isolate carbon in soils and biomass, along with initiatives to start reforestation to restore our carbon sinks and mitigate deforestation-related damage and destruction. This lowers food waste production while shifting toward plant-based diets, continuing the fight against carbonizing the food and agriculture system (Mondejar 2021; Frone and Frone 2015).

The primary aim is to transition into a low-carbon economy and develop a more adaptable and sustainable future using green process technologies, which is one method for achieving these goals. Nevertheless, this must go hand in hand with meaningful emission reductions, which would require collaborative efforts from all sectors across governments, industry, and individuals, among others, to prioritize innovations, investments, and policy support toward green technologies. We can embrace these solutions, hence mitigating climate change impacts for future generations and, after that, having a sustainable planet (Söderholm 2020).

3 Resource Efficiency and Conservation

Resource scarcity and environmental degradation caused by the overexploitation of natural resources and unnecessary waste generation have posed significant challenges to global sustainability (Singh and Singh 2016). In such times, resource efficiency and conservation have been critical in sustainable development. Green process technologies have been deeply rooted in the principles of the 4Rs', Reduce, Reuse, Recycle, and Recover, to address sustainability challenges. These implementations catalyze industries that are aiming toward transitioning into circular economy models. With cutting-edge technology and frequent integration in various industries, AI has the power to revolutionize industrial practices altogether. The numerous benefits include operation efficiency, resource allocation, optimized decision-making, etc. The synergy between AI and green process technology, facilitated by the AI-based Hybridized Intelligent Framework (AIHIF), promises enhanced resource efficiency and conservation. With the help of the AIHIF framework, industries can bring about more streamlined operations, and as a result, this will improve resource efficiency and conservation efforts (Kan 2021).

4 Principles of 4-R and AI Implementation

The principle of "Reduce" focuses mainly on waste generation and ways to minimize resource consumption through various enhancements and optimizations. "Reuse" is the term used to denote the extending lifespan of products or materials through restoration and remanufacturing, leading to minimizing the exploitation of virgin resources. "Recycling" involves using the waste streams to retrieve and reprocess constituents to create alternate products or raw materials. Lastly, "Recovery" highlights the extraction of energy or valuable resources from waste streams through anaerobic sludge digestion processes, composting, biofilm passages, or waste-to-energy conversion processes. The foundation of green process technology lies in these 4Rs frameworks, pioneering a roadmap in a circular economy setting, and from this perspective, integrating Artificial Intelligence with Green Process Technologies ushers in a new era of innovation and improvement. AI algorithms can analyze large amounts of data, apply big data analytics, and predict how clients will behave. It would also recognize some patterns and possibilities that are sometimes impossible for humans to comprehend. Companies may use AI to optimize production processes, predict changes in demand, and develop efficient and effective ways to reduce and recycle waste. The AI-based Hybridized Intelligent Framework (AIHIF) represents an inventive approach that uses the power of AI to improve decision-making and streamline operations in green process technologies. AIHIF applies the principles of machine learning algorithms, predictive analytics, and optimization techniques to facilitate real-time decision-making and improve performance in various industries (Kan 2021).

Further case studies across various industries are needed to comprehend the real-world applications of AIHIF in Green Process Technology. AIHIF, for instance, can be applied in the manufacturing industry to enhance production procedures, minimize waste, and improve energy efficiency

(Gu 2019). Continuous improvement can be achieved using new machine learning algorithms to analyze production data and detect trends and inefficiencies. Predictive AI-driven maintenance systems can also identify equipment breakdowns early, saving downtime and boosting operational dependability. By giving us real-time knowledge of soil health and assisting in our adaptation to pest and climate change, AIHIF has the potential to transform conventional crop management techniques in the agricultural sectors completely. Data from drones and satellite imagery can be collected, and with AIHIF's aid, irrigation schedules, pesticide usage, and crop yields can be maximized while minimizing environmental impact (He et al. 2019).

Additionally, in the waste management sector, AIHIF can enhance the waste collection routes by identifying recyclable materials, and new technologies that use the idea of waste-to-energy generation can be implemented. Preferred AI algorithms can also be applied to analyze waste composition and generation patterns. Municipalities can customize waste management strategies according to their locality, enhancing efficiency and reducing environmental burden (Heidari and Jabbarzadeh 2019). One such example that can be cited here is of the urban agglomeration sector of China City, a waste management enterprise pioneering the development of a sophisticated network dedicated to handling and distributing artificial intelligence waste, implementing the fundamentals of AIHIF. Currently, the network comprises 20 distinct waste generation sites, a central collection center for unwanted materials, a facility dedicated to waste resale, and a disposal site. The goal is to optimize the flow of artificial intelligence waste through the network, using prompts and effective handling at each stage of the waste management process. By implementing advanced technologies and strategic infrastructure development, the enterprise attempts to set new standards for waste management excellence in the region (Fujii and Managi 2019).

Integrating AI into green process technology offers many advantages, like proper decision-making, optimized resource sources, and increased operational efficiency. By leveraging AI algorithms, industries can extract actionable insights from vast datasets, anticipate future trends, and optimize processes with unprecedented precision. Moreover, AI-powered predictive analytics can identify inefficiencies in industrial processes, predict possible risks, and continuously improve industrial systems. However, it has a few challenges along with its advantages. Some of them are data privacy breaches, algorithm bias, and technical complexities that must be addressed. Nevertheless, powerful governance frameworks, collaboration among stakeholders, and initial investment in research and development are established. In that case, these challenges can be mitigated, unlocking the full potential of AI and eventually aiding in conserving the goals (Mao 2019).

5 Future Prospects

Looking into the future, integrating AI into green process technologies will significantly impact and transform various industrial processes. Moreover, advancements in explainable AI and ethical AI frameworks will introduce transparency, accountability, and trust in AI-powered systems, allowing widespread adoption and acceptance. Collaborations between academia, industry, and policymakers will play a significant role in innovations and driving policies that utilize AIHIF. By fostering these processes and promoting knowledge exchange, stakeholders can take advantage of the transformative potential of AI to accelerate progress toward fulfilling sustainable development goals (Kan 2021; Mao 2019).

5.1 Market Demand and Competitive Advantage

With recent advancements, integrating sustainability principles into green process technologies has become indispensable for organizations aiming to thrive in competitive markets (Ruiz-Mercado et al. 2016). It has become essential to align business strategies with sustainability objectives to capitalize on emerging market opportunities, enhance various operations, and meet customer demands. With the increasing focus on concerns and changing norms, consumers prioritize eco-friendly products and services more than traditional ones (Khan et al. 2021; Bhatia 2021). This shift has increased demand for long-term solutions across various sectors and industries. Market trends have been hugely influenced by regulatory requirements, government incentives, and industry standards promoting sustainability to reduce environmental impact. Organizations that meet this growing demand for green technologies effectively stand out to gain a competitive edge and can evenly capitalize on new business opportunities (Horbach et al. 2012). There are numerous competitive advantages, some of them being cost savings by efficiently utilizing the resources, a better brand reputation, and access to new markets on a global scale (Soni 2015). Companies can differentiate themselves from competitors by accepting sustainability practices and attracting environmentally conscious consumers, which helps to build stronger connections with different stakeholders.

5.2 Business-Related Sustainability Development

Business-related sustainability practices involve the principles strategically implemented into core practices like product design, manufacturing procedures, supply chain management, and stakeholder engagement (Khan et al. 2021). Adopting circular economy models and eco-friendly design principles are key strategies to achieve sustainability (Chen et al. 2020). Along with all these proper leadership commitments and employee empowerment, they are also responsible for establishing green and sustainable practices in the business sector. The vital role of investing in green technologies is environmental protection alongside stimulating economic growth, facilitating employment, and inducing innovation.

Organizations need help adopting green process technologies, including high upfront costs, technological limitations, and specific regulatory uncertainties—however, these challenges also present opportunities for innovation, collaboration, and partnership (Bhardwaj et al. 2015).

In the case of technological challenges, it has also been observed that positive IT managerial resources and implementation success have certain competitive advantages, as revealed in studies by Byrd and Turner (2001) and (Statistics 2017). Studies by Dehning and Stratopoulos (2003) also show that IT helps build strategic operations during technology implementation in business sectors, which eventually plays a positive role in sustainable competitive advantage. Therefore, both of them have essential roles in achieving sustainable competitive advantage. Better IT implementation and managerial skills, as a whole, social influence, increase the sustainable competitive advantage (Porter 1985; Chin 1998; McWilliams and Siegel 2011; Smith 2007).

Finally, collective action from businesses, governments, and civil society is essential for prioritizing sustainability by investing in green process technologies and thus conducting a genuine environmental administration.

5.3 Long-Term Economic Benefit

Green technology is an umbrella term that brings about the concept of energy efficiency, which can focus on reducing unnecessary energy expenditure and using eco-friendly modes of energy generation and usage. One pioneer aspect of Green Technology is its initial cost, which demands considerable capital to invest in infrastructure. Using solar heaters to replace nonrenewable energy sources for energy generation will require capital to be functional (Iravani et al. 2017).

Keeping this in mind, we must consider the long-term benefits associated with these methods. The primary issue that has already affected the populated sectors of the world is a worldwide energy shortage and increasing pollution that harms the ecosystem. The fauna and flora of a region have changed due to the adverse effects of pollution, which negatively affects sustainability (Mikhno et al. 2021). Various agricultural and artificial practices have increased the accumulation of xenobiotics in soil. Using back-dated technologies for bioremediation is not a solution as it will only slow down the process, giving undesirable results (Iravani et al. 2017).

Hence, the new importance of new methods is necessary. We need to keep in mind that the cost of the implementation of new energy needs to be affordable. Ease of use and affordability are among the primary aspects of green technology. This states that the cost of production of equipment should be less so that the selling price is lower, too. Developed countries spend a lump sum of their budget to exercise green technologies in various sectors.

5.4 Increased Productivity

Integrating green technology has become a beacon of positive change in the ever-evolving landscape of industry and technology. Beyond only helping the environment, adopting eco-friendly methods has sparked a new wave of higher production. This comprehensive movement is changing how we work, cooperate, and live—it is not just about using cleaner energy or leaving less carbon impact (Fuchs 2008).

5.5 Green Technology and Humans

At its core, implementing green technology is not just about machines and algorithms but about people. It is about realizing that every innovation results from a group working together to improve the world. This human-centric approach is evident in how green technology reshapes industries and empowers workers. One of the benefits is the availability of new job opportunities (Söderholm 2020). With time, industries would seek to hire professionals in renewable energy and energy management. This would address job issues and give people a sense of purpose in making the world greener.

The dynamic nature of the field requires continuous adaptation and skill development. This commitment to lifelong learning fosters a culture of personal and professional development. Employees are dynamic individuals thriving alongside the technology they use. Green technologies also encourage multi-dimensional collaboration (Zongyu et al. 2021). The ethos is collaborative, from brainstorming innovative solutions to implementing sustainable practices. This collective attitude fosters a sense of shared responsibility and improves creativity and problem-solving skills.

In our connected world, technology plays a vital role in fostering collaboration. Digital platforms and communication tools break down geographic barriers, enabling seamless communication. Interacting with like-minded people from all backgrounds broadens perspectives and creates a sense of community. It is an integrated device that works in unison; individuals work together to solve common problems and build a better future (Zongyu et al. 2021).

5.6 Challenges and Opportunities for Growth

Alongside various advantages, there are certain obstacles to its widespread use. Transitioning to sustainable practices may require significant initial expenditures, and not all businesses can execute these changes. Nonetheless, this difficulty offers creativity and cooperation opportunities (Awan et al. 2019). Communities, companies, and governments may collaborate to develop laws and other initiatives that encourage the use of green technology. Regulations, tax breaks, and financial assistance may all be reasonably practical in motivating companies to make the required adjustments. Facilitating cooperation between public and commercial entities can enable a more seamless and extensive shift toward sustainable practices.

5.7 Loopholes in Green Process Technology

A sustainable future is becoming increasingly likely due to the development of green technologies in the era of climate change and environmental consciousness (Ilhan 2021). Governments, businesses, and citizens are all investing in technology to reduce their environmental impact and carbon footprints. Although these projects have proper intentions, a closer study reveals that, even in green technology, there are areas for improvement that might jeopardize the objectives they seek (Ilhan 2021).

6 Greenwashing: A Mirage of Sustainability

One of the most pervasive loopholes in green technology is the phenomenon of "greenwashing" (de Freitas Netto et al. 2020). Companies often use this practice, presenting themselves as environmentally friendly without substantial proof to support their claims. This can lead to a false sense of security among consumers who believe they support eco-friendly products or services. The reality is that some companies manipulate marketing strategies to exploit the growing demand for green solutions without making significant strides in sustainability.

Superficial claims on eco-friendliness need to be checked by stringent regulations and certifications. Transparency and accountability are two main pillars to prevent greenwashing and are steps toward nurturing genuine progress in green technology (de Freitas Netto et al. 2020).

7 E-waste Management Challenges

The usage of electronic devices in our daily lives has increased considerably in the past decades, creating a new environmental challenge: electronic waste or e-waste (Needhidasan et al. 2014). Harmful materials may leak into the environment, harming ecosystems when e-waste is disposed of improperly. Although green technology focuses on eco-friendly ways of waste management, an increasing amount of e-waste is of great concern.

Addressing the e-waste loophole requires a comprehensive approach that includes responsible manufacturing practices, efficient recycling programs, and public awareness campaigns (Kumar and Singh 2013). Stringent rules need to be established by Governments and Corporations to help reduce the amount of E-waste generated and take a step toward a greener future.

8 Hidden Environmental Costs

While reducing direct environmental impact, some green technologies may inadvertently contribute to hidden ecological costs. For example, mining and processing rare earth minerals are required to develop numerous renewable energy sources, including solar panels and batteries for electric vehicles. The extraction and use of these minerals can adversely affect the environment, contaminating the soil and damaging the landscapes. Sustainable mining processes are expected to solve the problem of devastated landscapes and contaminated soil. Assessing technology's overall environmental impact, from raw material extraction to disposal, should be done by life cycle assessments (Hauschild 2005).

9 Limited Accessibility and Affordability

Accessibility of green technology, particularly in poor nations, is a significant weakness. Thus, a socio-economic divide is created, where only specific demographics can actively participate in the green revolution (Iravani et al. 2017).

Governments and international organizations can work toward making green technology more accessible through financial incentives. Collaboration between the public and private sectors is essential to bridge the affordability gap and ensure that sustainable solutions are available to a broader spectrum of society. While green technology holds immense promise for a more sustainable future, acknowledging and addressing these loopholes is crucial for its success. Learning about and addressing these gaps can pave the path for future generations to live greener, cleaner, and more sustainable lives (Iravani et al. 2017).

10 Conclusions

Green technology focuses on creating a healthy environment and sustainable development by reducing carbon-intensive processes that negatively harm the ecosystem. These technologies will be a revolutionary strategy for reducing carbon emissions and intensity (Jiao et al. 2020). The adoption of green technology denotes a fundamental transformation in our collective mindset, which is not solely a modification of instruments and procedures. It represents a shift from the long-held belief that resources are limitless (Rashid et al. 2013). Green technology forces us to reconsider our relationship with the environment; the ecosystem must be maintained for future generations, which requires our protection and care. A significant shift in the human perspective is represented here. It originates from an increasing understanding of how we interact with the world and how our activities have far-reaching effects, which is often overlooked. Green technology expresses our obligation to the next generation. It represents our dedication to leaving behind a planet home to biodiversity, sustainable resources, and one that is livable and habitable. One of the undeniable facets of the green technology revolution is its impact on economies worldwide. Green technology fosters job creation, which injects development and vitality into oral and global greenies. The recent shift toward renewable energy and sustainable agriculture has generated a demand for diverse skilled workers. Immediate economic concerns, a cycle of continuous learning, and professional development are addressed in this economic surge. Empowering local communities is one of the most promising aspects of the green revolution. The implementation of renewable energy projects or sustainable practices often occurs at the grassroots level, which creates a cascade of effects that are far beyond immediate concerns in a highly positive way (Sianipar et al. 2013; Ehresman and Okereke 2014).

Green technology's impact extends beyond the boundaries of national borders, creating a web that connects us to the worldwide effects of global consequences. Climate change, deforestation, and pollution recognize no boundaries, emphasizing the urgency of a collective, international response. The human touch in this global narrative is reflected in our shared responsibility toward the planet and each other (Khoday and Perch 2012; Ajmone-Marsan et al. 2012). In a fast-paced world where development and technologies are inseparable from humans, the future needs to be safe and sustainable to procreate life, and we must keep moving toward achieving more and making the Earth a better place to stay. Overall, the benefits of implementing these have been discussed, and we need to catch up to the desired outputs. But this is the beginning of a new journey with a different aim for the betterment of all. Green technology will be the key to a greener, calmer, cleaner, better future (Rose et al. 2017; Leigh 2005; Mintzer and Michel 2001; Aithal and Rao 2016).

References

Aithal PS, Rao P (2016) Green education concepts & strategies in higher education model. Int J Sci Res Mod Educ (IJSRME) ISSN 1(1):793–802

Ajmone-Marsan M et al (2012) The emerging energy web. Eur Phys J Spec Top 214:547–569

Anastas PT, Warner JC (1998) Green chemistry: theory and practice. Oxford University Press, New York; Horvath I, Anastas PT (2007) Chem Rev 107:2167

Anastas PT, Williamson TC (1996) In green chemistry: designing chemistry for the environment. American Chemical Series Books, Washington, DC, pp 1–20

Awan U, Sroufe R, Kraslawski A (2019) Creativity enables sustainable development: supplier engagement as a boundary condition for the positive effect on green innovation. J Clean Prod 226:172–185

Banerjee S et al (2014) Advantages of green technology. Recent Res Sci Technol 6(1):97–100

Bhardwaj M et al (2015) The advantages and disadvantages of green technology. J Basic Appl Eng Res 2(22):1957–1960

Bhatia MS (2021) Green process innovation and operational performance: the role of proactive environment strategy, technological capabilities, and organizational learning. Bus Strateg Environ 30(7):2845–2857

Brundtland GH (1987) For world commission on environment and development, our common future. Oxford University Press, Oxford, UK

Byrd TA, Turner DE (2001) An exploratory analysis of the value of the skills of IT personnel: their relationship to IS infrastructure and competitive advantage. Decis Sci 32:21–54

Chen TL et al (2020) Implementation of green chemistry principles in the circular economy system towards sustainable development goals: challenges and perspectives. Sci Total Environ 716:136998

Chin WW (1998) The partial least squares approach to structural equation modeling. Mod Methods Bus Res 295:295–336

de Freitas Netto SV et al (2020) Concepts and forms of greenwashing: a systematic review. Environ Sci Eur 32:1–12

Dehning B, Stratopoulos T (2003) Determinants of a sustainable competitive advantage due to an IT-enabled strategy. J Strateg Inf Syst 12:7–28

Ehresman TG, Okereke C (2014) Environmental justice and conceptions of the green economy. Int Environ Agreem: Politics Law Econ 15:13–27

Fischhendler I et al (2021) The impact of community split on the acceptance of wind turbines. Sol Energy 220:51–62

Frone DF, Frone S (2015) Resource efficiency objectives and issues for a green economy. Sci Pap Ser Manag Econ Eng Agric Rural Dev 15:133

Fuchs C (2008) The implications of new information and communication technologies for sustainability. Environ Dev Sustain 10:291–309

Fujii H, Managi S (2019) Decomposition analysis of sustainable green technology inventions in China. Technol Forecast Soc Chang 139:10–16

Grajzl P, Murrell P (2021) A machine-learning history of English case-law and legal ideas prior to the industrial revolution II: applications. J Inst Econ 17(2):201–216
Gu F (2019) Exploring "Internet+Recycling": mass balance and life cycle assessment of a waste management system associated with a mobile application. Sci Total Environ 649:172–185
Habibi S, Obonyo EA, Memari AM (2020) Design and development of energy-efficient re-roofing solutions. Renew Energy 151: 1209–1219
Hauschild MZ (2005) Assessing environmental impacts in a life-cycle perspective
He K, Zhang J, Zeng Y (2019) Knowledge domain and emerging trends of agricultural waste management in social science: a scientometric review. Sci Total Environ 670:236–244
Heidari R, Yazdanparast R, Jabbarzadeh A (2019) Sustainable design of a municipal solid waste management system considering waste separators: A REAL-world application. Sustain Cities Soc 47: 101457
Hessel V, Kralisch D, Krtschil U (2008) Sustainability through green processing – novel process windows intensify micro and milli process technologies. Energy Environ Sci 1:467–478
Horbach J, Rammer C, Rennings K (2012) Determinants of eco-innovations by type of environmental impact — the role of regulatory push/pull, technology push, and market pull. Ecol Econ 78:112–122
Ilhan UD (2021) Business strategy for sustainable development in the digital era: green management
Iravani A, Hasan Akbari M, Zohoori M (2017) Advantages and disadvantages of green technology; goals, challenges and strengths. Int J Sci Eng Appl 6(9):ISSN-2319–7560
Jiao J, Chen C, Bai Y (2020) Are green technology vertical spillovers more significant in mitigating carbon intensity? Evidence from Chinese industries. J Clean Prod 257:120354
Kan HY (2021) Environmental planning based on reduce, reuse, recycle and recover using artificial intelligence. Environ Impact Assess Rev 86:106492
Khan SJ et al (2021) Green process innovation: where we are and where we are going. Bus Strateg Environ 30(7):2845–3423
Khoday K, Perch L (2012) Green equity: environmental justice for more inclusive growth, vol 19. The International Policy Centre for Inclusive Growth
Klein C et al (2015) Social equity and the probability of success of biodiversity conservation. Glob Environ Chang 35:299–306
Kumar U, Singh DN (2013) E – waste management through regulations. Int J Eng Invent 3(2):e-ISSN: 2278–7461, p-ISSN: 2319–6491
Leigh P (2005) The ecological crisis, the human condition, and community-based restoration as an instrument for its cure. Ethics Sci Environ Politics 2005:3–15
Liu M, Shi Y, Fang F (2014) Combined cooling, heating and power systems: a survey. Renew Sust Energ Rev 35:1–22
Mao S (2019) Opportunities and challenges of artificial intelligence for green manufacturing in the process industry. Engineering 5(6):995–1002
McWilliams A, Siegel DS (2011) Creating and capturing value: strategic corporate social responsibility, resource-based theory, and sustainable competitive advantage. J Manag 37:1480–1495
Mehmood S et al (2024) The role of green industrial transformation in mitigating carbon emissions: exploring technological innovation channels and environmental regulation. Energy Built Environ 5(3):464–479
Mikhno I, Koval V, Shvets G, Garmatiuk O, Tamošiūnienė R (2021) Green economy in sustainable development and improvement of resource efficiency
Mintzer I, Michel D (2001) Climate change, rights of future generations and intergenerational equity: an in-expert exploration of a dark and cloudy path. Int J Glob Environ Issues 1:203–222
Mondejar ME (2021) Digitalization to achieve sustainable development goals: steps towards a smart green planet. Sci Total Environ 794:148539
Needhidasan S, Samuel M, Chidambaram R (2014) Electronic waste – an emerging threat to the environment of urban India. J Environ Health Sci Eng 12(36)
Pandit MK (2021) Climate change, urbanization and impact on natural environment: the Indian scenario. In: Urban growth and environmental issues in India. Springer Singapore, pp 173–187
Porter ME (1985) Competitive advantage: creating and sustaining superior performance, vol 43. Free Press, New York, p 214
Pradhan RP et al (2017) Is there a link between economic growth and insurance and banking sector activities in the G-20 countries? Rev Financ Econ 33:12–28
Rashid A et al (2013) Resource conservative manufacturing: an essential change in business and technology paradigm for sustainable manufacturing. J Clean Prod 57(15):166–177
Ribbing S, Xydis G (2021) Renewable energy at home: a look into purchasing a wind turbine for home use—the cost of blindly relying on one tool in decision making. Clean Technol 3:299–310
Rose SK et al (2017) The Paris Agreement and next steps in limiting global warming. Clim Change 142:255–270
Ruiz-Mercado GJ, Carvalho A, Cabezas H (2016) Using green chemistry and engineering principles to design, assess, and retrofit chemical processes for sustainability. ACS Sustain Chem Eng 4(11):6208–6221
Sianipar CPM et al (2013) Community empowerment through appropriate technology: sustaining sustainable development. Procedia Environ Sci 17:1007–1016
Singh RL, Singh PK (2016) Global environmental problems. Appl Environ Sci Eng Sustain Future (AESE):13–41
Smith AD (2007) Making the case for the competitive advantage of corporate social responsibility. Bus Strateg Ser 8:186–195
Söderholm P (2020) The green economy transition: the challenges of technological change for sustainability. Sustain Earth 6(3)
Soni GD (2015) Advantages of green technology. Soc Issues Environ Probl 3(9):ISSN-2350–0530(O) ISSN-2394–3629(P)
Sovacool BK et al (2020) Differences in carbon emissions reduction between countries pursuing renewable electricity versus nuclear power. Nat Energy 5:928–935
Sovacool BK, Griffiths S, Kim J, Bazilian M (2021) Climate change and industrial F-gases: a critical and systematic review of developments, sociotechnical systems, and policy options for reducing synthetic greenhouse gas emissions. Renew Sust Energ Rev 141:110759
Statistics (2017) Percentage share of SMEs DGP and Malaysian GDP in 2017. Department of Statistics; West Group, Eagan
United Nations Conference on Environment and Development (UNCED) (1992) Rio de Janeiro, June 3–14
Wu Y, Zhang L (2017) Can the development of electric vehicles reduce the emission of air pollutants and greenhouse gases in developing countries? Transport Res Part D: Transport Environ 51:129–145
Zongyu M, Zheng Y, Sun H (2021) Cooperative green technology innovation of an e-commerce sales channel in a two-stage supply chain. Sustainability 13(13)

Role of Organic Products, Clean Water, and Green Energy in Life Sustainability

Trashi Agrah Singh and Payal Basu

Abstract

Air, water, and food along with healthy and hygienic place to stay are basic requirements to sustain life. Human progression in every field has resulted in polluting all these sources drastically along with depleting them in various ways. Using chemicals and pesticides for farming tops the list to get good quantity yield, which along with soil and eatables pollutes water bodies too. Now to go green, a number of research and works are going on for being eco-friendly. The use of organic products (eatables grown using organic fertilizers), water purified using natural ways, and generation of energy from green renewable sources has become a major priority. These are most important to combat a lot of health issues caused by overuse of chemicals all around. Organic agriculture has reduced energy use of nonrenewable sources by decreasing agrochemical needs, which had played a very important part in reducing soil and water pollution and has helped in enhancing the quality of agricultural products. This chapter deals with all the modern developments that aim at going green in all aspects, and probability of increasing life sustainability.

Keywords

Eco-friendly · Nonrenewable sources · Green · Sustainability · Agrochemicals

T. A. Singh (✉)
Bhagyoday Tirth Institute of Paramedical Sciences, Sagar, Madhya Pradesh, India

P. Basu
Tropical Forest Research Institute, Jabalpur, Madhya Pradesh, India

1 Introduction

In order to sustain a healthy life, food and water are the main basic requirements. Along with these two, other needs include energy resources which we use in our day-to-day life. Mainly we can conclude that these three basics are the foundation needed for a proper sustainable life.

2 Organic Products

The term organic products refers to implementation of techniques that require the least or minimal number of artificial and chemicals that are toxic to nature and mankind. Major part of this contributes to foods that are consumed by us (Tilman et al. 2011). Many products that are being consumed today have been claimed to be sustainable but when tested, they have in many cases proven to be far from all the claims (Rockström et al. 2009). In many countries, agriculture and farming techniques are at stake due to economic resilience (Godfray and Garnett 2014). These challenges can be easily addressed by distinguishing three fundamentally different strategies that include efficiency, consistency, and sufficiency.

The efficiency strategy aims at optimizing the relationship between the negative impacts of generated output (product).

The consistency strategy tries to lead the production strategy closer to natural principles or systems. Finally, the sufficiency strategy focuses on reduction of negative impact on resources that are being produced (Schaltegger et al. 2003).

Organic practices may lead to economic resilience because of their alternative approach toward sustainability of nature (Schader et al. 2012), but debates continue and many research studies and field observations suggest that organic methods decrease the production, impacting Global Hunger Index (Schader et al. 2012).

J. A. Parray et al. (eds.), *Environmental Landscape and Sustainable Biodiversity for Healthy Green Growth*, Sustainable Landscape Planning and Natural Resources Management, https://doi.org/10.1007/978-3-031-75405-0_3

Sustainability of a product can be accessed by concentrating on three levels, namely, operator level, product level, and policy level.

According to Bystrický et al. (2014); the operator level deals with the area of operation with some indicator sets. These determine how functioning of the company or farm may affect surroundings or associated population along with the neighboring area. The indicators are based on SAFA guidelines (FAO 2013), which include all parameters and conditions for good governance and practices related to sustainability. Also, the guidelines may vary with the product or industry dealing with different products. For instance, guidelines for a dairy industry would be different as compared to organic farming products or other herbal organic consumables.

The product level assessments, as described by Finkbeiner et al. (2010), deal with assessing effects of product and its production on environment using policies by LCA (Life Cycle Assessment) like eutrophication, acidification, global warming potential, etc. Also, sustainability of the product can be accessed by checking fact that how much it is able to fulfil the energy needs along with influence of socioeconomic factors on local areas of production.

3 Clean Water

Water is most basic necessity of life, because a person can survive for some time without food but can't without water. Water constitutes around 75% of human weight. According to Water, Sanitation & Hygiene-Action Against Hunger (2016), the most alarming fact is that clean water for consumption is not available to around 2 billion people around the globe. Water has a number of physiological and metabolic functions in human body so it plays a vital role in human health. Normally 20–50 litres of clean water is needed for survival by a human being. Cholera, typhoid, Urinary Tract Infections, etc. are major diseases and infections that are caused by consumption of contaminated water. According to article entitled "Safe Drinking Water is Essential—Why is Safe Water Essential" (2016), diarrhoeal diseases from cholera are responsible for around 1.8 million deaths all-round the globe. This fact highlights the importance of clean water for consumption in daily life. So access to clean water has been labeled as human right by the United Nations.

One of the factors that affect availability and access to clean water is population and demographic location. The countries with large population have more issues as some of the population has easy access to clean and potable water but some part of population fails to obtain it. Also, demographic location also hinders access to potable water as many parts of the globe face shortage of water. Populations in these areas have to struggle to get potable and clean water. Also, if a river is source of water and it flows through different countries, then it has maximum chances of getting contaminated. For example, in Matlab, Bangladesh, people with low socio-economic status are more prone to cholera outbreaks (Root et al. 2013).

According to World Vision New Zealand Water Topic Sheet (2017), access to clean water depends upon its ease of availability to humans and the source through which it is being provided. For instance, water from creeks and stagnant water bodies can easily get contaminated by animals and birds along with other dumping. In some circumstances, nature also plays a part in contaminating water bodies. Availability of clean water also depends on socio-economic factors like rich persons have more ease to access to clean water as compared to poor ones. Other conditions include geopolitical boundaries which limit access to water by borders of their countries or areas. Contaminated water not only affects health but also leads to loss of productivity (adults are unable to work properly due to poor health conditions), reduced production of food, migration of people in search of clean and easy access to water, and last but not the least, conflicts related to accessibility to water.

World Vision New Zealand Water Topic Sheet (2017) states that to overcome these aspects, United Nations General Assembly has set Sustainable Development Goals for every country round the globe by ending all forms of poverty, fight inequality and injustice and protect environment by 2030. To achieve the above goals, six aims are targeted to focus on. These are:

1. Making water safe to drink
2. Reducing water pollution
3. Increasing water use efficiency and reducing water scarcity
4. Supporting developing countries to improve water and sanitation
5. Integrating international water resource management
6. Helping local communities get involved

According to Goal 6, Clean Water and Sanitation, around 1.8 billion people globally use a source of drinking water that is contaminated with fecal matter. Some 2.4 billion people lack access to basic sanitation services, such as toilets or latrines. Water- and sanitation-related diseases remain among the major causes of death in children under five; more than 800 children die every day from diarrhoeal diseases linked to poor hygiene. By managing our water sustainably, we are also able to better manage our production of food and energy and contribute to decent work and economic growth. Worldwide, more than 2 million people die every year from diarrhoeal diseases. Poor hygiene and unsafe water are responsible for nearly 90% of these deaths and mostly affect children. The economic impact of not investing in water and sanitation costs 4.3% of Sub-Saharan African GDP. The World Bank estimates that 6.4% of India's GDP is lost due to adverse economic impacts and costs of inadequate sanitation.

So access to clean and potable water is one of the main and necessary aspects of human survival.

4 Green Energy

The term green has become a synonym for eco-friendly and healthy production of any product. It has caught attention of everyone in the past decade as most available sources of energy are noneverlasting. Also, they had a negative impact on the environment. All the sources of energy generating lower environmental impact as compared to other conventional sources are considered as green energy sources (Bhowmik et al. 2017). Energy in any form is basic need of life. Without energy, sustaining life will not be possible today. The sources that generate those energy had negative impacts on environment mainly pollution, global warming, etc. So, obtaining energy from greenways or sources (Green energy) is quite a new concept which is being worked on for quite long.

Energy is a basic requirement without which imagining life is not even possible. So when we look forward for greener alternatives, major point of concern remains on the fact that novel sources should be efficient enough as compared to the previous source. Secondly, emissions produced by them should be either negligible or less as compared to available conventional sources. Thirdly, it should be affordable to the people who have utility of that particular source.

4.1 Why Green Energy?

The first question that strikes the mind is why there is a need to shift to green sources from conventional sources? What is the main reason of this transition? Transition toward green energy from natural available sources is known as "premeditated product." This became essential because exploitation of natural sources in abundance leads to their depletion as they are mostly nonrenewable sources of energy and are also not eco-friendly. This induces serious environmental problems and imposes negative effects on human health. This leads to increase in the concentration of Greenhouse gases in the air. They also leave behind lot of residual wastes which play as an add on in polluting environment. This increase in concentration of greenhouse gases leads to global warming which had played an important role in increasing temperature, sea level, and exhaustion of many species around the globe. According to Bernstein et al. (2007) and Lau et al. (2010), by 2100, the existence of half of the world's population would be in danger because of increasing global temperature on an average of 6.4 °C from 1.10 °C and rising the sea level by about 16.5 to 53.8 cm, respectively. This will mainly affect populations living in or nearby coastal areas.

Green energy has potential of reducing level of emissions along with minimizing undesirable effects on environment. Green energy has an upper hand among available sources of energy. Also, it is considered to be competitive enough as compared to natural sources. With the rising concerns about climate change and financial crisis around the globe, green energy is being considered as potential for lowering the above negative aspects on society and environment (Bekhet and Harun 2017; Dogan and Seker 2016; Mundaca et al. 2016).

4.2 Barriers in Development

Using green energy as perspective for growth and development for environmental sustainability faces lot of barriers. It is not a highway track that can reach target smoothly in a single go. Before implementation, it faced many barriers which include economic barriers, lack of information and awareness, socio-cultural issues, and market barriers.

Economic barriers include analysis of social costs and benefits. This can be understood as every industry works on principle of the lowest investment and long-term high output. Major issue being faced by green technology is this source is still very much underrated. This has led to minimal number of investors investing in it. Also, raw goods required for production bear high cost so the output is not cost-effective. Since investment cost is high and future in the field is still not clear, it is still not certain how it will affect the economy. So financial risks are also high. These are first-step barriers or economic barriers that are being faced.

According to Ucar and Balo 2009; Akdag and Güler 2010 second barrier is of lack of awareness and information. In simple words, it is lack of available presentable data about a particular source. For instance, wind and solar energy are considered as most favorable renewable sources of energy. But how much of energy they can generate and what percentage of it can be utilised is most important question. Main demerit of wind energy is that amount of energy generated depends on wind speed which is not constant. So, amount of energy generated remains in direct proportion of the speed of wind. Similarly, intensity and availability of sunlight are proportional to amount of energy generated. Again, here investment cost is high and output results obtained are not directly available to the public which may be helpful to make them aware about the particular source.

Socio-cultural barriers are conditions where one particular source can be utilized for generation of other sources of energy, but people of the particular area may raise conflicts about the same. Also, it may have beneficial effects for one set of population but not be that fruitful to others. Like an agricultural land can be used for farming or can be used for making hydel power plants or can be used to install windmills or solar panels but not everyone would be convinced for such changes. The other aspect which may arise is it might open doors for one source of energy but may deplete a source of food or dairy production (Sovacool 2009; West et al. 2010).

Last but not the least the market implementation of energy source or product is another important factor. Since this is rising new field, there is relatively very less competition among companies. If there is any monopoly for a particular product, then valuation and pricing may increase which can have adverse effect on market cap of the particular green product.

These barriers need to be crossed for successful implementation of green energy all around the world.

5 Conclusion

To sustain a healthy life, the basic requirements need to be available at affordable cost and ease. They include healthy food, water, and energy sources. To increase productivity and quantity of products, and to prevent them from contaminants and various parasites, lots of agrochemicals, chemical fertilizers, etc. are used. These have adverse effects on human health and on soil, water, and air as well. So organic farming and production of organic goods have become important. If chemicals are not used in farming, one source of water pollution automatically decreases which is the second major issue. Potable fresh water is not available to everyone with ease which is a necessity. Economically weaker sections are the most affected individuals as they are ones who struggle most to obtain clean water for consumption. They suffer from most of the waterborne infections caused by consumption of contaminated and infected water. Apart from food and water the energy sources, they use are also responsible for increase in global warming. Green energy generated from nongreen sources cannot be considered green. There are lots of sectors that need to be checked for full implementation of green energy as a source worldwide.

6 Future Prospects

The source from which energy is being generated and amount of energy generated are interrelated. If the source is green, only then energy generated can be considred as green. For instance today, electric vehicles are being considered as eco-friendly as they do not require fossil fuel as a source of energy and also they do not emit harmful gases in output. But recent studies indicate that these produce other emissions like burnt rubber particles and road particles in air as they are much heavier as compared to conventional vehicles because of heavy batteries that are being installed in them as source of energy. Also, how those batteries generate the power is other factor and how these batteries would be disposed after they completely run out? What effects will it have on the environment? These are the questions which are yet to be answered.

Organic farming and making clean water available are the task that needs to be completed earliest as even today not each and everyone is getting clean water and healthy food for consumption. Production of organic products is little costly as it requires more land and resources but it is eco-friendly. Many research data reveal that organic products cannot fulfil the required needs, but it is not impossible to make ends meet in this case as many research studies have developed many tools and techniques where production can be increased in the less available land area without using harmful chemicals and fertilizers.

Future seems bright with these new technologies and it may take time but sooner or later they will make life more easy and more sustainable for mankind.

References

Akdag SA, Güler Ö (2010) Evaluation of wind energy investment interest and electricity generation cost analysis for Turkey. Appl Energy, Elsevier 87(8):2574–2580

Bekhet HA, Harun NH (2017) Elasticity and Causality among electricity generation from renewable energy and its determinants in Malaysia. Int J Energy Econ Policy 7(2):202–216

Bernstein L, Bosch P, Canziani O, Chen Z, Christ R, Davidson O, Yohe G (2007) Climate change 2007: summary for policymakers. Hemisphere 57(8):267–269

Bhowmik C, Bhowmik S, Ray A, Pandey KM (2017) optimal green energy planning for sustainable development: a review. Renew Sust Energ Rev 71:796–813

Bystrický P, Machová E, Čížová A (2014) Effect of carboxymethylation on antioxidant properties and radical degradation of mannans and glucans. Carbohydr Polym 112:603–607. ISSN 0144-8617

Dogan E, Seker F (2016) Determinants of CO2 Emissions in the European Union: the role of renewable and non-renewable energy. Renew Energy 94:429–439

FAO (2013) Sustainability Assessment of Food and Agriculture systems (SAFA), Rome. Available at http://www.fao.org/nr/sustainability/sustainability-assessments-safa/en/

Finkbeiner M, Schau E, Lehmann A, Traverso M (2010) Towards life cycle sustainability assessment. Sustainability 2(10):3309–3322

Godfray HCJ, Garnett T (2014) Food security and sustainable intensification. Philos Trans R Soc B Biol Sci 369:20120273

Lau KM, Kim KM, Sud YC, Walker GL (2010) Response of the water cycle of West Africa and Atlantic to Radiative Forcing by Saharan Dust. In: AGU fall meeting abstracts

Mundaca L, Neij L, Markandya A, Hennicke P, Jinyue Y (2016) Towards a green energy economy? Assessing POLICY CHOICES, STRATEGIES AND TRANSITIONAL PATHWAYS. Appl Energy 179:1283–1292

Rockström J, Steffen W, Noone K, Persson Å, Chapin FS, Lambin EF, Lenton TM, Scheffer M, Folke C, Schellnhuber HJ, Nykvist B, de Wit CA, Hughes T, van der Leeuw S, Rodhe H, Sörlin S, Snyder PK, Costanza R, Svedin U, Falkenmark M, Karlberg L, Corell RW, Fabry VJ, Hansen J, Walker B, Liverman D, Richardson K, Crutzen P, Foley JA (2009) A safe operating space for humanity. Nature 461:472–475

Root E, Rodd J, Yunus M, Emch M (2013) The role of socioeconomic status in longitudinal trends of Cholera in Matlab, Bangladesh. PLoS Negl Trop Dis 7(1):1993–2007

Schader C, Stolze M, Gattinger A (2012) Environmental performance of organic farming. In: Boye JI, Arcand Y (eds) Green technologies in food production and processing, pp 183–210

Schaltegger S, Burritt R, Petersen H (2003) An introduction to corporate environmental management: striving for sustainability. Greenleaf Publishing

Sovacool BK (2009) The cultural barriers to renewable energy and energy efficiency in the United States. Technol Soc 31(4):365–373

Tilman D, Balzer C, Hill J, Befort BL (2011) Global food demand and the sustainable intensification of agriculture. Proc Natl Acad Sci 108:20260–20264

Ucar A, Balo F (2009) Evaluation of wind energy potential and electricity generation at six locations in Turkey. Appl Energy 86(10):1864–1872

West J, Bailey I, Winter M (2010) Renewable energy policy and public perceptions of renewable energy: a cultural theory approach. Energy Policy 38(10):5739–5748

A Roadmap to Environmental Sustainability and Global Partnership

Chamin Chimyang, Isha Pokhrel, Sentisuba,
N. Anthony Baite, and Dawa Dolma Bhutia

Abstract

Economic Growth and development are the top priorities for any nation, and they often tend to dominate and pressure the global environmental crisis. The world's booming population and the fast-paced changing technology need to be fueled with a large capacity of energy, 81% of which is realized by fossil fuels. The study will discuss the various goals for achieving environmental sustainability, a combined effort of the public, the state, and nations, and it will involve relationships between multiple countries. Apart from that, individual behavior towards our Environment plays a vital role in achieving a healthy society will also be elaborated. The goal of a healthy environment is long and requires calculated steps of every nation since the climate is typical to all. The chapter will also document the role of media and other communication sectors that must be strengthened to make ordinary citizens aware of climate change, increasing global temperature, and its consequences and possible solutions.

Keywords

Economic growth · Environment · Fossil fuels · Global partnership · sustainability

C. Chimyang
Department of Genetics and Plant Breeding, Institute of Agricultural Sciences, Banaras Hindu University, Varanasi, India

I. Pokhrel · D. D. Bhutia (✉)
Department of Mycology and Plant Pathology, Institute of Agricultural Sciences, Banaras Hindu University, Varanasi, India

Sentisuba
Department of Genetics and Plant Breeding, School of Agricultural Sciences, Nagaland University, Medziphema, India

N. A. Baite
Department of Agronomy, Institute of Agricultural Sciences, Banaras Hindu University, Varanasi, India

1 Introduction

Economic growth and development are the top priorities for any nation, and they often tend to dominate and pressure the global environmental crisis. The world's booming population and the fast-paced changing technology need to be fueled with a large capacity of energy, 81% of which is realized by fossil fuels (Xu et al. 2010). Furthermore, the combustion of fossil fuels results in the emission of high amounts of greenhouse gases (GHGs) (Ideris et al. 2021), which mainly consist of water vapor (H_2O), carbon dioxide (CO_2), methane (CH_4), nitrous oxide (N_2O), and ozone (O_3). Though these GHGs have a minor concentration in the atmosphere, they play a vital role in trapping solar radiation and keeping the planet warm (Antolini et al. 2019; Mokhta et al. 2020). Otherwise, the surface temperature would have been 15 °C colder than the average of 14 °C (Tuckett 2019). However, a rise in CO_2 emissions due to human activities is leading to a subsequent increase in the temperature, which we know today as global warming, which is part of a more significant phenomenon of climate change.

Earlier environmental activists associated climate change with the depletion of the ozone layer, which was first discovered in the Antarctic region in 1987 (Solomon 1988; Newman et al. 2006; Abbasi and Abbasi 2017; Kuttippurath and Nair 2017). However, a heat wave and a droughty summer witnessed by North America in 1998 changed the climate change perspective (Trenberth et al. 1988; Trenberth and Branstator 1992; Smoyer-Tomic et al. 2003). In the same year, the United Nations Environment Programme (UNEP) and the World Meteorological Organization (WMO) formed the Intergovernmental Panel on Climate Change (IPCC) to take scientific measures to tackle this pressing issue. However, a question remains of how to limit the degrading Environment without hampering a nation's economy.

Many environmental scientists believe that the best approach is to handle it with sustenance in which neither further Environment is degraded in the development process;

J. A. Parray et al. (eds.), *Environmental Landscape and Sustainable Biodiversity for Healthy Green Growth*, Sustainable Landscape Planning and Natural Resources Management, https://doi.org/10.1007/978-3-031-75405-0_4

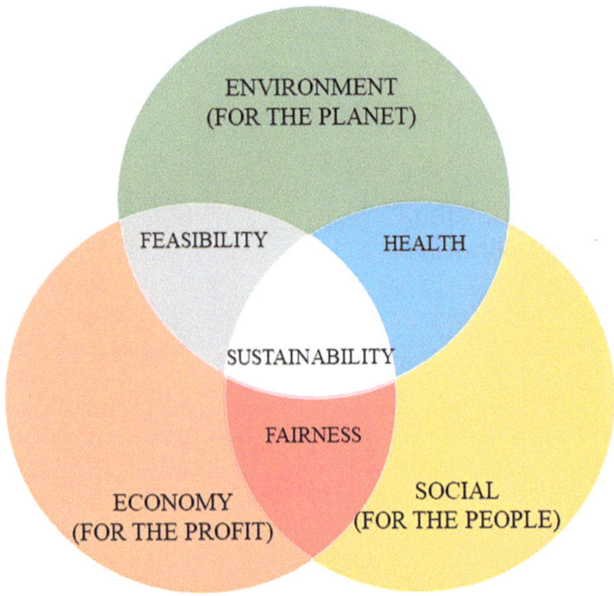

Fig. 1 A typical diagram of sustainability showcasing its three pillars, which are Economy, Environment, and Social Sustainability

this strategy is known as environment sustainability (ES). The latter is one of the pillars of sustainability, of which the other two are economic and social sustainability and are typically represented by three overlapping circles (Purvis et al. 2019; Boussemart et al. 2020) and is illustrated in Fig. 1. A discussion of ES will always comprise key role players such as the corporations, the government, and the nongovernmental organizations (NGOs) at domestic levels and their contribution to lower carbon emission, limiting the dependency on natural resources and further conversion of energy sources from nonrenewable to renewable source and maintenance in resource consumption (Shrivastava and Guimarães-Costa 2017). However, the twenty-first century is a jet age, which means the world has transformed into a global village (Eluwole et al. 2020). The attempt to curb CO_2 emissions and safeguard our Environment has evolved to a transnational level from a domestic level.

Conference of Parties (COP) is a transnational-level product where nations have joined hands with the United Nations at the frontline. Egypt hosted the most recent COP27 in November 2022, where the collaborators concluded upon a low-carbon economy with a required investment of approximately $6 trillion annually and additionally to follow uptight actions to avoid exceeding the global warming temperature of 1.5 °C (Akram et al. 2023). Another issue nexus to CO_2 emission is the depletion of natural resources, which has received much attention. The heavy reliance on the global production of natural resources contributes to economic Growth. In addition to that, the inhabitants of this planet depend upon natural resources as well, which makes it a significant topic of discussion.

Therefore, it can be concluded that the depletion of natural resources and CO_2 emissions and its further consequences are the two ends of the rope and between which various regulating factors are present to narrow the degradation of the Environment, such as green energy, eco-innovation, environmental policies, and carbon taxation which are of significant prominence and will be discussed further in this chapter. Other international organizations such as G7 countries, though their role is scanty but evolving, which include Canada, France, Germany, Italy, Japan, United Kingdom, and the United States, are also on the frontline of global campaigning to enact carbon neutrality and lead the way for other nations.

2 Understanding Environmental Sustainability

A human population has two points for the flow of energy; one is the source or production site. At the same time, the other is consumption or the sink, but both are connected to the Environment to help the human pool grow and develop. The source, the input, or the production site may be food, water sources, air, and energy, while the output or the sink can be the waste. The Environment is the support system from which the resources are drawn, and the consumed product is dumped back (Fig. 2). The more the human population grows, the more resources will be drawn out of the Environment, resulting in more significant waste that goes back into the Environment.

On the source side, the depletion of natural resources can be seen. Conversely, environmental pollution can be seen irrespective of land mass, water body, or the air we breathe. It has been reported that in developing countries, about 92% of deaths are due to pollution (Landrigan et al. 2018), and both air pollution and noncommunicable diseases are responsible for increasing the rate of health care and consumption about 10% of the Gross Domestic Product (Arora et al. 2018a, b). Plastic, which is a ubiquitous and notorious artificial pollutant, is said to be dumped at 79,000 tons annually in the Pacific alone, and the Great Pacific Garbage Patch is a credible product of human activity that can be seen floating over the ocean and is said to be three times the size of France (Pettit 2018). Additionally, the deposition of heavy metals has been reported in both soil and water; the main reason is mega industrialization and rapid urbanization (Yan et al. 2018).

In addition to that, global air pollution is a menace and is a controversial topic. It has been stated that the GHGs are at their peak in human history, and even if emission is to be stopped immediately, the effect, viz. climate change, might continue for many years after that. It is said that the rate of CO_2 emission is highly correlated with climate

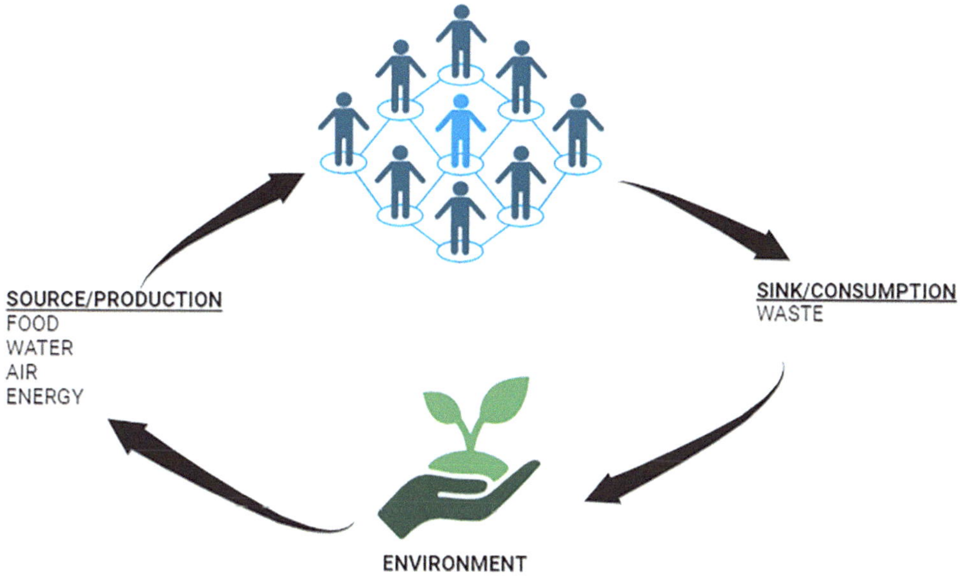

Fig. 2 A typical representation of the flow of energy where the Environment is the support system from where natural resources are drawn

change. From 1750 to 2012, the concentration of GHGs has skyrocketed immensely; CO_2 has increased by 280 ppm to 395 ppm, while CH_4 has increased from 715 ppb to 1882 ppb, and the same for N_2O that has increased from 227 ppb to 323 ppb (Arora et al. 2018a, b; Kumar et al. 2019, 2022). This climate variation is the main factor behind warmer global temperatures, including the polar regions, melting of glaciers, alarming rise in sea levels, bleaching of coral reefs, heavy and erratic precipitation patterns, longer dry spells, and variation in the distribution of flora and fauna and natural disaster (Loehman 2010; Kraaijenbrink et al. 2017). This increased rate of global carbon emission is credited to fossil fuels combusting engines (392 billion metric tons (BMT)) and coal-based power plants (279 BMT) (Gurney et al. 2009), cement producing plants (113 BMT) (Worrell et al. 2001), Transportation (191 BMT) (Lu et al. 2009), Industries (178 BMT), and land use pattern (13 BMT) (Davis et al. 2010) which are all due to the act of human development and Growth. Furthermore, the minor contribution is due to nonhuman activity, which is respiration by plants, animals, and humans (7 BMT) (Cai et al. 2018), air exchange in oceans (7 BMT) (Ramanathan 1981), soil-inhabiting microbes (1.54 BMT) (Oertel et al. 2016), and volcanic eruption (0.15 BMT) (Ritchie et al. 2020). Sector-wise, global contributors to CO_2 emission are presented in Fig. 3. However, Bhutan is the only country in the world with harmful carbon emissions, and the reason for this credible role is due to high forest coverage (The Climate Council 2017).

Land degradation resulting in agricultural constraints by reducing soil fertility and decreasing productivity is another by-product of climate change. It has directly impacted the lives of about 1.5 billion people due to the loss of 15 billion tons of fertile soil (Hamdy and Aly 2014), with an economic loss of about 300 billion USD (Hossain et al. 2020). Variations in climate have also led to frequent drought and flood cycles with longer spells of water scarcity, and land degradation is interconnected to drought. In a report by FAO, it has been stated that with every temperature rise, grain is declining by 5%. (FAO 2016).

Climate change has not only influenced environmental degradation but also has led to habitat loss and biodiversity loss. The biological heterogeneity of both territorial and aquatic flora and fauna has declined. Even though extinction is a natural process, anthropogenic activity has accelerated the process. Overexploitation, environmental pollution, invasive diseases and alien species, and fishing and hunting are other explanations for biodiversity loss (Beumer and Martens 2016; Keziah and Devi 2017; Arora 2018).

Human-induced biodiversity degradation has also been seen in micro-level disbalancing of the overall ecosystem by excessive usage of nitrogen-based fertilizers, making the soil acidic and further losing soil organic matter. The best example cited is a decreased rate of legume nodulation due to pollution and salinity (Arora 2018)

Additionally, the capacity of the Environment to regenerate is slower than the conversion rate of input to output due to the fast-growing population, which further implies that with time, the quality of input or resources drawn has degraded. This resulted in a massive gap between the poor and the rich, creating social disparity. Still, it should always be remembered that the Environment is an unbiased entity for every living being. Every living person has the right to quality resources, current and down the generation, which can only be achieved by playing the game of sustainability

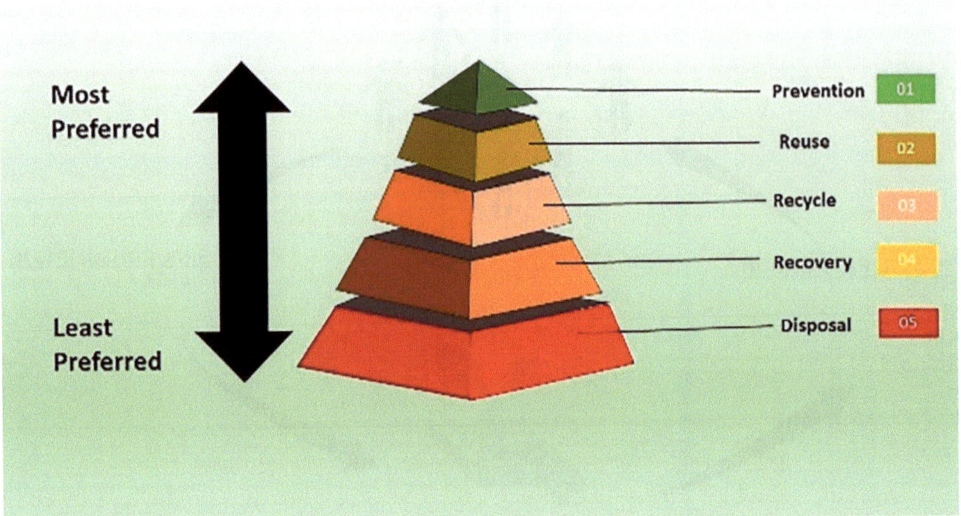

Fig. 3 Hierarchy system in EU for waste management

while the actual cost of rehabilitation of our Environment is yet to be worked out and calculated.

Every nation has its strategies in terms of combating the global environmental crisis. Therefore, the steps a nation takes may vary from one country to another to achieve its goal. Environmental sustainability is not the final goal, but maintaining it is. This chapter primarily discusses strategies available to attain the former for safeguarding our Environment and future generations.

3 Green Technology and Renewable Energy

New technology and innovation are the essential appliances of this century. These tools have offered many beneficial results to humankind. Technological advancements and industrial revolutions have also led to economic and environmental developments, including higher living standards, rising economies, and increased use of fossil fuels. On the one hand, this advancement is improving countries' economies and finances; on the other hand, the same technology threatens environmental sustainability. Economic expansion leads to increased energy demand. More energy demand means the overuse of natural resources. Amplified utilization of renewable and nonrenewable sources causes ecological *deterioration* due to carbon dioxide emissions.

CO_2 emissions affect the Earth in various ways, including sea-level oscillations, global warming, pollution, unpredictable weather, and changes in the ecosystem. Environmental degradation has sparked concern among researchers, scientists, and decision-makers. Substantial actions must be implemented to mitigate the negative impact on the Environment.

Cheng et al. (2019) signify the vital role of green energy from renewable sources such as wind, biofuel, sun, tides, waves, and geothermal heat in reducing environmental degradation. Kirikkaleli and Sowah also recognized the benefits of green energy for economic and ecological Growth (Kirikkaleli and Sowah 2020). Among many other technologies, innovative green technologies can help battle ecological degradation. Green technology is becoming more popular every day. Green technology, one of the eco-friendly or clean technologies, is developing and applying products, services, and processes that aim to conserve natural resources, reduce environmental impact, and promote sustainability. Green technology (GT) helps to reduce environmental challenges like pollution, climate change, and ecological imbalance (Bilal et al. 2021). GT covers various topics such as renewable energy, energy saving, sustainable agriculture and Transportation, waste and water management, etc. Green technology encompasses solar panels, electric vehicles, energy-efficient lighting, green roofs, wind turbines, and water conservation devices (TECAM 2023).

3.1 Innovations in Renewable-Based Green Technology

With every passing day, the field of green technology introduces new products and services that are environmentally safe. Innovation is essential since we must rely on something other than traditional methods in this period. Some of this field's exciting and innovative outcomes include eco-vehicle, waste recycling, energy reduction, wind power, sustainable smartphones, vertical farming, and many more.

3.1.1 Solar Energy

Solar energy contributes significantly to a sustainable environment in a variety of ways. Solar power is renewable and always available as long as the sun shines. Unlike fossil fuels, which are finite and pollute the Environment when harvested and used, solar energy is abundant and sustainable. Solar power generation emits far fewer greenhouse gases than fossil fuels. Solar energy may lessen our reliance on carbon-intensive sources such as coal, oil, and natural gas, limiting climate change and its negative consequences.

Solar desalting or desalination is one of the emerging solar energy-based green technologies. Using solar thermal energy to produce clean water from seawater or brackish water makes it suitable for human consumption, agriculture, or industrial use. This approach provides a long-term solution to water scarcity issues in areas with plenty of sunlight but limited access to freshwater resources.

3.1.2 GT-Based Sustainable Architecture

Green architecture, often known as sustainable architecture or eco-friendly architecture, is the process of designing buildings and structures to minimize environmental effects and resource consumption throughout their existence. Green design seeks to develop energy-efficient, environmentally safe, and healthy structures for users. The principle of green architecture mainly focuses on protecting resources, maintaining water quality, and eco-friendly design of the site (Ragheb et al. 2016).

3.1.3 Wind Energy

After a decade of unprecedented increase in wind power capacity worldwide, the sustainable use of wind energy resources has become critical. It possesses enormous capacity as a renewable resource. Wind generating capacity has grown significantly during the past 20 years, from 6.1 GW in (1996) to 539.6 GW (2017), including onshore and offshore facilities (Nikitas et al. 2019). With advancements in windmill technology, we can employ the power of the wind to produce clean electricity on a bigger scale.

3.1.4 Electrical Transportation

Transportation is a significant source of environmental pollution, mainly because fossil fuel-powered cars generate carbon dioxide into the atmosphere. However, the transition to green vehicles now gives new rays of a clean environment. Electrical vehicles (EVs) are considered the most viable alternative to internal combustion engine vehicles (Casals et al. 2016). EVs have zero tailpipe emissions since they run on electricity rather than petrol or diesel, ultimately lowering overall CO_2 emissions from Transportation.

3.1.5 Sustainable Mobile Phone

A sustainable mobile phone is designed, manufactured, used, and disposed of to minimize its environmental impact while considering social and economic aspects. Sustainable phones attempt to reduce their carbon impact throughout their entire life cycle, from production to disposal. This involves using renewable energy sources in manufacturing and lowering greenhouse gas emissions from Transportation and distribution. Manufacturers of sustainable phones provide recycling programs to ensure that outdated handsets are appropriately disposed of and recycled at the end of their lifespan (van Der Velden 2016).

3.1.6 Electronic Waste (E-Waste) Recycling

E-waste is one of the fastest-increasing solid waste streams, expanding at 3%–5% annually (Liu et al. 2023). E-waste recycling is recovering valuable materials from discarded electronic devices such as computers, smartphones, televisions, and appliances while assuring the safe disposal of dangerous components. Recycling e-waste uses less energy than mining and processing raw materials. For example, recycling aluminum from electronic equipment requires substantially less power than mining bauxite ore and refining it into aluminum. The cost of recycling metal from e-waste is significantly below that of mining unrefined ore, implying that e-waste recycling is an energy-saving and environmentally friendly strategy (Liu et al. 2023).

3.1.7 Carbon Pickup and Storage

Carbon pickup, also known as carbon capture and storage (CCS), is a crucial green technology that aims to mitigate climate change by lowering carbon dioxide (CO_2) emissions from industrial operations and electricity generation (Anderson and Newell 2004). CCS entails absorbing CO_2 emissions from primary point sources such as power plants, industrial sites, and natural gas processing plants, transferring the captured CO_2 to an appropriate storage location, and safely storing it underground to avoid release into the atmosphere. CCS is essential in reaching carbon neutrality and harmful emissions by collecting CO_2 from the atmosphere and storing it underground. When paired with bioenergy production (BECCS), which uses biomass to collect CO_2 during Growth and stores it through CCS, the outcome is a net removal of CO_2 from the atmosphere, which aids in balancing emissions from other sectors and achieving climate targets (Don et al. 2012).

3.1.8 GT-Based Biofuels

Green technology-based biofuels provide a sustainable and ecologically friendly alternative to traditional fossil fuels, helping to improve energy security, reduce greenhouse gas emissions, and accelerate the transition to a low-carbon economy (Solomon 2010). Bio diesel, cellulosic Ethanol,

Biogas, algae biofuels, and pyrolysis oil are some key examples of biofuels that offer a promising alternative to conventional fossil fuels.

3.1.9 Tidal Energy

Tidal and wave energy-producing devices can be a significant renewable energy source as global energy demand continues climbing (Frid et al. 2012). Tidal and wave energy are renewable energy sources powered by natural phenomena like the moon's gravitational pull and the energy of ocean waves. Unlike fossil fuels, which are finite and contribute to climate change, tidal and wave energy may be harnessed continually, offering a dependable and long-term source of electricity. Tidal and wave energy projects that are appropriately designed and positioned can cause little impact on marine ecosystems, wildlife habitats, and coastal landscapes, hence protecting biodiversity and ecosystem services.

3.2 A Tripartite Importance

Green technology, renewable energy, and sustainability are all interconnected concepts that help to address environmental issues, reduce carbon emissions, and promote a more sustainable future. Sustainability refers to the long-term balance of ecological, social, and economic elements that meet the requirements of current and future generations without jeopardizing ecosystem integrity or depleting natural resources. To achieve sustainability, conservation, efficiency, equity, and resilience principles must be integrated into societal decision-making processes. Sustainable practices include using renewable energy, resource conservation, waste reduction, sustainable agriculture, green urban planning, responsible consumerism, and social equality programs. Communities prioritizing sustainability can reduce climate change, protect Biodiversity, improve quality of life, and promote inclusive economic development. Green technology and renewable energy help to achieve the United Nations' Sustainable Development Goals (SDGs) by addressing interconnected issues like poverty reduction, climate action, clean energy access, environmental conservation, and social fairness. Integrating green technology, renewable energy, and sustainability principles is critical for tackling major ecological concerns, accelerating socioeconomic progress, and guaranteeing a more egalitarian and resilient future for all.

4 Conservation of Biodiversity

Biodiversity indicates the diversity of life on Earth at all levels. It includes all living species, including plants, animals, fungi, and microbes, as well as the ecosystems and ecological processes in which they sustain (American Museum of Natural History). During the 1980s, "biodiversity" was first introduced by a group of naturalists concerned about environmental deterioration (Lévêque and Mounolou 2004). Biodiversity provides multiple ecological services vital for human well-being and all the tiny creatures in the Environment. Clean air and water, pollination of crops, climate regulation, nutrient cycling, and cultural and recreational benefits are some of the essential aspects of this subject. Biodiversity is frequently described on three levels, i.e., genetic diversity (varieties of genes within a species), species diversity (varieties of species), and ecosystem diversity. Each of these diversities has its composition and importance (Pathak and Abido 2014).

Talking about the importance of Biodiversity, the food we eat, and the air we breathe cannot go unnoticed. These basic human needs depend on Biodiversity. Biodiversity is interconnected with every part of the ecosystem. In the world of Biodiversity, every species has the right to exist and survive, even if they have little or no valuable impact on human beings. From scientific research to the country's economic Growth, Biodiversity is harnessed. In addition to its inherent value, Biodiversity provides critical ecosystem services for the sustainability of natural ecological processes (Alho 2008).

Ecosystems regulate the Earth's climate. Forests, wetlands, and oceans absorb and store carbon dioxide, a primary greenhouse gas. Biodiversity is the basis for genetic variation. This genetic diversity is critical to agriculture, health, and industry breeding programs. Wild species frequently have genetic characteristics that might enhance crop resistance to pests, diseases, and changing environmental conditions. Biodiverse ecosystems are more resistant to disruption. They are more resistant to environmental disturbances such as temperature variations and invasive species. If one species is impacted, another can balance its ecological role. Any loss or deterioration of Biodiversity can jeopardize all these qualities and affect human well-being (Biodiversity et al.).

The global loss of Biodiversity has reached an alarming stage. If the measures to recover the loss are not taken up and brought into action, then it might lead to the collapse of the ecosystem. The species once lost cannot be brought back into existence; it will be forever extinct, as the loss of Biodiversity is permanent, i.e., it is irreversible (National Research Council 1992). The pattern of Biodiversity has changed due to unwanted human interference. Overuse of Biodiversity has resulted in the loss of natural habitats for many species. In earlier times, as humans gradually started changing small areas of the Earth's surface, the momentum of local extinctions was hardly noticeable from the natural rate. Currently, the loss of species might have been accelerated at the speed of 1000 to 10,000 times (Wilson 1988). Although the immediate reasons behind biodiversity decline are biological, the fundamental causes involve societal and economic mechanisms that exert influence on a worldwide level. Using Biodiversity and its components is necessary for a country's development, but at the same time, humans need to adopt measures that slightly disturb the species. We need to change our actions judiciously in using Biodiversity and halt the speeding of environmental deterioration and species loss.

4.1 Conservation Strategies

The approaches, methods, and regulations regarding biodiversity conservation have evolved over the years and experienced different success levels. Conventional methods of biodiversity conservation, like protected areas, environmental rehabilitation, wildlife reserves, and national parks (in-situ and ex-situ), have also transformed. It is first necessary to take initiative from local participation so that people in both rural and urban areas can understand the need for conservation (Rands et al. 2010).

Conservation approaches should first be directed toward species management, such as habitat restoration, eradication of invasive species, captive propagation, and reintegration programs. Between 1994 and 2004, such measures also saved the extinction of 16 vulnerable bird species (Butchart et al. 2006). Since 1992, the worldwide system of protected areas has consistently expanded, with an average annual increase of 2.5% in total area and 1.4% in the number of sites. By 2006, this network encompassed over 24 million square kilometers across approximately 133,000 designated sites (Butchart et al. 2010).

Among various approaches, landscape strategies for mitigating biodiversity decline have gained greater significance over time. In Landscape approaches, conservation efforts beyond designated areas should encompass (Rands et al. 2010).

(i) Establishment of protected zones within areas designated for production forests
(ii) Creation of buffers around aquatic ecosystems
(iii) Strategic planning and positioning of road networks
(iv) Meticulous arrangement of harvesting zones in terms of space and time
(v) Implementation of suitable practices for managing wildfires

It also includes cross-border conservation, sustainable farming, and the creation and restoration of extensive habitats. Habitat management should be holistic, large, and interconnected rather than isolated.

Biodiversity conservation depends on habitat management; therefore, only some principles should be followed to meet the objective (Lindenmayer et al. 2006).

(i) Management of the relationship between habitats, communities, and ecological processes.
(ii) Conservation of aquatic systems through the continual support of geomorphological and hydrological processes.
(iii) Management of landscape and its heterogeneity.
(iv) Applying natural disturbance patterns allows the species to be adjusted to the disturbance under which they have evolved.

For endangered species, preserving them in their natural habitat (in situ conservation) might be preferable, but this alone might not guarantee their survival in all cases. Apart from conventional techniques like seed banking and the establishment of living collections, biotechnology tools like *in-vitro* methods can serve as supplementary backup repositories and offer alternative means for propagating and conserving species. As extinction is on the rise, it is essential to prioritize the vulnerable species and implement strategies, including the *in vitro* methods to preserve those species from extinction (Reed et al. 2011). To safeguard Biodiversity in the forthcoming decades, conservation strategies must consider the significance of sites for species at present and in the future (Zhu et al. 2023).

4.2 Protected Areas and Reforestation

Protected areas are those locations managed through legal and other norms where human intervention and manipulation of resources are limited. Protected areas provide long-term Biodiversity conservation since the exploitation is up to some range. Species and other ecological parameters are protected in their natural habitat. Protected areas, encompassing national parks, wilderness areas, community-managed zones, nature reserves, and similar designations, are fundamental to preserving Biodiversity. Additionally, they play a crucial role in supporting local livelihoods. Although 13% of Earth's land is designated as protected areas, there are still some voids in their representation of global Biodiversity. Therefore, strategies need to expand the international network of protected areas. "Irreplaceability" has been widely employed in identifying prospective new protected areas. The concept of irreplaceability signifies the capacity of sites to obtain the desired objectives or the degree to which opportunities for achieving those objectives diminish if the site is compromised or lost.

Effectively managed protected areas can offer essential ecosystem services, including water purification and retention, erosion mitigation, and the mitigation of flooding and human-induced wildfires. Protected regions act as a protective barrier for human communities against various environmental threats and dangers. Furthermore, they bolster food and health security by preserving crop diversity and economic or subsistence significance species. Protected areas are crucial in strategies based on ecosystems for adapting to climate change. They also reduce its impact by storing and capturing carbon (United et al. on Biodiversity).

Forests are those parts of Biodiversity that have been exploited continuously for human needs. Although forestry helps to recover the country's economic aspect, it should be noted that while over-harnessing these trees, species residing in tree habitats lose their homes, cannot move elsewhere, and consequently face extinction. Deforestation is a significant contributor to species extinction, and reforestation is one of the recognized solutions for combating this decline (Kemppinen et al. 2020). Species do not become extinct immediately soon after their habitat loss as numerous species persist in the remaining forest part. This delay offers a chance for conservation efforts, mainly through reforestation. Introducing new forest areas can bolster population sizes by enlarging habitats and facilitating immigration from source populations. The significant environmental challenges that can be addressed by reforestation are reducing the impact of climate change, protecting and preserving various species and ecosystems, maintaining soil health and preventing degradation caused by erosion, sustainable management of water resources and the ecosystems that support them, reducing pollutants and enhance the quality of the air (Di Sacco et al. 2021). Before initiating reforestation, it is essential to explore methods that protect current forests; it is better to protect the existing forest first. Reforestation is not a one-person task, so it is vital to involve all the stakeholders, including local participation from planning to delivery. The aim of reforestation should meet other objectives like carbon sequestration, ecosystem services, and economic benefits, and the most essential thing should be considered in selecting appropriate sites for reforestation. The earlier nonforested lands should be avoided, and the previous land used for forestation should be reused along with expanding those lands or connecting more lands. The knowledge of ongoing activities of deforestation should also be monitored.

5 A Sustainable Approach to Agriculture

The agricultural sector, the backbone of every nation's economy, already has a constraint on undulating topography, climate, and soil characteristics. To make the scenario worse, the growing population with rapid urbanization has diminished the arable lands. Emerging countries such as China, India, Mexico, and others will likely face much more significant consequences because of population inclination, and urbanization will rapidly accelerate by 2050 (Khan et al. 2021). On top of that, global warming and climate change with intermittent natural disasters have created a bottleneck between demand and supply of food grains.

The sector is both a driving force to global environmental changes and the industry most affected by the said challenges. It is also the largest user of freshwater, with 70% of freshwater being withdrawn annually. However, it is said to be the key to a UN Sustainable Development Goal that might help to eradicate hunger and ensure food for 9–10 billion of mouth by 2050, also to ensure the livelihood of 2.5 billion small landholders who are solely dependent on farming (Rockström et al. 2017).

Current agriculture is only based on the intensification of crop productivity and is less sensitive to the changing Environment. Therefore, a paradigm shift is required that is not lopsided but creates a balance between producing food grain to feed the growing population and negating the pressure on the Environment, otherwise known as sustainable intensification of agriculture (SIA).

(i) *Greenhouse technology (GT)*: One such strategic approach to intensify crop production in a minimal land area is that of a greenhouse. A greenhouse is an artificial structure that employs solar radiation to regulate and create a favorable microclimate for crop plants (Syed and Hachem 2019). This reduces the overall running cost of the greenhouse, and, more importantly, off-season vegetable cultivation can also be done. The interception of solar radiation depends upon the geometry, orientation, and covering material employed, which ranges from glass to low-density polyethylene (LDP) to high-density polyethylene (HDP). Depending upon the material used, heat retention and loss are regulated inside the greenhouse along with uniform light distribution. Other factors that affect the productivity of greenhouses are ventilation and shading systems; the latter is used in hot regions to reduce the intensity of solar radiation and convert it into latent heat (Achour et al. 2021). Apart from these, employment of renewable energy sources has been seen to be a recent feat. The primary source is solar radiation, which meets the electricity load, making it economically feasible and sustainable (Kabeel et al. 2017). A photovoltaic system is used to convert the energy into electricity. In addition, solar thermal collectors are used to collect maximum solar radiation. Another heat-generating source used in greenhouses is geothermal energy. A fluid is circulated that uniformly distributes the heat inside the structure. Geothermal energy can be employed efficiently only if properly architected inside the greenhouse. It is a great way to reduce CO_2 emissions (Yousefi et al. 2018) and decrease the rate of NO_2 and SO_2 emissions (Yousefi et al. 2015), thus considered an Environment-friendly system.

(ii) *Use of plant microbiomes*: Plant microbiomes constituting both endophytic and epiphytic microbes help plants grow and develop by releasing growth factors and enhancing the nutrient cycle. These plant growth promoting (PGP) microbes range from different species and encourage plant growth and development directly or indirectly by releasing growth regulators, solubilization of phosphorus (P), potash (K), and zinc (Zn), fixing atmospheric nitrogen, production of siderophores, ammonia (NH_3), hydrogen cyanide (HCN), and other secondary metabolites which might be antagonist to plant pathogens. PGP microbes can be used as biofertilizers or bioinoculants to replace chemical-based fertilizers, which can be as effective and Environmentally friendly (Yadav 2020).

(iii) *De-novo-domestication to unlock new alleles*: Some researchers also have discussed designing crop plants based on beneficial alleles with desirable traits such as increment of yield, improvement of nutritional quality, input use efficiency, and tolerance to both abiotic and biotic stresses. This can be achieved by de-novo-domestication. During evolution, some crops, such as millet, were semi-domesticated, and they were abandoned as the domestication progressed for significant crops such as rice, wheat, maize, sorghum, and other crops. Such semi-domesticated crops are known as "orphan crops." Though these orphan crops have low productivity coupled with low cultivated area, they significantly harbor essential traits such as high nutritional value and tolerance to abiotic and biotic stress, which might help with the current challenges the agricultural sector has faced in recent decades. Another concept that comes under de-novo-domestication is that of domestication of perennial crops because the latter need not have to be sown annually, have more extended growing seasons with long and sturdy root systems, have increased water use efficiency (WUE), much more tolerant to stresses as compared to annual crops and have higher carbon storage capacity in soil. Generation of such diverse perennial crops may be agriculturally sustainable by outcrossing annual cultivars with their related wild species. Re-domestication of current cultivated crops is another strategy for domesticating existing or ancestral crops as per the desired allele (Tian et al. 2021). For example, re-domestication of current high-oil-content soybean to high-protein-content can be answerable for shortage in livestock feeds (Patil et al. 2017), while re-domestication of maize for its oil-containing oleic acid (Zheng et al. 2008).

(iv) *Unlocking new food sources*: Cultivation of fungi (mushrooms) as a new crop species requires low input and generates limited waste, thus sustainable and, at the same time, can be tailored according to consumer preferences, which can be achieved at domestic and community levels and even has the potential to reach the multi-commercial levels. About 85% of cultivated mushrooms belong to only five genera: *Agaricus*, *Lentinula*, *Pleurotus*, *Auricularia*, and *Flammulina*. These edible mushrooms are a rich source of fiber, selenium (Se), potassium (K), copper (Cu), zinc (Zn), and vitamin B. Mushrooms exposed to sunlight and UV radiation during their growth period and post-harvest have higher concentrations of vitamin D_2 (Ulian et al. 2020). Even inoculation of mycorrhiza can increase the nutritive value of plant edible parts, as mycorrhizal inoculation in strawberries increases the concentration of anthocyanin and phenols. In contrast, in tomatoes, P, N, Cu, and flavoring compounds increased (Torres et al. 2018). China is the most diverse in my-agriculture (Zhang et al. 2014; Fang et al. 2018)

(v) *At the genetic level*, Growth and productivity depend upon the shoot architecture of a crop. For example, teosinte (*Zea mays parviglumis*), an ancestor of maize from Central America, had highly branched shoots. After the domestication of a favorable trait, a single stem with a singular tassel and a lateral ear was formed. A mutation in the single gene encoding *Teosinte-branched 1/Cycloidea/Proliferating* (TCP) transcription factor encoding gene *TEOSINTE BRANCHED1* (TB1) had led to these significant changes (Doebley et al. 1997; Wang et al. 1999; Studer et al. 2011). Likewise, deactivation in GA3 has led to the shortening of internodes in rice and wheat, causing a revolution in the agricultural sector known as the green revolution.

(vi) Similarly, inflorescence structure is an essential trait for yield, determined by flowering time and floral meristem. Such as the genes involved in the flowering time of Arabidopsis *TERMINAL FLOWER1*, its ortholog *RICE CENTRORADIALIS* in rice, and *SELF-PRUNING* in tomatoes maintain shoot inflorescence identity. Thus, knowing the key regulators involved in a plant architecture can help in designing our future crops according to human requirements, which can be achieved by marker-assisted breeding (MAS), significant gene pyramiding, and genome editing to improve the quality and quantity of a crop.

(vii) For example, a maize cultivar was made feasible for high-density planting (HDP) by inserting a gene from a wild *Upright Plant Architecture2* allele into a modern cultivar (Tian et al. 2019). Recent developments in gene editing tools have opened new doors toward increasing plant productivity with tolerance to various stresses, otherwise crop designing. Therefore, it would be advantageous if such editing tools are used in the improvement of orphan crops, which also have been demonstrated in Solanaceae crops (Rodríguez-Leal et al. 2017; Lemmon et al. 2018; Li et al. 2018; Zsögön et al. 2018). Usually, variation in agriculturally important traits is due to single nucleotide polymorphism (SNP) in the noncoding intergenic regions; otherwise, a point mutation in the coding sequence has provided an opportunity for CRISPR-based crop engineering (Li et al. 2020). However, editing a chloroplast genome has been challenging since the organelle has been the center of biological metabolism (Maliga and Bock 2011; Bock 2015; Fuentes et al. 2018). However, the induction of targeted mutation in mitochondria via interbacterial cytidine deaminase toxin specific for dsDNA has recently been achieved, which has further paved the way for the chloroplast genome. Another bottleneck of plant biotechnology in plant tissue culture is the lack of a universal plant transformation protocol, leading to limited transformations in major crop cultivars (Atkins and Voytas 2020). For instance, WUCHEL2 and BABY BOOM in maize have increased transformational rates but are limited to only a few cereals. However, a recent development of de-novo induction of reagents that can edit genes along with the delivery of totipotent genes into the somatic cells of dicots crops such as potatoes, tomatoes, tobacco, and grapes has skipped the restriction in transformation (Maher et al. 2020). Highly efficient virus-based transformations have also been successful in the Solanaceae family (Ellison et al. 2020; Ma et al. 2020). Transgenic crops, skipping the tissue culture operation has also been achieved via nanotechnology, such as in pollen magnetofection in cotton plants (Zhao et al. 2017).

(viii) Moreover, all these gene editing and transformation can be achieved if a proper prior database is present. The first plant genome sequencing was done for Arabidopsis in the year 2000. With the development of next-generation sequencing (NGS) technology, a vast database has already been collected in rice, maize, and wheat; even population-level gene evolution can now be understood. Also, the Earth BioGenome Project (EBP) (Lewin et al. 2018; Exposito-Alonso et al. 2020) will generate the eukaryotes' reference genome within the next decade.

6 Waste Management

In recent years, waste management has become an essential concern around the globe. Waste is generated at both production and the end of the consumption of a product. The consequences of irresponsible waste disposal can lead to health hazards, especially in poorer sectors of society, increased GHGs, and depletion of natural resources. Due to the acceleration of urbanization, there has been an alarming increase in their amount with variation in their diversity. Municipal solid waste (MSW) is one such result of urbanization, which is responsible for air, water, and soil pollution, resulting in the deteriorating living conditions of people (Sanjeevi and Shahabudeen 2015). According to the World Bank, MSW will rise to 2.2 billion tons by 2050 for two main reasons: the booming population and urbanization to accommodate this population (World Bank 2018). Emission of GHGs, depositions of heavy metals such as cadmium, explosion hazards, odor, acidification of air, and eutrophication are some other damages that can be caused by the mismanagement of waste (Abila and Kantola 2019).

Previously, waste management leaned more toward aesthetics, health, resources, land use, and economic purpose. Still, this has been balanced recently, and environmental health has also been looked out for due to climate change. Waste handling contributes 5% of GHG emissions (Gautam and Agrawal 2021). Waste management has become a critical measure of healthy human and environmental well-being.

Sustainable waste management can be described as any practice and strategy that is goal-specific that controls, monitors, and regulates the production of waste and evaluates and makes policies as seen fit to lessen the pressure on the Environment; it differs from developed to developing countries and is more important in later due to incomplete infrastructure involved in its management.

Waste management is goal-specific in developed countries, and every state has its strategy, but it is expected to be nationwide adoption. For example, the Swiss waste management policy minimizes the pollutants at production and finished goods and undergoes recycling. The remaining pollutant is made environment-compatible before release. The European Union (EU) has a hierarchy system for waste management based on the three Rs, i.e., reduce, reuse, and recycle, and are as follows.

(i) *Prevention*: This involves minimization of the generation of pollutants both at the production or manufacturing level and at the end of product consumption. The applicable policy at the former end might be avoiding heavy packaging material, while at the former end, it might buy reusable and recyclable material.

(ii) *Reuse*: The process involves the further use of material in its original form itself when its prior use end is met. It usually occurs at the generation of waste.

(iii) *Recycle*: This process involves the collection, sorting, and reprocessing of used or unused material, which is transformed into a completely new form or its original self. It is an excellent means to mitigate any form of pollution.

(iv) *Recovery*: It is a concept of "garbage-to-energy" that involves energy extraction from waste material by any means, such as combustion, anaerobic digestion, pyrolysis, and gasification (USEPA 2015). The energy thus generated can be used to power internal combustion (IC) engines, turbines, or biogas that can generate electricity (Zaman et al. 2022).

(v) *Disposal*: It is the most common and accessible process but, at the same time, not a very sustainable approach. The area where waste is dumped is known as a landfill. According to the EU hierarchy of waste management (Fig. 3), it should be limited to the minimum and under the guidance of Directives of 1999/31/EC (European Commission 2015).

However, developing countries still need to adopt a proper waste management hierarchy, which generally differs from nation to nation. The most preferred approach to handling waste is the best technology, which will be the most accessible but simultaneously at a sustainable cost (Ferrari et al. 2016). For example, in Botswana, recycling and reuse are the preferred options in waste management (Mmereki 2018).

Another example of managing waste is from a developing country such as Singapore. Singapore manages the waste generated from its state through three means: limited waste production, incineration, and landfills. The Semakau landfill, 8 km away from the mainland, is constructed strategically over the sea between two islands. However, there are two other landfills at Lim Chu Kang and Lorong Halus; the former is the only sanitary and offshore landfill. The construction of the Semakau landfill is based on civil and landscape engineering (Bai and Sutanto 2002), and it receives 2300 tons of incinerated and nonincinerated waste daily. To avoid leaching from the landfill to the sea, impermeable bunds are made, and leachates are usually treated at on-site plants as per national standards before being released into the sea. There are also plans to convert this landfill into a recreation park after it has achieved its maximum capacity (Chan 2016).

7 Public Awareness and Behavioral Change Toward Carbon Footprint

William Reese and Matthias Wackernagel introduced the term carbon footprint in the twentieth century. It refers to the amount of CO_2 and its equivalent gases released in the atmosphere in time by an individual, organizations, or various institutes. Carbon footprint can be classified into different types viz individual or personal carbon footprint, products or services or events carbon footprint and institute or organization and companies carbon footprint. However, in this section, we will mainly discuss personal carbon footprint. It has been reported that the amount of CO_2 emission is 4 tons per capita per year, though the ideal emission rate is 2 tons per year.

In a study conducted in the Biology Department, Faculty of Science, El-Mergib University, Al-Khums in Libya, 34% of the students had previously heard about personal carbon footprint. In comparison, 77% of the students came across the term while taking an ecology course, while 22.3% listened to it via friends, social media, and various other sources. This study was relevant to show how students have come across this term by multiple means. It was also reported that the students who had taken up the course in ecology were more aware practically of minimal waste generation and conservative electricity usage (Ashokri and Abuzririq 2023).

In another study conducted on employees of Ankara province on general awareness of carbon footprint and renewable, it was found that 77.6% of the employees knew about carbon footprint and 59% preferred electric vehicles. In comparison, 51% retorted on the costly installation of renewable energy (Ersöz and Bulbul 2023).

There is also a mobile-based app, Smiling Earth, that integrates energy and transport usage by an individual and provides feedback in the form of visualized data about an individual's daily activity. This app aims to motivate users to reduce self-CO_2 emissions and explore how Information and Communication Technologies have tried to educate individuals (Petersen et al. 2020).

8 International Collaboration and Partnerships

The like-mindedness and overall participation of different countries in their effort to tackle global issues and problems under the discipline of an overseeing general body can always present itself as a more efficient and effective way to deal with matters of international concern. Proper and thorough estimation of the problem at hand and then laying common objectives over paneled discussion of the representatives sharing the same vision can pave commendable progress in dealing with the subject. Transparency and trust among the members sharing the same goal and addressing all proceedings respectfully is a sign of a positive partnership. Global collaborations can not only be confined to trade, education, and technology; they also include every other aspect of the different streams of interests, including the Environment and its sustainability. Global collaborations provide a window to share and exchange ideas and tools, allowing an all-round perception and assessment of the problems in discussion under diplomatic agreement. Furthermore, the cooperation among various interested groups, agencies, organizations, and nations imparts a scope to finding solutions that can be harnessed in more dimensions than one, given the potential of a broader spectrum in the discernment of skill and knowledge pool from every party committed to the cause of setting a better and continued progress in addressing the objectives and goals.

International partnerships can be either governmental organizations or nongovernmental organizations depending on the representation of the members affiliated with the government of the member countries, where governmental organizations officially represent the government. In contrast, nongovernmental organizations do not officially represent the government. The partnerships' norms, rules, objectives, and visions are conferred through diplomacy and institutionalized in a framework of proper understanding. There has been an increase in interest in collaborative works toward environmental sustainability among policymakers, academicians, various nonprofit organizations, state representatives, and nations. These partnerships also create a network among the members that can be used as a tool to shed light on the scenario of the Environment currently in a widespread manner that will affect other spheres like industry, transport, communication, and development, which in turn will favor a more thought-out approach toward environmental sustainability. There is always something new to take away from such exchanges, and a progressive step can be very beneficial. The logistics of a task of this magnitude require integrating all available resources and a mindset of collective motives. The positivity of such collaborations brings certain drawbacks and criticisms that are indefinitely a part of any work of such magnitude. The major shortcoming of such endeavors can be the discontinuous nature of disseminating awareness given the size of the work and how sometimes it becomes difficult to reach every nook and corner due to it being a colossal task.

Furthermore, it is sometimes seen to be challenging to manage the pace of work, and the degree of impact sometimes shows itself to be less than what the actual work plan intended at the start. Collaboration typically occurs among those entities with the necessary resources and socio-political clout, such as government or industry, whose interests reinforce society's status quo and political structure. Collaboration tends to marginalize politically weaker groups, such as the poor, unemployed, less developed countries, future generations, or radical environmentalists (Levy 1997). However, the benefit of the doubt lies in the fact that initiating such missions with a shared purpose ignites some potential advances in a case that needs more attention than given. So, the overall verdict in such collaborations and partnerships can be drawn out as a step toward sustainability that always, with its pace and scheme, works toward resolving underlying drawbacks and continuously shows an openness to refinement.

Government and nongovernment organizations work toward environmental sustainability and spreading awareness. The functionality and work process of the two differ, but the aim is similar in the case of environmental sustainability. Government organizations can be seen as bodies that decide policies and regulations under the law and strive toward achieving the goals by enforcing said policies. They see to it that every agenda follows the weight of the laws and serves as an entity to maintain order in the action of plans for environmental sustainability. Their interest lies in forming structures of official rules that are fair and ethical and maintain a balance so that no acting body under its umbrella should go out of hand. Government organizations belong to the governments of the countries to which they belong, and they act as representatives of their governments.

On the other hand, nongovernmental organizations do not directly represent their government. Still, they must advocate their interests and beliefs under the guidance of the rules and laws set up by the government and government organizations. Nongovernmental organizations are confined to a country and can be international and operate independently. The government can also provide funding, but its functioning is dictated by independence. Under the UN, UNEP is an intergovernmental body that works extensively toward global climate change, ecosystem management, and the effect of waste and chemicals. The Ministry of Environment and Forest, Central Pollution Control Board, and Indian Board for Wildlife are some front-running government organizations that work toward environmental sustainability in India.

Generally, for most countries, the government sets a separate ministry to take up the works about the Environment and its preservation, and the government overlooks the functioning of these organizations. For nongovernment organizations, Greenpeace and WWF are two giants that work for the Environment in terms of their platform on the international scale. The working members of Greenpeace are comprised of scientists, government officials, and activists from business backgrounds, coming together with a shared motive to safeguard and preserve the Environment, pooling tools, intellect, and capital for the cause. World Wide Fund for Nature advocates similarly to Greenpeace and negotiates with agencies under other organizations for proper dialog and work relating to environmental conservation. Appropriate investment toward a resolve to counter anti-environmental actions and form a decorum to assess current impacts and find measures and solutions to mitigate such impacts, thus forming a continued chain of projects and tasks. Environmentalist Foundation of India is one such nongovernment organization in India that works for environmental sustainability functioning in 18 states of the country. Contrary to modern beliefs of pessimism toward such movements and their effectiveness, there is also growing enlightenment in people's minds about the changes that these organizations bring about confidence in such activities and movements.

The United Nations headlines the key bodies to work and administer toward global environmental sustainability, development, and awareness. UNEP (United et al.) is the lead program under the UN and is dedicated to the Environment, with the United Nations Environment Assembly as its governing body, the UN, which was established in 1972. The UN Environment Assembly is the UNEP's policy-making organ that works out laws and policies for the course of action to be taken. There are also four significant bodies functioning under the United Nations dedicated to Environment and climate change:

(i) High-level Political Forum on Sustainable Development
(ii) Commission on Sustainable Development (CSD)
(iii) Intergovernmental Panel on Climate Change (IPCC)
(iv) UN Forum on Forests

The Intergovernmental Panel on Climate Change (IPCC) was also established by the United Nations Environment Programme (UNEP) and World Meteorological Organization (WMO) in 1988. The establishment of the IPCC was endorsed to provide governments at all levels with scientific innovations and information to develop strategies regarding the climate. The IPCC, in its synthesis report at the 52nd-panel session, agreed upon the introduction of three main timeframe sections viz., the first section (Current Status and Trends) covers the historical and present period, the second section (Long et al.) sets itself for the projected futures upto 2100 and beyond and finally the third section (Near term Responses in a Changing Climate) that considers the current international policy timeframes and the time between now and 2030–2040. The UN has also introduced specific goals to be attained under the banner Sustainable Development Goals 17 with 17 distinct goals, which also look toward the Environment and sustainability, all tailored toward a brighter future, keeping in mind the partnerships and cooperations of all nations.

Certain treaties have been signed for environment sustainability under Multilateral Environment Agreements, viz., the Convention on Biological Diversity (1992) and the Convention for the Conservation of Antarctic Marine Living Resources (1980). Other treaties like the Stockholm Declaration (1972), Kyoto Protocol (1997), and Paris Agreement (2015) have all garnered support for environmental sustainability.

The debate always persists about the Environment and how sometimes data and estimation for the consistency of improvement in projects attached to their sustainability are unstable; however, there is evidence of the successes of said projects in tremendous fashion if we look closely. Awareness, participation, and a sense of responsibility are always required for such missions, and only with this understanding can it gain steam. Volunteering and being dedicated to saving what we have in nature can fuel the pursuit of a better and more sustainable environment. Other factors like geopolitics, economy, and socio-cultural views of different countries and organizations can sometimes slow down the development of issues such as these. Still, it is only with hope that we can find a sense of inclination toward realizing the importance of environmental sustainability and its protection in this day and age. Without a visionary mindset, it sometimes can succumb to a null.

9 Conclusion

The goal to achieve environmental sustainability is a combined effort of the public, the state, and nations and involves relationships between various countries. Apart from that, individual behavior toward our environment plays a vital role in achieving a healthy society. The goal of a healthy environment is long and requires calculated steps of every nation since the climate is typical to all. In the recent COP26 event held at Glasglow, the UK agreed to minimize the use of coal, which is the most significant contributor to climate change. Furthermore, 140 countries pledged for net-zero emission, including 90% of world GDP, and more than a hundred countries pledged for reforestation. However, there is a massive gap between ambition and goal realization, and the stakeholders have emphasized those problems more, which is pulling down the sustainable goals. Also, the role of media and other communication sectors must be strengthened to make ordinary citizens aware of climate change, increasing global temperature, and its consequences and possible solutions.

References

Abbasi SA, Abbasi T (2017) Ozone hole: past, present, future. Springer

Abila B, Kantola J (2019) Waste management: relevance to environmental sustainability. Int J Environ Waste Manag 23(4):337–351

Achour Y, Ouammi A, Zejli D (2021) Technological progress in modern sustainable greenhouse cultivation is the path toward precision agriculture. Renew Sust Energ Rev 147:111251

Akram R, Ibrahim RL, Wang Z, Adebayo TS, Irfan M (2023) Neutralizing the surging emissions amidst natural resource dependence, eco-innovation, and green energy in G7 countries: insights for global environmental sustainability. J Environ Manag 344:118560

Alho CJR (2008) The value of biodiversity. Braz J Biol 68:1115–1118

American Museum of Natural History

Anderson S, Newell R (2004) Prospects for carbon capture and storage technologies. Annu Rev Environ Resour 29:109–142

Antolini D, Ail SS, Patuzzi F, Grigiante M, Baratieri M (2019) Experimental investigations of air-CO2 biomass gasification in reversed downdraft gasifier. Fuel 253:1473–1481

Arora NK (2018) Biodiversity conservation for a sustainable future. Environ Sustain 1(2):109–111

Arora NK, Fatima T, Mishra I, Verma M, Mishra J, Mishra V (2018a) Environmental sustainability: challenges and viable solutions. Environ Sustain 1:309–340

Arora NK, Khare E, Singh S, Tewari S (2018b) Phenetic, genetic diversity and symbiotic compatibility of rhizobial strains nodulating pigeon pea in Northern India. 3 Biotech 8:1–11

Ashokri HAA, Abuzririq MAK (2023) The impact of environmental awareness on personal carbon footprint values of biology department students, Faculty of Science, El-Mergib University, Al-Khums, Libya. Acta Biology Forum V02i02 18:22

Atkins PA, Voytas DF (2020) Overcoming bottlenecks in plant gene editing. Curr Opin Plant Biol 54:79–84

Bai R, Sutanto M (2002) The practice and challenges of solid waste management in Singapore. Waste Manag 22(5):557–567

Beumer C, Martens P (2016) BIMBY's first steps: a pilot study on the contribution of residential front-yards in Phoenix and Maastricht to Biodiversity, ecosystem services and urban sustainability. Urban Ecosyst 19(1):45–76

Bilal A, Li X, Zhu N, Sharma R, Jahanger A (2021) Green technology innovation, globalization, and CO2 emissions: recent insights from the OBOR economies. Sustainability 14(1):236

Biodiversity Conservation Trust

Bock R (2015) Engineering plastid genomes: methods, tools, and applications in basic research and biotechnology. Annu Rev Plant Biol 66:211–241

Boussemart JP, Leleu H, Shen Z, Valdmanis V (2020) Performance analysis for three pillars of sustainability. J Prod Anal 53:305–320

Britannica, T. Editors of Encyclopaedia (2024a, February 17) WWF. Encyclopedia Britannica

Britannica, T. Editors of Encyclopaedia (2024b, February 9) Greenpeace. Encyclopedia Britannica

Butchart SH, Stattersfield AJ, Collar NJ (2006) How many bird extinctions have we prevented? Oryx 40(3):266–278

Butchart SH, Walpole M, Collen B, Van Strien A, Scharlemann JP, Almond RE et al (2010) Global biodiversity: indicators of recent declines. Science 328(5982):1164–1168

Cai Q, Yan X, Li Y, Wang L (2018) Global patterns of human and livestock respiration. Sci Rep 8(1):9278

Casals LC, Martinez-Laserna E, García BA, Nieto N (2016) Sustainability analysis of the electric vehicle use in Europe for CO2 emissions reduction. J Clean Prod 127:425–437

Chan JKH (2016) The ethics of working with wicked urban waste problems: the case of Singapore's Semakau Landfill. Landsc Urban Plan 154:123–131

Cheng L, Zhu J, Abraham J, Trenberth KE, Fasullo JT, Zhang B, Song X (2019) 2018 continues record global ocean warming

Davis SJ, Caldeira K, Matthews HD (2010) Future CO2 emissions and climate change from existing energy infrastructure. Science 329(5997):1330–1333

Di Sacco A, Hardwick KA, Blakesley D, Brancalion PH, Breman E, Cecilio Rebola L et al (2021) Ten golden rules for reforestation to optimize carbon sequestration, biodiversity recovery and livelihood benefits. Glob Chang Biol 27(7):1328–1348

Doebley J, Stec A, Hubbard L (1997) The evolution of apical dominance in maize. Nature 386(6624):485–488

Don A, Osborne B, Hastings A, Skiba U, Carter MS, Drewer J et al (2012) Land-use change to bioenergy production in Europe: implications for the greenhouse gas balance and soil carbon. GCB Bioenergy 4(4):372–391

Ellison EE, Nagalakshmi U, Gamo ME, Huang PJ, Dinesh-Kumar S, Voytas DF (2020) Multiplexed heritable gene editing using RNA viruses and mobile single guide RNAs. Nature Plants 6(6):620–624

Eluwole KK, Saint Akadiri S, Alola AA, Etokakpan MU (2020) Does the interaction between growth determinants a drive for global environmental sustainability? Evidence from world top 10 pollutant emissions countries. Sci Total Environ 705:135972

Ersöz B, Bulbul HI (2023) A qualitative study on employees carbon footprint and renewable energy awareness. Int J Renew Energy Res (IJRER) 13(2):699–706

European Commission (2015) Waste [online] http://ec.europa.eu/environment/waste/landfill/index.htm. Accessed 19 Aug 2015

Exposito-Alonso M, Drost HG, Burbano HA, Weigel D (2020) The Earth BioGenome project: opportunities and challenges for plant genomics and conservation. Plant J 102(2):222–229

Fang R, Kirk P, Wei J, Li Y, Cai L, Fan L et al (2018) Country focus: China. In: Willis KJ (ed) State of the world's fungi, pp 48–55

FAO (2016) The state of food and agriculture: climate change, agriculture and food security. FAO, Rome

Ferrari K, Gamberini R, Rimini B (2016) The waste hierarchy: a strategic, tactical and operational approach for developing countries. The case study of Mozambique. Int J Sustain Dev Plan 11(5):759–770

Frid C, Andonegi E, Depestele J, Judd A, Rihan D, Rogers SI, Kenchington E (2012) The environmental interactions of tidal and wave energy generation devices. Environ Impact Assess Rev 32(1):133–139

Fuentes P, Armarego-Marriott T, Bock R (2018) Plastid transformation and its application in metabolic engineering. Curr Opin Biotechnol 49:10–15

Gautam M, Agrawal M (2021) Greenhouse gas emissions from municipal solid waste management: a review of global scenario. In: Carbon footprint case studies: municipal solid waste management, sustainable road transport and carbon sequestration, pp 123–160

Gurney KR, Mendoza DL, Zhou Y, Fischer ML, Miller CC, Geethakumar S, de la Rue du Can, S. (2009) High-resolution fossil fuel combustion CO2 emission fluxes for the United States. Environ Sci Technol 43(14):5535–5541

Hamdy A, Aly A (2014) Land degradation, agriculture productivity and food security. In: Proceedings of the fifth international scientific agricultural symposium, AGROSYM, pp 708–717

Hossain A, Krupnik TJ, Timsina J, Mahboob MG, Chaki AK, Farooq M, Hasanuzzaman M (2020) Agricultural land degradation: processes and problems undermining future food security. In: Environment, climate, plant and vegetation growth, pp 17–61

Ideris F, Nomanbhay S, Kusumo F, Silitonga AS, Ong MY, Ong HC, Mahlia TMI (2021) Optimization of ultrasound-assisted oil extraction from Canarium odontophyllum kernel as a novel biodiesel feedstock. J Clean Prod 288:125563

Kabeel AE, Hamed MH, Omara ZM, Kandeal AW (2017) Solar air heaters: design configurations, improvement methods and applications – a detailed review. Renew Sust Energ Rev 70:1189–1206

Kemppinen KM, Collins PM, Hole DG, Wolf C, Ripple WJ, Gerber LR (2020) Global reforestation and biodiversity conservation. Conserv Biol 34(5):1221–1228

Keziah SM, Devi CS (2017) Essentials of conservation biotechnology: a mini-review. In: IOP conference series: materials science and engineering, vol 263, No. 2. IOP Publishing, p 022047

Khan N, Ray RL, Sargani GR, Ihtisham M, Khayyam M, Ismail S (2021) Current progress and future prospects of agriculture technology: gateway to sustainable agriculture. Sustainability 13(9):4883

Kirikkaleli D, Sowah JK (2020) A wavelet coherence analysis: nexus between urbanization and environmental sustainability. Environ Sci Pollut Res 27:30295–30305

Kraaijenbrink PD, Bierkens MF, Lutz AF, Immerzeel WW (2017) Impact of a global temperature rise of 1.5 degrees Celsius on Asia's glaciers. Nature 549(7671):257–260

Kumar R, Kumar P, Kumar P (2019) Understanding the evidence of climate change influencing Indian agriculture. Innov Agric Environ Health Res Ecol Restor 2(4):20

Kumar SS, Wani OA, Krishna JR, Hussain N (2022) Impact of climate change on soil health. Int J Environ Sci 7:70–90

Kuttippurath J, Nair PJ (2017) The signs of Antarctic ozone hole recovery. Sci Rep 7(1):585

Landrigan PJ, Fuller R, Acosta NJ, Adeyi O, Arnold R, Baldé AB, Zhong M (2018) The Lancet Commission on pollution and health. Lancet 391(10119):462–512

Lemmon ZH, Reem NT, Dalrymple J, Soyk S, Swartwood KE, Rodriguez-Leal D et al (2018) Rapid improvement of domestication traits in an orphan crop by genome editing. Nature Plants 4(10):766–770

Lévêque C, Mounolou JC (2004) Biodiversity. Wiley

Levy DL (1997) Environmental management as political sustainability. Organ Environ 10(2):126–147

Lewin HA, Robinson GE, Kress WJ, Baker WJ, Coddington J, Crandall KA et al (2018) Earth BioGenome Project: sequencing life for the future of life. Proc Natl Acad Sci 115(17):4325–4333

Li T, Yang X, Yu Y, Si X, Zhai X, Zhang H et al (2018) Domestication of wild tomato is accelerated by genome editing. Nat Biotechnol 36(12):1160–1163

Li C, Zhang R, Meng X, Chen S, Zong Y, Lu C et al (2020) Targeted, random mutagenesis of plant genes with dual cytosine and adenine base editors. Nat Biotechnol 38(7):875–882

Lindenmayer DB, Franklin JF, Fischer J (2006) General management principles and a checklist of strategies to guide forest biodiversity conservation. Biol Conserv 131(3):433–445

Liu K, Tan Q, Yu J, Wang M (2023) A global perspective on e-waste recycling. Circ Econ:100028

Loehman RA (2010) Understanding the science of climate change: talking points-impacts to Arid Lands

Lu IJ, Lewis C, Lin SJ (2009) The forecast of motor vehicle, energy demand and CO2 emission from Taiwan's road transportation sector. Energy Policy 37(8):2952–2961

Ma X, Zhang X, Liu H, Li Z (2020) Highly efficient DNA-free plant genome editing using virally delivered CRISPR–Cas9. Nature Plants 6(7):773–779

Maher MF, Nasti RA, Vollbrecht M, Starker CG, Clark MD, Voytas DF (2020) Plant gene editing through de novo induction of meristems. Nat Biotechnol 38(1):84–89

Maliga P, Bock R (2011) Plastid biotechnology: food, fuel, and medicine for the 21st century. Plant Physiol 155(4):1501–1510

Mmereki D (2018) Current status of waste management in Botswana: a mini-review. Waste Manag Res 36(7):555–576

Mokhta ZM, Ong MY, Salman B, Nomanbhay S, Salleh SF, Chew KW, Chen WH (2020) Simulation studies on microwave-assisted pyrolysis of biomass for bioenergy production with special attention on waveguide number and location. Energy 190:116474

National Research Council (US) Panel on Biodiversity Research Priorities (1992) Conserving biodiversity: a research agenda for development agencies. National Academies Press, Washington, DC, 1. Biodiversity and Development

Newman PA, Nash ER, Kawa SR, Montzka SA, Schauffler SM (2006) When will the Antarctic ozone hole recover? Geophys Res Lett 33(12)

Nikitas G, Bhattacharya S, Vimalan N, Demirci HE, Nikitas N, Kumar P (2019) Wind power: a sustainable way to limit climate change. Manag Glob Warm:333–364

Oertel C, Matschullat J, Zurba K, Zimmermann F, Erasmi S (2016) Greenhouse gas emissions from soils – a review. Geochemistry 76(3):327–352

Pathak MR, Abido MS (2014) The role of biotechnology in the conservation of Biodiversity. J Exp Biol 2:352–363

Patil G, Mian R, Vuong T, Pantalone V, Song Q, Chen P et al (2017) Molecular mapping and genomics of soybean seed protein: a review and perspective for the future. Theor Appl Genet 130:1975–1991

Petersen SA, Petersen I, Ahcin P (2020) Smiling Earth—raising awareness among citizens for behaviour change to reduce carbon footprint. Energies 13(22):5932

Pettit H (2018) The 'great Pacific garbage patch' is bigger than thought: toxic area is three times the size of FRANCE and contains 79,000 tonnes of plastic

Purvis B, Mao Y, Robinson D (2019) Three pillars of sustainability: in search of conceptual origins. Sustain Sci 14:681–695

Ragheb A, El-Shimy H, Ragheb G (2016) Green architecture: a concept of sustainability. Procedia Soc Behav Sci 216:778–787

Ramanathan V (1981) The role of ocean-atmosphere interactions in the CO2 climate problem. J Atmos Sci 38(5):918–930

Rands MR, Adams WM, Bennun L, Butchart SH, Clements A, Coomes D et al (2010) Biodiversity conservation: challenges beyond 2010. Science 329(5997):1298–1303

Reed BM, Sarasan V, Kane M, Bunn E, Pence VC (2011) Biodiversity conservation and conservation biotechnology tools. In Vitro Cell Dev Biol Plant 47:1–4

Ritchie H, Roser M, Rosado P (2020) CO_2 and greenhouse gas emissions. Our World in Data

Rockström J, Williams J, Daily G, Noble A, Matthews N, Gordon L et al (2017) Sustainable intensification of agriculture for human prosperity and global sustainability. Ambio 46:4 17

Rodríguez-Leal D, Lemmon ZH, Man J, Bartlett ME, Lippman ZB (2017) Engineering quantitative trait variation for crop improvement by genome editing. Cell 171(2):470–480

Sanjeevi V, Shahabudeen P (2015) Development of performance indicators for municipal solid waste management (PIMS): a review. Waste Manag Res 33(12):1052–1065

Shrivastava P, Guimarães-Costa N (2017) Achieving environmental sustainability: the case for multi-layered collaboration across disciplines and players. Technol Forecast Soc Chang 116:340–346

Smoyer-Tomic KE, Kuhn R, Hudson A (2003) Heatwave hazards: an overview of heat wave impacts in Canada. Nat Hazards 28:465–486

Solomon S (1988) The mystery of the Antarctica ozone "hole". Rev Geophys 26(1):131–148

Solomon BD (2010) Biofuels and sustainability. Ann N Y Acad Sci 1185(1):119–134

Studer A, Zhao Q, Ross-Ibarra J, Doebley J (2011) Identification of a functional transposon insertion in the maize domestication gene tb1. Nat Genet 43(11):1160–1163

Syed AM, Hachem C (2019) Review of construction; geometry; heating, ventilation, and air-conditioning; and indoor climate requirements of agricultural greenhouses. J Biosyst Eng 44:18–27

TECAM (2023) 10 examples of green technology. TECAM

The Climate Council (2017) Bhutan is the world's only carbon negative country, so how did they do it. The Climate Council

Tian Z, Wang JW, Li J, Han B (2021) Designing future crops: challenges and strategies for sustainable agriculture. Plant J 105(5):1165–1178

Tian J, Wang C, Xia J, Wu L, Xu G, Wu W, Li D, Qin W, Han X, Chen Q, Jin W (2019) Teosinte ligule allele narrows plant architecture and enhances high-density maize yields. Science 365(6454):658–664

Torres N, Antolín MC, Goicoechea N (2018) Arbuscular mycorrhizal symbiosis as a promising resource for improving berry quality in grapevines under changing environments. Front Plant Sci 9:897

Trenberth KE, Branstator GW (1992) Issues in establishing causes of the 1988 drought over North America. J Clim 5(2):159–172

Trenberth KE, Branstator GW, Arkin PA (1988) Origins of the 1988 North American drought. Science 242(4886):1640–1645

Tuckett R (2019) Greenhouse gases. In: Encyclopedia of analytical science. Elsevier, pp 362–372

Ulian T, Diazgranados M, Pironon S, Padulosi S, Liu U, Davies L et al (2020) Unlocking plant resources to support food security and promote sustainable agriculture. Plants People Planet 2(5):421–445

United Nations Decade on Biodiversity

US Environmental Protection Agency (USEPA) (2015) Non-hazardous Waste Management Hierarchy [online] http://www.epa.gov/wastes/nonhaz/municipal/hierarchy.htm. Accessed 26 Aug 2015

van Der Velden M (2016) Design as regulation: opportunities and limitations for sustainable mobile phone design. In: Culture, technology, communication. Common world, different futures: 10th IFIP WG 13.8 international conference, CaTaC 2016, London, UK, June 15–17, 2016. Revised selected papers 10. Springer, pp 32–54
Wang RL, Stec A, Hey J, Lukens L, Doebley J (1999) The limits of selection during maize domestication. Nature 398(6724):236–239
Wilson EO (1988) The current state of biological diversity. Biodiversity 521(1):3–18
World Bank (2018) Solid Waste Management [online] http://www.worldbank.org/en/topic/urbandevelopment/brief/soild-waste-management. Accessed 18 Apr 2018
Worrell E, Price L, Martin N, Hendriks C, Meida LO (2001) Carbon dioxide emissions from the global cement industry. Annu Energy Environ Rev 26(1):303–329
Xu M, Crittenden JC, Chen Y, Thomas VM, Noonan DS, Desroches R, French SP (2010) Gigaton problems need gigaton solutions
Yadav AN (2020) Plant microbiomes for sustainable agriculture: current research and future challenges. Springer, pp 475–482
Yan X, Liu M, Zhong J, Guo J, Wu W (2018) How human activities affect heavy metal contamination of soil and sediment in a long-term reclaimed area of the Liaohe River Delta, North China. Sustainability 10(2):338
Yousefi H, Noorollahi Y, Abedi S, Panahian K, MirAbadi AH, Abedi S (2015) Economic and environmental feasibility study of greenhouse heating and cooling using geothermal heat pump in northwest Iran. In: Proceedings World Geothermal Congress, Melbourne, Australia
Yousefi H, Ármannsson H, Roumi S, Tabasi S, Mansoori H, Hosseinzadeh M (2018) Feasibility study and economical evaluations of geothermal heat pumps in Iran. Geothermics 72:64–73
Zaman K, Malik M, Awan U, Handayani W, Jabor MK, Asif M (2022) Environmental effects of bio-waste recycling on industrial circular economy and eco-sustainability. Recycling 7(4):60
Zhang Y, Geng W, Shen Y, Wang Y, Dai YC (2014) Edible mushroom cultivation for food security and rural development in China: bio-innovation, technological dissemination and marketing. Sustainability 6(5):2961–2973
Zhao X, Meng Z, Wang Y, Chen W, Sun C, Cui B et al (2017) Pollen magnetofection for genetic modification with magnetic nanoparticles as gene carriers. Nature Plants 3(12):956–964
Zheng P, Allen WB, Roesler K, Williams ME, Zhang S, Li J et al (2008) A phenylalanine in DGAT is a key determinant of oil content and composition in maize. Nat Genet 40(3):367–372
Zhu G, Giam X, Armsworth PR, Cho SH, Papeş M (2023) Biodiversity conservation adaptation to climate change: protecting the actors or the stage. Ecol Appl 33(2):e2765
Zsögön A, Čermák T, Naves ER, Notini MM, Edel KH, Weinl S et al (2018) De novo domestication of wild tomato using genome editing. Nat Biotechnol 36(12):1211–1216

Revolutionizing Sustainable Development Policies to Reduce Inequality Among Countries

Rubiya Dar, Tanvirul-Hassan-Dar, Hina Mushtaq, and Bilal Ahmad Wani

Abstract

The origins of inequality may be traced back to one's parents and place of birth, which determine differences in the opportunities and resources individuals have access to. Early life social and economic injustices and disadvantages restrict prospects for full potential realization and the capacity to exercise one's rights. Discrimination based on age, gender, race or ethnicity, handicap status, sexual orientation, immigration status, place of residence, or other variables causes disadvantage to certain people in numerous, often undetectable ways throughout their lives. The next generations inherit these patterns of inequity. Thus, the key to ending intergenerational poverty and inequality is early interventions and investing in all children, especially the most disadvantaged and marginalized ones. Sustainable planning promotes social equity by addressing disparities, reducing poverty, and fostering social cohesion. It entails laws and initiatives that guarantee fair distribution of development advantages, empower underserved populations, and advance equitable access to resources and services. Equitable resource distribution, investments in education and skill development, the implementation of social protection measures, the fight against discrimination, the support of marginalized groups, and international cooperation for fair trade and financial systems are all necessary to reduce inequality within and between countries.

Keywords

Inequality · Marginalization · Empowerment · Skill development · Economic justice

1 Introduction

Sustainable development can be defined as an approach to economic development that meets the needs of the present without compromising the ability of future generations to meet their own needs. Today's most influential global framework for international cooperation, the 2030 Agenda for Sustainable Development and its Sustainable Development Goals (SDGs), built on sustainability, adopted by all United Nations Member States in 2015, provides a shared draft for peace and prosperity for people and the planet, now and into the future. At its heart, the 17 Sustainable Development Goals (SDGs) are an urgent call for action by all countries, developed and developing in a global partnership. They recognize that ending poverty and other deprivations must go hand-in-hand with strategies that improve health and education, reduce inequality, and spur economic growth—all while tackling climate change and working to preserve our oceans and forests. Global environmental leaders gathered in June 2022 to assess our current state and future directions and to create momentum for the U.N. Decade of Action to achieve the Sustainable Development Goals (SDGs). These conferences mark significant turning points in the governance of sustainable development. However, the work done daily by everyone from small communities and people to multinational organizations and beyond serves as their actual test. The concept of sustainable development holds that human civilizations cannot exist and satisfy their needs at the expense of future generations' capacity to do the same. First

R. Dar (✉) · B. A. Wani
Centre of Research for Development (CORD), University of Kashmir, Srinagar, Jammu and Kashmir, India
e-mail: rubiyadar@uok.edu.in

Tanvirul-Hassan-Dar
Department of Biotechnology, School of Biosciences and Biotechnology, BGSB University, Rajouri, Jammu and Kashmir, India

H. Mushtaq
Government Degree College for Women, Baramulla, Jammu and Kashmir, India

J. A. Parray et al. (eds.), *Environmental Landscape and Sustainable Biodiversity for Healthy Green Growth*, Sustainable Landscape Planning and Natural Resources Management, https://doi.org/10.1007/978-3-031-75405-0_5

outlined in the 1987 Brundtland Report, sustainable development gained popularity during the Industrial Revolution when Western cultures realized the significant impact of their industrial and commercial activities on the environment and social balance. This entails considering the immediate and long-term imperatives, such as maintaining the environment and natural resources or social and economic equity and increasing awareness of the need for a more sustainable paradigm. Some global economic and social catastrophes in the twentieth century include the American banking crisis of 1907, the American hyperinflation crisis that began in 1923, and the financial crisis of the 1930s.

Furthermore, there were global demonstrations against bureaucracy in 1968, followed by oil shocks in 1973 and 1979, and the debt shock of emerging nations in 1982. Ecological catastrophes during this period include nuclear fallout from Rongelap in 1954; the Torrey Canyon oil leak in 1957; the Seveso catastrophe in 1976; the Bhopal disaster in 1984; the Chornobyl nuclear disaster in 1986; the Exxon Valdez oil spill in 1989; the Erika disaster in 1999; and the Mercury crisis of Minamata in 1956. Ongoing issues with the ozone layer, global warming, air pollution, and biodiversity loss exist. In 1972, Meadows et al. conducted a computer simulation at the request of the Club of Rome to forecast potential outcomes in a world with finite resources. The study examined the interplay among five dimensions: global population expansion, industrialization, pollution generation, food production, and the depletion of nonrenewable resources, the conclusion of which suggested that without restrictions on growth, there would be a social and economic collapse by the end of the twenty-first century. After more than 40 years of this study, these forecasts about pollution and its effects, which jeopardize sustainable development, appeared accurate.

The following are some crucial tactics that can support sustainable development.

2 Environmental Sustainability

Sustainable development strategies should prioritize resource management and environmental protection to protect the welfare of present and future generations. This includes supporting sustainable farming methods, renewable energy sources, and the preservation of ecosystems and biodiversity. The first significant gathering of world leaders to address the relationship between economic progress and environmental degradation was the United Nations Conference on the Environment, held in Stockholm in 1972. A primary objective of this assembly was to identify a shared perspective and guiding ideals to motivate and direct the global populace in protecting the "human environment." The assembly aimed to establish common guiding principles to encourage global efforts in safeguarding the "human environment." The annual Human Development Index (HDI) report is used by the U.N. Development Programme to rate nations and measure global development progress. It functions as a recurring method of tracking nations' degrees of progress. The HDI evaluates nations' economic and social progress by considering mobility, human security, financial flows, health, and education factors. In an ideal world, humanity would live below the maximum ecological footprint per person and at least reach the minimal HDI. Residing over the minimal HDI would ensure that fundamental requirements, including health and education, are met.

The most significant amount of consumption that any individual may have while still maintaining Earth's ecological capability is known as an environmental footprint. Because the earth can replenish itself, living underneath it wouldn't endanger future generations. We are headed toward a sustainable future if we maintain our ecological footprint per capita below the maximum (which is dropping as the human population grows) and above the minimal HDI. However, Earth Overshoot Day occurs earlier each year, marking the point at which humanity consumes more natural resources than the Earth can replenish in the same year. This creates a deficit as we use more resources than the Earth can sustainably provide and must adequately manage our waste. Unlike nature's cyclical processes, our linear consumption patterns are, therefore, threatening sustainable development. The U.N. Development Programme and the World Meteorological Organisation established the International Panel on Climate Change in response to growing awareness of the potential effects of climate change on the environment and human existence. Its goal was and continues to be to increase public understanding of how human activity affects climate change. It also seeks to investigate the origins, effects, and strategies for halting climate change. Gases like CO_2 and methane are necessary for the Earth to maintain its perfect temperature and support life as we know it. However, the planet's temperature rises due to the overproduction of these gases. This occurs because some of the Earth's heat emitted and would otherwise escape into space is trapped. The concept of sustainable development is thus based on the idea of the triple bottom line—coined by John Elkington, the founder of a sustainability consulting firm. This concept suggests businesses should consider their financial performance and social and environmental impact. In other words, companies should evaluate their social responsibility across their entire value chain and quantify their global environmental impact.

2.1 Promoting Renewable Energy Sources

Renewable energy is a key player in pursuing the Sustainable Development Goals (SDGs). Its potential to provide clean and affordable energy, mitigate climate change, drive economic growth, create jobs, and enhance the quality of life for urban and rural communities alike makes it indispensable for achieving the global vision of sustainable development. However, realizing this potential requires concerted efforts from governments, businesses, and individuals to invest in renewable energy infrastructure, research, and technology development. By harnessing the power of renewable energy, the world can move closer to a future where the SDGs are not just aspirational but achievable. The transition to renewable energy requires collaboration among governments, the private sector, civil society, and international organizations. Public–private partnerships can drive investment in renewable energy projects, improve energy infrastructure, and facilitate technology transfer to developing countries. International cooperation is crucial for advancing renewable energy adoption, as many renewable technologies require significant research and development efforts. Collaborative efforts can accelerate the deployment of renewable energy solutions and ensure that no country is left behind in transitioning to a sustainable energy future. Sustainable development has emerged as a global priority in the twenty-first century, with the United Nations Sustainable Development Goals (SDGs) serving as a comprehensive framework to address pressing global challenges (Gusmao et al. 2018). These goals include ending poverty, ensuring access to clean water, reducing inequality, and combating climate change. One key driver of sustainable development is the transition to renewable energy sources. Approximately 789 million people worldwide lack access to electricity, with even more relying on dirty and inefficient energy sources like kerosene and biomass. Renewable energy technologies like solar, wind, and hydroelectric power can bring clean and reasonable energy to remote and underserved communities. Solar energy, in particular, has witnessed remarkable growth in off-grid and decentralized applications. Solar panels can be installed in rural areas to power homes, schools, and healthcare facilities, bridging the energy divide and improving the quality of life for those whom traditional grid expansion efforts have left behind. Moreover, renewable energy technologies are often scalable and adaptable to the needs of communities, making them a resourceful tool for addressing energy poverty. SDG 13 centers on urgent action to combat climate change and its impacts. Burning fossil fuels for energy production is a principal driver of greenhouse gas emissions, the primary cause of global warming and climate change. Transitioning to renewable energy sources is essential in extenuating the adverse effects of climate change and achieving the targets set by SDG 13. Renewable energy technologies produce little to no greenhouse gas emissions during operation, making them a cleaner substitute for fossil fuels. SDG 8 promotes sustained, comprehensive, sustainable economic growth, full and productive employment, and decent work. Renewable energy has become a significant driver of economic growth and employment in many regions. As funds for renewable energy projects increase, they create employment opportunities in manufacturing, installation, maintenance, and research and development. Countries that invest in renewable energy technologies can experience a "green jobs" boom, as workers are needed to manufacture solar panels, construct wind farms, and maintain renewable energy infrastructure. These jobs are often locally based, providing opportunities for communities to thrive. Additionally, the renewable energy sector can attract investment and foster innovation, further contributing to economic development. SDG 7 also emphasizes the importance of affordable energy. Renewable energy sources offer a unique advantage in this regard. While initial costs are associated with installing solar panels, wind turbines, and other renewable technologies, the ongoing equipment costs are significantly lower than those of fossil fuel-based power generation. Once installed, renewable energy systems can produce electricity for decades with minimal maintenance, reducing long-term energy expenses for individuals, businesses, and governments. Furthermore, decentralizing energy production through renewable sources can enhance energy security. Communities and nations that rely on a diverse portfolio of renewable energy resources are less vulnerable to disruptions in energy supply and price volatility associated with fossil fuels. SDG 11 focuses on making cities and human settlements inclusive, secure, resilient, and sustainable. Cities are hubs of energy consumption, and their energy choices substantially impact environmental sustainability. Renewable energy is crucial in making cities more sustainable by reducing air pollution, lowering carbon emissions, and improving urban residents' overall quality of life. Renewable energy technologies, such as rooftop solar panels, can be integrated into urban landscapes, generating clean energy at the point of consumption and reducing the strain on centralized power grids (IRENA 2020). Additionally, electric vehicles, which can be charged using renewable energy sources, suggest a sustainable transportation solution for urban areas, reducing air pollution and promoting cleaner, more livable cities. SDGs 14 and 15 address life below water and life on land, respectively, aiming to protect and sustainably manage these ecosystems. The transition to renewable energy sources is also critical in achieving these goals. International cooperation is crucial for advancing renewable energy adoption, as many renewable technologies require significant research and development efforts. Collaborative efforts can accelerate the deployment of renewable energy solutions and ensure no country is left behind in the changeover to a sustainable energy future.

2.2 Sustainable Agriculture Practices and Their Management

The importance of sustainable agriculture is closely emphasized today because people strongly rely on animal welfare concerns and human health risks. Sustainable agriculture aims to address the growing need for food, fiber, and livestock fodder due to the planet's population growth and the potential scarcity of resources. Sustainable agriculture demands a limited use of nonrenewable resources to protect the environment and future generations. The strategy promotes reducing land usage, transitioning to renewable energy sources, and eliminating natural pollutants. Farmers look for effective executive solutions, and data for the most precise and reliable analytics is provided by remote sensing technology in sustainable agriculture. Sustainable agriculture is a system of farming that strives to provide the resources necessary for present human populations while conserving the planet's ability to sustain future generations. Despite its efficiency, modern industrial agriculture has several drawbacks, including the degradation of ecosystems and the related biodiversity loss and loss of crop diversity. Such issues prioritize "planetary health," i.e., the planet's stability determines human well-being. Its basic canon includes promoting socioeconomic equity, earning profit, and maintaining ecosystem health. Because modern agriculture has played a substantial role in precipitating a mass extinction of plant and animal species on Earth, sustainable agriculture actively endeavors to protect and support biodiversity.

2.3 Crop Production

Sustainable agriculture emphasizes planting diverse crops, including heirloom plants, often suited to a region's climate. Rather than relying on a single crop in industrial monoculture, sustainable agriculture advocates using polyculture, where multiple crops are grown together. Although polyculture is frequently more labor-intensive than industrial monoculture, polyculture can reduce the need for chemical pesticides and fertilizers and improve soil quality. Similarly, crop rotation can help preserve soil productivity and lessen the need for agricultural chemicals for fertilization and pest control. Nitrogen-fixing cover crops, mother crops, and green manures can help restore soils and reduce wearing away. Composting crop residues and other agricultural wastes helps recycle nutrients to the farmland.

2.4 Animal Agriculture

Animal agriculture is a critical area that sustainable agriculture seeks to develop. Livestock production is responsible for a large proportion of the greenhouse gases driving global warming. Sustainably managing manure and implementing animal feed additives can reduce methane emissions, a potent greenhouse gas. Other methods of improving animal agriculture's sustainability focus on maintaining livestock health. Intensive animal agriculture can spur health crises. For example, diseases, including the Nipah virus and swine flu, have emerged from crowded factory farms. Avian influenza can be transmitted from wild birds to poultry farms and vice versa, and the disease has been responsible for the culling of millions of chickens and other poultry worldwide. Reducing animal crowding and increasing farm hygiene standards are sustainable agriculture measures that reduce the environmental health risks of cultivating livestock. Animal and crop production are often combined on smaller-scale farms to form interconnected systems that minimize waste.

2.5 Water Pollution and Conservation

Water conservation is a significant facet of sustainable agriculture. Globally, about 70% of all available freshwater resources are used for agriculture. Methods of reducing water waste can involve improving water storage practices to prevent evaporation losses and seepage and planting drought-resistant crops or crops that are appropriate for the climate. Many agricultural areas rely on simple flooding or surface irrigation as the principal means of irrigation. However, flooding often inundates fields with more water than crops require, and significant amounts of water are lost to evaporation or during transportation from the water source. Some sustainable farmers seek to implement reduced-volume irrigation, which provides slow streams of water to meet the water needs of specific crops while lessening water. Sustainable agriculture also aims to address the contamination of surface water and groundwater. Large-scale agriculture often produces pollutants, such as agrochemical runoff and pathogen-laden animal waste that seep into bodies of water and damage the surrounding environment, affecting wildlife and humans. Soil erosion also degrades water quality, and the loss of productive topsoil reduces crop yields and the total land available for agriculture. Farmers can reduce the frequency and intensity of tillage or practice no-till methods to mitigate these impacts. Organic or synthetic fertilizers and pesticides should be applied only sparingly and during dry conditions to minimize runoff; the judicious use of agricultural chemicals can reduce air pollution caused by airborne drift. Some farmers use buffer plants near waterways to absorb polluting nutrients before they can leach into bodies of water.

2.6 Conservation and Sustainable Use of Biodiversity

The present rate of biodiversity loss is up to 1000 times faster than the natural rate. Referred to by some scientists as "Earth's sixth mass extinction," the crisis is similar to the previous excellent extinction catastrophe that occurred 65 million years ago. These irrevocable extinctions pose a significant risk to human health and welfare. The foundation of biodiversity protection is the designation and administration of protected areas. Nevertheless, biodiversity is still declining even if there are more protected places in the globe overall. An integrated landscape approach to conservation planning is critical in ensuring suitable habitats for species. However, many protected areas need to be fixed as effectively as intended due to limited resources to maintain these areas and enforce relevant legal frameworks. In addition, current protected area networks may need to be re-aligned to account for climate change. Efforts to preserve biodiversity must consider the physical environment and social and economic systems well connected to biodiversity and ecosystem services.

Protective Measures

(a) While sustainable use and management of land outside protected areas remain a top concern, expanding protected areas in many nations contributes to preserving alternatives for the future.
(b) If the environmental state of conservation areas is to be improved, landscape-scale techniques and related measures are desperately needed if their efficiency is to be increased.
(c) Insufficient technical, financial, and capacity resources might hinder the expansion of creative solutions, underscoring the necessity of regional and subregional collaboration.
(d) Building capacity is essential for preventing and minimizing land degradation and well-informed restoration.
(e) Three levels of government should be involved in addressing capacity development needs: national, provincial, and local.
(f) To allow nongovernmental sources to advise pertinent departments and policies on biodiversity, capacity training is required.

UNESCO works on conserving biodiversity and the sustainable use of its components through UNESCO-designated sites, including biosphere reserves, World Heritage Sites, and UNESCO Global Geoparks. In 2018, UNESCO designated sites protected over 10 million km^2, an area equivalent to the size of China. These conservation instruments have adopted policies and strategies to conserve these sites while supporting the broader objectives of sustainable development. One such example is the policy on integrating a sustainable development perspective into the processes of the World Heritage Convention. UNESCO is also the depository of the Convention on Wetlands of International Importance, and countless species of plants and animals depend on these delicate habitats for survival.

2.7 Climate Change and Mitigation

Mitigating climate change entails lowering the heat-trapping greenhouse gases released into the atmosphere. This can be done in two ways: either by reducing the sources of these gases (such as burning fossil fuels for transportation, electricity, or heat) or by improving the "sinks" that absorb and store these gases (such as the oceans, forests, and soil). Avoiding significant human influence on Earth's climate is the aim of mitigation, which aims to "stabilize greenhouse gas levels in a timeframe sufficient to allow ecosystems to adapt naturally to climate change, ensure that food production is not threatened, and enable economic development to proceed sustainably." Reducing human exposure to the negative consequences of climate change—such as rising sea levels, more severe extreme weather events, or food insecurity—is the aim of mitigation.

Making the most of any favorable chances climate change brings also falls under this category (e.g., extended growing seasons or enhanced yields in particular places). Another mitigation aspect is adapting to a changing environment and the current or anticipated climate. It will be increasingly challenging the faster the environment changes. Although it affects the entire world, climate change is perceived locally. As a result, local governments are leading the adaptation charge. Addressing their climate-related issues has been a global priority for cities and local communities. Better-draining pavements are being installed to deal with floods and stormwater, plans for heat waves and greater temperatures are being made, flood defense is being built, and water storage and usage are being improved.

3 Investment in Education and Healthcare

Ensuring access to high-quality healthcare and education is essential for reducing inequality. This involves investing in healthcare facilities, schools, and vocational training programs, especially in underprivileged areas (Suffian et al.2015). Education for Sustainable Development (ESD), also known as Education for Sustainability (EfS), emphasizes the importance of human coexistence with the environment. The U.N. promotes ESD as a worldwide educational goal. Japan originally introduced the idea of ESD (Fig. 1) at the World Summit on Sustainable Development in 2002. Since then, UNESCO, the principal UN agency for ESD, has assumed the lead in international initiatives based on the international frameworks, the "Global Action Programme on Education for Sustainable Development (ESD) (GAP)" (2015–2019) adopted at the 37th UNESCO General Conference in 2013 and the "U.N. Decade of Education for Sustainable Development (DESD)" (2005–2014) adopted at the 57th Session of the General Assembly of the United Nations in 2002. According to the U.N., education for sustainable development (ESD) "empowers people to change their thoughts and work toward a sustainable future."

The globe is dealing with several issues brought on by human development activities, including resource depletion, biodiversity loss, climate change, and poverty. Accepting these problems of modern society as our own and tackling the issues in our immediate surroundings (think globally, act locally) are the first steps toward creating alternative values and transformational acts that lead to problem-solving and the realization of a sustainable society. ESD is made up of educational and learning activities that aim to create these kinds of values and transformational acts. This will guarantee that future generations may live in abundance because of humanity. ESD is the kind of education that helps people build sustainable societies. The education sector's response from UNESCO to the planet's pressing and significant concerns is called Education for Sustainable Development (ESD). Human activity has changed the earth's ecosystems to the point that our very survival is in jeopardy due to everyday changes that are harder to undo. As a worldwide spokesperson and working to improve the government's ability to deliver high-quality Climate Change Education (CCE), UNESCO develops and disseminates research, offers technical assistance and policy recommendations to its member states, and carries out on-the-ground programs. Through collaborations, networking, and the media, UNESCO promotes new ideas and improves non-formal education initiatives (Fig. 1).

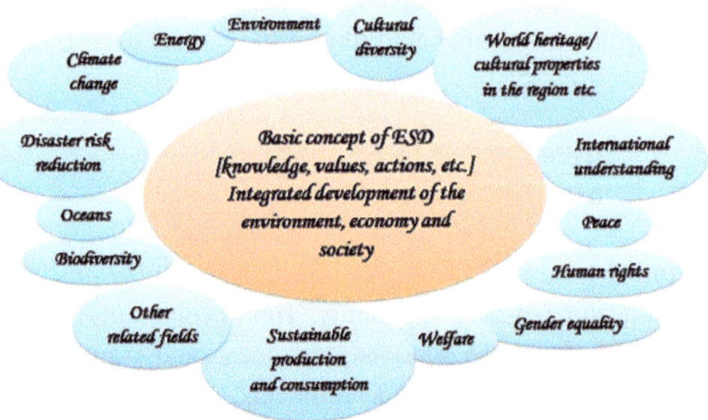

Fig. 1 Basic concept of education for sustainable development

ESD is also a part of Goal 4 of the Sustainable Development Goals set forth by the U.N., which is committed to achieving and accomplishing these goals by 2030. ESD for 2030 seeks to achieve all 17 SDGs and increase ESD implementation to create a more equitable and sustainable world. After ESD for 2030 was adopted, UNESCO released a roadmap outlining the precise steps that needed to be followed by this framework. The plan comprises five critical areas of action: (1) Policy advancement, (2) Learning environment transformation, (3) Educator capacity building, (4) Activating and empowering young people, and (5) Quickening actions at the local level and six crucial implementation areas: (1) Putting ESD for 2030 into practice at the national level, (2) Making the most of cooperation and partnerships, (3) Communicating to take action, (4) Tracking problems and trends, (5) Organizing resources, and (6) Keeping an eye on developments.

ESD Practices Continuous improvement of pedagogies and learning is vital, focusing on "proactive learning, interactive learning, and in-depth learning." This entails emphasizing the exploratory learning process—placing problem-solving learning in the proper context, enhancing opportunities for proactive, learner-centered learning, and incorporating experiences and activities—and carefully analyzing which learning process steps would be most successful and where to place them. Additionally, being conscious of the viewpoint of "building a sustainable society" might change students' values. To promote ESD effectively, the implementation of ESD must be positioned in the school management strategy, the internal school organization must be streamlined, and the school as a whole must systematically handle ESD. ESD is appropriately placed in the teaching plans, the perspective of collaboration with the local community, universities, and businesses is incorporated, and students' communication and reflection on learning outcomes are appropriately carried out.

4 Promotion of Gender Equality

Empowering women and girls through policies that promote gender equality in education, employment, and political participation can help reduce inequality. This includes equal pay for equal work, combating gender-based violence, and providing access to reproductive health services. Throughout the world, there are still disparities in wealth, sex, age, handicap, sexual orientation, color, class, ethnicity, religion, and opportunity. Inequality undermines people's sense of fulfillment and self-worth, endangers long-term social and economic growth, and damages efforts to reduce poverty. This can then lead to increased crime, illness, and environmental damage. If individuals are denied the opportunity to live better lives, we will never be able to accomplish sustainable development and improve the earth for all living things. Without access to healthcare, countless numbers of women and children perish every day from avoidable illnesses like T.B. and measles or during childbirth. Globally, prejudice and limited possibilities for older people, migrants, and refugees are problems that must be addressed. Five out of ten people said they had experienced discrimination based on at least one illegal factor under international human rights law. Issues and difficulties such as poverty, global warming, migration, and financial crises are never limited to a single nation or area. There are still areas in the wealthiest nations where people live in extreme poverty.

Even the oldest democracies struggle with religious intolerance, homophobia, transphobia, and racism. All of us are impacted by global inequality, regardless of our identities or origins. Ensuring a dignified existence for every individual is a feasible and worthy goal. Social, political, and economic policies must be universal and give special consideration to the needs of underprivileged and marginalized groups. Enacting radical change is necessary to reduce inequality. Extraordinary measures are required to end severe poverty and hunger and to increase funding for social safety, health, education, and suitable employment, particularly for young people, immigrants and refugees, and other vulnerable populations. Empowering people and fostering inclusive social and economic prosperity are crucial in nations. We can guarantee equal opportunity and lessen wealth disparities by eliminating discriminatory laws and regulations. For solutions to be more effective, trustworthy, and responsible, we must ensure that developing nations are more fully represented in the countries that make decisions on international matters. Governments and other stakeholders can also promote safe, regular, and responsible migration, including through planned and well-managed policies, for the millions of people who have left their homes seeking better lives due to war, discrimination, poverty, lack of opportunity, and other drivers of migration (Carroll 1991).

- The incomes of the poorest 40% of the population had been growing faster than the national average in most countries. However, emerging yet inconclusive evidence suggests that COVID-19 may have dented this positive trend of falling within-country inequality. The pandemic has also caused the most significant rise in between-country inequality in three decades.
- One in six people worldwide has experienced discrimination in some form, with women and people with disabilities disproportionately affected.
- The year 2022 witnessed the highest number of refugees (34.6 million people) ever documented. This year is also a deadly one for migrants, with nearly 7000 deaths recorded globally.
- Reducing within and between-country inequality requires equitable resource distribution, investment in education and skills development, social protection measures, combating discrimination, support for marginalized groups, and fostering international cooperation for fair trade and financial systems.

5 International Cooperation to Address Economic Inequality and Promote Sustainable Development

Economic inequality is a pervasive issue that has plagued societies worldwide for centuries. The gap between the rich and the poor has widened over the years, leading to numerous social and economic challenges. The impact of economic inequality on sustainable development cannot be ignored, as it threatens social stability, economic growth, and the environment. International cooperation is crucial in addressing economic inequality and promoting sustainable development across the globe. Economic inequality is a complex issue that affects various aspects of social and economic development. It refers to unequal wealth, income, and resource distribution among individuals and communities. Economic inequality can lead to poverty, poor health outcomes, limited access to education, and limited social mobility. A recent report by Oxfam International indicates that the world's richest 1% possess more than twice the wealth of 6.9 billion people. This inequality has far-reaching consequences, including poverty, poor health outcomes, limited access to education, and limited social mobility. Sustainable development addresses these challenges by promoting economic growth, reducing poverty, and ensuring social and environmental sustainability.

International cooperation is essential in addressing economic inequality and promoting sustainable development. The global nature of economic inequality requires collective action and collaboration among nations and international organizations. Another significant effect of economic inequality is the limited access to healthcare. People from low-income backgrounds are often unable to afford quality healthcare and are more likely to suffer from chronic illnesses. This burdens healthcare systems unequally and ultimately affects the population's overall health. Economic inequality thus threatens sustainable development by hindering economic growth and development. Countries with high levels of economic inequality often experience lower economic growth, as wealth and resources are concentrated in the hands of a few. This can make it difficult for countries to invest in infrastructure, social services, and other initiatives that promote sustainable development.

International cooperation is crucial in addressing economic inequality because it is a global issue that requires a collective response. Countries must work together to address the root causes of economic inequality, develop policies that promote economic equality, and ensure sustainable development. International cooperation allows countries to share resources, knowledge, and best practices to promote economic equality and sustainable development. The World Bank, International Monetary Fund, and United Nations (1987) are examples of international organizations that significantly promote economic equality and sustainable development. The World Bank provides financing, technical assistance, and policy advice to developing countries to promote economic growth and poverty reduction. The International Monetary Fund promotes global financial stability and assists needy countries financially. The United Nations plays a critical role in promoting sustainable development by implementing the Sustainable Development Goals (SDGs), which aim to end poverty, protect the planet, and ensure prosperity for all.

6 Fairtrade and Economic Policies

Many developing countries face significant debt burdens and are vulnerable to unfair trade practices. Addressing debt sustainability through debt relief initiatives and advocating for fair trade policies can help mitigate economic inequalities and promote more equitable global economic relations. Fair Trade (FT) is a societal movement that aims to support poor and vulnerable producers in developing nations to attain improved trading conditions with direct links to consumers and excluding mediators in the trading chain (Barratt 1993; Young and Utting 2005). The concept of FT, in general, indicates that it has, at its core, two main factors or sights. The first sight is to provide an effective model of international trade that differentiates between consumers' and producers' involvement. The second and more important aspect is to defy the accepted view in business practices, not just through publicity but as an "instrument to revise the influential economic model" (Renard 2003). Therefore, FT allows poor producers to be part of a trading organization that ensures a fair and steady price for their products. It also provides them and their systems with different levels of banking facilities and stimulates a sustainable environment. Fair Trade (FT) is being measured as an alternative to traditional trade and is gradually gaining popularity worldwide (Reed 2009). The concept is related to the movement that seeks to sustain equity to a greater extent in international trade partnerships between developing countries through dialogue, transparency, and respect (Charter 2009). Therefore, a secondary goal of moving toward FT is to raise consumers' consciousness of unjust and unfair global trade practices. So, FT is more than just trading; it demonstrates that justice is possible at the greatest level in world trade. It focuses on the necessity of changing the procedures and practices of traditional trade and presents how an effective trade can place people first. It plays a perceptible role in competing against poverty, economic crisis, and climate change. In addition, it enhances sustainable development (SD) by proposing better environments for trading and obtaining the freedom and rights of producers in developing countries that are marginalized (Cremonas and Duran 2012). Charter (1992) defined SD as "development that meets the needs of the present without compromising the ability of future generations to meet their own needs. The concept is focused on economic development, social development, and environmental protection for future generations." Moreover, three fundamental views ensure movement, as well as the contribution of FT to SD (Strong 1997): First, establishing a more coherent association between a group of consumers and producers in the two worlds and a better level of perception between consumers of the need of producers to assist their independent improvement. Second, traded goods fairly stimulate a pattern of production and consumption that decreases pressure on the environment and meets the needs of humankind. Lastly, establishing FT practices throughout consumer relationships helps decrease environmental degradation and poverty. Although the contribution of FT to SD is observable, there still is a knowledge gap about the status quo in the so-called changeover economies. The Concept of Fair Trade, the fierce competition for limited resources, and the globalization process that began in the second half of the twenty-first century have led to controversy over the side effects of these processes, including natural environmental degradation and increasing income inequality between developed and developing nations. FT helps to create a model of economic development in the least developed countries for the expansion of capital and the usage of natural resources. FT is the best course of action to meet the social needs of a community based on its specific environmental circumstances. FT provides a platform for alleviating poverty and environmental problems (Huybrechts 2010). The provision of FT practices also plays a role in building a consumer-producer relationship (Strong 1997). Therefore, FT creates wealth and plays a creative role in achieving economic sustainability by rewarding all stakeholders fairly and equally (Jacobs 1991; Pearcy and Schrock 2014). Also, the principle of FT incorporates noneconomic standards of equity, i.e., ethical aspects, environmental standards, and standard of life, into the distribution of profits and trading facilities. Thus, the goals of FT differ from conventional economic goals and are not based on purely financial goals but on fair price (i.e., an FT premium or an FT minimum price), sustainable production costs, sustainable use of natural resources, and rebuilding community profits for health care and education. As per the Brundtland report, FT's relationship with the environmental dimension is possible by maintaining "the needs of the present generations without compromising the ability of future generations to meet their own needs." Organizations should try to attain environmental sustainability within ecological limits (Ketola 2009). The principle of FT thus combines preservation assessment and collaboration with good working quality and environment. In this context, it deals with the widespread utilization of mineral and forest resources to improve the quality of life and access to a greater number of consumer products and services by a portion of humanity. Ketola (2009) explained using renewable energy, recyclable or renewable natural resources, and zero emissions, such as the environmental dimension of SD As a strategic part of climate change, FT has developed FT climate standards to support smallholders and rural populations in becoming more prosperous in the effects of climate change internationally (FLO 2015). The focal points are minimized and protected usage of agrochemicals, safe and proper waste management, maintaining the fertility of soil and water resources, prohibited use of heritably modified bacteria, and pursuing Sustainable Development Goals (WFTO 2008, 2013).

6.1 Barriers to Fair Trade and Way Forward as a Whole

FT movement supports SDG by eliminating poverty, ending hunger, achieving gender equality, promoting decent work for all, ensuring sustainable consumption and production patterns, taking urgent action to combat climate change, promoting peaceful and inclusive societies for SD, providing access to justice for all, and strengthening the means of implementation and revitalize the global partnership for SD (Ronchi (2002), Jaffee (2007), Calo and Wise (2005)). The FT initiatives have significantly improved the product quality of smallscale coffee farmers through training and improved skills, as well as the well-being of their families, especially with credit facilities and external funding (Murray et al. 2003). FT farmers were more successful in expanding their production, gaining better pricing experience for their crops, and seeing improved monthly household food intake and livelihoods, significantly reducing infant mortality (Becchetti and Constantino 2008). FT can make an authentic contribution to SD by reducing poverty and sustaining development in developing nations (Boto and Peaccerell 2008). The FT movement and the trade and development constituencies can support ensuring that this contribution is going forward. So, at last, it remarks us that FT mainly serves two objectives for developing countries, namely, it offers real prospects for the improvement of smallscale workers and producers in developing nations, and it has had a positive impact on the worldwide trading system and private sectors, encouraging them to be more equitable, socially inclusive, and supportive of SD.

7 Capacity Building and Knowledge Sharing

Building institutional capacity and promoting knowledge sharing among countries can effectively enhance their ability to implement sustainable development policies. This involves providing technical assistance, promoting South-South cooperation, and facilitating technology transfer and innovation. Global efforts to foster sustainable development came together at the Rio "Earth Summit" UNCED conference in 1992. At Rio, the agreement on the Rio Declaration and Agenda 21 sought to define principles and an "action plan" for global sustainable development. It was aimed at fostering sustainable development through domestic and international actions. The following are of p relevance: "States should cooperate to strengthen endogenous capacity-building for sustainable development through exchanges of scientific and technological knowledge and by enhancing the development, adaptation, diffusion, and transfer of technologies," reads Principle 9 of the Rio Declaration, which addresses technology transfer explicitly. The Declaration further stated that "States should cooperate to promote a supportive and open international economic system" as well as that "States should effectively cooperate to discourage or prevent the relocation and transfer to other States of any activities and substances that cause severe environmental degradation or are found to be harmful to human health." Several other Rio Declaration principles addressed requirements for states to develop domestic policies supportive of sustainable development, including participation, environmental legislation, liability and compensation for ecological damage, internalization of environmental costs, environmental impact assessments, and international cooperation. Environmentally sound technologies should be available to developing countries, including expertise and related issues. Developing countries should have access to relevant information on technological choices, and international information exchange systems and clearing houses should be developed. Access to and transfer of environmentally sound technology should be promoted "on favorable terms, including on concessional and preferential terms, as mutually agreed, taking into account the need to protect intellectual property rights and the special needs of developing countries for the implementation of Agenda 21". Specific measures should include policies and programs to encourage public and private technology transfer, regulatory measures, including subsidies and tax policies, and appropriate mechanisms for improved access and transfer of relevant technologies. National capacities, particularly of developing countries, should be built to develop and manage environmentally sound technologies, including human resource development, and strengthen research and development capacities. The development of indigenous technology and technology assessment should be promoted, "a collaborative network of research centers" should be established, and "programs of cooperation and assistance" should be strengthened. The importance of technology transfer through business commerce is recognized, and while the availability to developing countries is of concern, "fair incentives to innovators" should also be provided; here, the role of patent and property rights should be examined. Long-term partnerships between holders and users of environmentally sound technologies, companies in developed and developing countries, and joint ventures should be promoted. The Montreal Protocol on Substances that Deplete the Ozone Layer is one of the most successful international treaties related to global environmental problems. It provides an excellent example of an international agreement continuously amended in response to new scientific and technical assessments.

The Convention on Biological Diversity was adopted at the Rio Summit in 1992. It addresses one of the most pressing issues of the modern era: sustainable management of the planet's living resources. It adopts a comprehensive rather than a sectoral approach to conserving the planet's biodiversity and sustainable use of its biological resources for the first time. In the four sessions of the COPs, the Parties developed a mechanism for the normative development and implementation of the Convention. The Convention provides "access to and transfer of technology" in Article 16. When this is read with Article 19, dealing with "handling of biotechnology and distribution of its benefits," it reflects the years of North-South debate in other fora over the issue of technology transfer and the issue of intellectual property rights Glowka et al. (1994). The closely interlinked provisions along with Articles 16 and 19 are Articles 12 (research and training), 17 (exchange of information), and 18 (technical and scientific cooperation). Together, they seek to "provide and facilitate access for and transfer to other Contracting Parties" of technologies relevant to the conservation of biological diversity, technologies pertinent to the sustainable use of its components, or technologies that use genetic resources.

CBD includes provisions for biodiversity-rich developing countries to regulate access to and use of these resources. This approach is now carried out in access to genetic resources and benefit-sharing agreements worldwide Reid et al. (1993). A vital component of the CBD supporting such mechanisms was the recognition of national sovereign rights over genetic resources. The CBD explicitly addresses access to and transfer of technology relevant to the conservation and sustainable use of biological diversity, including biotechnology. Such access and transfer to developing countries should be facilitated under fair and favorable terms, including concessional and preferential terms where mutually agreed upon and by the financial mechanism (GEF). The situation calls for a careful balance between technology needs, financial requirements, appropriate regulatory frameworks, and cooperation among countries. In this context—and for any other international agreement with technology transfer provisions—national strategies to determine technological needs play an essential role. The CBD also established that transfer should be under terms that recognize and are consistent with intellectual property rights over these technologies. Since the CBD is concerned directly with access to and use of genetic resources, IPRs have played a more central role than likely in the context of climate change. The debate has also been highly contentious, with some developing countries arguing that TRIP provisions may need to be amended or set aside in the CBD's context and several Western governments and industry groups vigorously contesting this position. Technology transfer in the case of technologies related to genetic resources, particularly biotechnology, which is primarily developed by the private sector, presents specific differences concerning climate change-related technology transfer.

8 Infrastructure Development

Investing in infrastructure such as roads, bridges, ports, and energy systems is essential for promoting economic development and reducing inequality. Improving infrastructure connectivity in rural areas can facilitate access to markets, healthcare, and education while attracting investment and creating jobs (Karamchandini et al. 2009). Sustainable infrastructure systems are planned, designed, constructed, operated, and decommissioned to ensure economic, financial, social, environmental (including climate resilience), and institutional sustainability over the entire infrastructure life cycle. Sustainable infrastructure can include built, natural, or hybrid infrastructure containing both elements. Significant investment in sustainable and resilient infrastructure is required to achieve the Sustainable Development Goals by 2030 and net zero emissions by 2050. The Organisation for Economic Cooperation and Development (OECD) estimates that USD 6.9 trillion per year is needed up to 2050 for investment in infrastructure to meet development goals and create a low-carbon, climate-resilient future. According to the Global Infrastructure Hub, however, there is currently a multi-trillion-dollar gap in these required investments, and evidence shows that most investments continue to go into "business-as-usual" infrastructure. A resolution on sustainable and resilient infrastructure was approved by member states of the U.N. Environment Assembly (UNEA) at its fifth session, which convened in Nairobi in March (UNEA/E.A.5/L.15). By encouraging the Member States to implement the International Good Practice Principles for Sustainable Infrastructure, promote investment in natural infrastructure, resilient infrastructure, and nature-based solutions, cooperate internationally to strengthen frameworks, including financing, and offer opportunities for engaging relevant stakeholders, this new resolution expands on a 2019 UNEA resolution (UNEP/EA.4/L.5). The Sustainable Infrastructure Partnership (SIP), which was established in 2018, provides a forum for the advancement and assistance of integrated methods for the design and implementation of sustainable infrastructure. As acknowledged by SIP, infrastructure has a crucial role in the 2030 Agenda for Sustainable Development. SIP approaches its work with an interdependent lens by considering the connections between various infrastructure systems, sectors, stages, and governance structures. The social, environmental, and financial benefits of infrastructure development may be maximized with this strategy.

9 Innovation and Entrepreneurship

The SDGs can only be realized with strong global partnerships and cooperation. The intersection of entrepreneurship, innovation management, and sustainable development has become increasingly important in an era characterized by rapid environmental and social changes. Entrepreneurs and innovators play a pivotal role in shaping the future as they create new business and innovator ventures, especially start-ups, products, and technologies that have the potential to drive positive change while addressing critical global challenges (Apostolopoulos et al. 2018; Iglesias-Sánchez and Jambrino 2024). The world is more interconnected than ever. Improving access to technology and knowledge is an important way to share ideas and foster innovation. Coordinating policies to help developing countries manage their debt and promoting investment for the least developed is vital for sustainable growth and development. The U.N. Sustainable Development Goals address numerous development issues. The 169 goals that make up the 17 U.N. Sustainable Development Goals (SDGs) address various issues, such as eradicating hunger and poverty and providing access to inexpensive, clean energy, clean water, and hygienic conditions. Numerous approaches have been employed to tackle these global development issues. These consist of loans from development banks, donor grants, and business concepts. Based on the past performance of these various strategies, the donor aid and development bank lending models have shown some degree of promise. However, the loan and aid approaches have come under fire for fostering reliance on them, especially in African nations (Moyo 2009). Several authors have promoted the entrepreneurial approach (Polack and Warwick 2013; Prahalad 2010). Sustainability and scalability are important factors in entrepreneurial models but are not as emphasized in the other two strategies. The funds given or lent for a particular project under the donor assistance and bank loan models often support the development activity for a limited time or scope. This is helpful and undoubtedly has a place in certain infrastructure-related projects. However, more entrepreneurial venture techniques are needed, given the scope and urgency of the demands in this field (Lokman and Chahine 2021). Additionally, entrepreneurial techniques provide chances for sustainability, long-term growth, and new revenue streams. Specifically, this section will outline the common themes found in ineffective business models that have been studied in the field. By identifying these successful models, the opportunity to apply them to other U.N. SDGs may be explored. The professional association for electrical, computer, and electronic engineers is called the Institute of Electrical and Electronic Engineers (IEEE). It has more than 450,000 members globally, making it the biggest professional association in the world. The IEEE Smart Village initiative is a flagship project of the IEEE Foundation, which supports humanitarian causes inside IEEE. About providing power access to the nearly 800 million people who lack it globally, this program assists entrepreneurs in bringing technology-based solutions to development difficulties in the field. To pilot a solution, the program first receives seed financing. It then receives matching funds to expand and scale the initiative to provide one million individuals with access to energy. Paras Loomba, Managing Director of Global Himalayan Expeditions (GHE), is one of the beneficiaries of the IEEE Smart Village financing, which has expanded his entrepreneurial enterprise. The initiative began with funding to electrify five isolated villages in Ladakh, India and has expanded to include 131 communities. The communities in this area now have more income sources because of the business strategy used here. Goals UN SDG 7—Affordable and Clean Energy—as well as UN SDG 8—Decent Effort and Economic Growth—and UN SDG 13—Climate Action—are all met as a consequence of this effort. The Energy Sector Management Assistance Programme (ESMAP) of the World Bank provides analytical and consultancy services in the renewable energy sector to low- and middle-income countries (LMICs). It is an international program providing technical support and information. "Mini Grids for Half a Billion People," a technical report published by ESMAP in September 2022, describes how investments in renewable energy might significantly boost the economies of LMICs (Energy Sector Management Assistance Programme 2022). This report's executive summary, released in 2019, highlights business strategies that use energy-efficient equipment and profitable ways to use power to boost local economies.The UN SDG 8 goal of decent work and economic growth, as well as the UN SDG 7 goal of clean and affordable energy, are addressed by this strategy. Sun King offers a range of products, starting with the entry-level Sun King Pico Plus solar LED light[://sunking.com/solar-lanterns/sun-king-pico-plus, 2023]. At 1169 Kenyan Shillings (less than $10), this solar light costs less than $10 in Kenya and has five times the brightness of a kerosene lamp. Solar LED lights are far more economical, healthier, and cleaner than kerosene lamps, which can be costly and expose children to harmful fumes and fire threats. Additionally, Sun King provides updated products, such as a solar house system that can recharge cell phones, several lights, and a larger solar panel. With additional funding of $70 million from Leap Frog Investments, Sun King recently increased its Series D investment to $330 million. The Sun blazer unit is produced as part of the IEEE Smart Village program (Singh et al. 2013; Singh 2022). With this modular design, batteries are housed in the frame of the towable trailer, where solar panels are attached. This system, which usually starts at 1 kW, can be towed to isolated areas and configured into a scalable system to power a hamlet. Larger-scale systems have recently been installed in villages by Renewable Energy Innovators Cameroon (REIC) to supply electricity for operating businesses. In addition, REIC is the first certified manufacturer of Sunblazer units in Africa. Tiny enterprises like barbershops, tiny workshops, refrigerators that serve cold drinks, and so on are supported by these multi-kilowatt systems. Once more, this gives small companies in developing communities access to new sources of income. Pay-as-you-go (PAYG) financing has been a major paradigm shift in recent years for supplying access to energy in low- and middle-income countries (LMICs). Households that cannot afford the upfront cost of a solar electric system can receive energy access in modest amounts, much like low-income households can purchase cell phone credits in small amounts to gain access to mobile phone service. Such schemes are provided by several organizations, such as Sun King [https://sunking.com/], delight [https://www.dlight.com/], and M-Kopa in Kenya [https://m-kopa.com/]. A small solar electric system (usually approximately 100–200 W) is installed in the client's home by the company for which the client pays a down payment. The corporation then finances the remainder of the system. A box that can connect and disconnect the electricity supply is located between the home and the solar electric system. The client can pay for an increment of electricity, e.g., 5 kWh, through mobile payments. The company can then unlock the box to access this chunk of electricity. Once consumed, the client can send a mobile payment to the company to purchase another chunk of electrical energy (Wuebben et al. 2020).

10 Global Partnerships

Implementing the United Nations Sustainable Development Goals (SDGs) has become increasingly important, as the COVID-19 pandemic has exacerbated many existing inequalities. These growing inequalities, especially those observed between industrialized and developing countries, hamper equitable, sustainable development in both social and environmental terms. For this reason, it is essential to recognize the importance of SDG 17—which focuses on partnerships—as necessary in meeting the other SDGs (Walter et al. 2022). SDG 17 advocates for greater cooperation between public, private, and third-sector organizations, regardless of their origin and size, to implement sustainable development, particularly with developing countries. It is of fundamental importance for the continuous development of these associations and for understanding their role. The complex challenges society faces on top of climate change, such as the COVID-19 pandemic, remind us of the constant and growing demands and the urgent need for equitable international partnerships. With good governance, partnerships among international organizations and local and regional associations can help address these growing deficits (Baimenov and Liverakos 2019). Nevertheless, there are many challenges to the successful implementation of SDG 17. Achieving the Sustainable Development Goals (SDGs) requires strong international collaboration and partnerships. In 2017, official development assistance totaled US$147 billion, remaining consistent but falling short of the goal. There is a continued need for increased funding and help to address humanitarian emergencies caused by war or natural disasters. Many countries also require official development assistance to promote trade and economic progress. Improving access to technology and information is crucial for promoting innovation and sharing ideas in today's interconnected world. Sustainable growth and development rely on coordinating policies to support investments in the least developed countries and help emerging nations manage their debt. The objective is to support national programs in achieving all the targets to enhance cooperation between developed and developing nations. Achieving a global trading system that is open, transparent, and benefits everyone involves promoting commerce internationally and assisting underdeveloped nations in growing their exports. International partnerships play different roles in achieving the SDGs depending on the organizations involved, the project requirements, and the setting. To achieve the SDGs faster, more effectively, and more fairly than can be accomplished by operating in silos, the overall objective is to forge synergies among the knowledge and resource bases of international organizations, nongovernmental organizations (NGOs), governments, and private sector institutions (Detomasi 2014). Ultimately, these synergies are intended to produce win–win circumstances in which each partner accomplishes their objectives and the shared ones (Beisheim and Simon 2018). There are five main categories of international partnerships, according to Bull and McNeill (2019a, b): market-based operations, policy formulation and lobbying, advocacy, resource mobilization, and local implementation. International partnerships can advance the agenda of SDGs in addition to these significant roles. These include advancing favorable regulations, creating cross-sectoral connections at the policy level, modifying individual and organizational behavior, and empowering marginalized and vulnerable people (Stott and Murphy 2020). In reality, partnerships take many forms depending on the unique set of abilities each partner brings to the table and the problems they aim to solve. Partnerships unquestionably help society's commitment to global sustainability by promoting economic progress, empowering individuals, and concurrently concentrating on sustainable development. The urgency of the issues these partnerships will attempt to solve is rising along with their necessity. Prominent challenges reinforcing each other's causes and remedies include economic inequality, political polarization, COVID-19 pandemic recovery and preparedness, climate change, and wars. Even though they are widespread and can have gradual beginnings, dealing with them frequently necessitates quick responses. To solve these issues and improve society, it is essential to use long-term, sustainable international collaborations as change agents (El-Jardali et al. 2018). SDG 17 presents many implementation issues, which are made worse by the SDGs' broad scope and lack of defined accountability and duty (Persson et al. 2016). Economics presents one of the most significant obstacles. The pursuit of the SDGs at the national policy level is frequently subordinated to economic development and stabilization, which has a detrimental impact on private enterprises' participation in SDG collaborations (Leal Filho et al. 2020; 2022).

Business has cooperated with multilateral organizations through public-private partnerships (PPPs) since the late 1990s. With their adoption of sustainable development goals (SDGs), multilateral institutions increasingly consider partnerships as a means to achieve their goals, given their limited implementation capacity. However, the global economic order has changed significantly since the first expansion of PPPs, mainly due to growing participation by non-western states and companies (Cheng et al. 2018). PPPs involving companies from different varieties of capitalism—private, market-based forms, and state-led forms of capitalism have been studied. It has been found that companies and other actors from Western countries still dominate PPPs. Moreover, businesses participate more in US- and Canadian-led partnerships than others. It has also been found that there are strong differences regarding what category of PPPs companies from different backgrounds engage in and discuss the linkages between varieties of capitalism and PPP participation (Bull and McNeill 2019a, b).

11 Conclusion

Inequality threatens social and economic development, harms poverty reduction, and destroys people's sense of fulfillment. Inequalities based on income, sex, age, disability, sexual orientation, race, class, ethnicity, religion, and opportunity continue to persist worldwide, leading to crime, disease, and environmental degradation. Discrimination has many forms, pointing to the urgent need for measures to tackle discriminatory practices and hate speech. Regardless of our background, poverty, climate change, migration, and economic crises affect us all. Reducing inequality requires transformative change, including eradicating extreme poverty, investing in health, education, social protection, decent jobs for vulnerable communities, and promoting inclusive growth. Eliminating discriminatory laws, policies, and practices can ensure equal opportunity and reduce income inequalities.

References

Apostolopoulos N, Al-Dajani H, Holt D, Jones P (eds) (2018) Entrepreneurship and the UN Sustainable Development Goals. R. Newbery Emerald Publishing

Baimenov A, Liverakos P (2019) Public Service excellence in the 21st century Ch, vol 10, pp 319–339

Barratt Brown M (1993) Fair Trade. Reforming the international trading system. Zed Books, London

Becchetti L, Constantino M (2008) The effects of fair trade on affiliated producers: an impact analysis on Kenyan farmers. World Dev 36(5):823–842

Beisheim M, Simon N (2018) Multistake holder partnerships for the SDGs: actors' views on UN metagovernance. Glob Gov 24:497 515

Boto I, La Peccerella C (2008) Does fair trade contribute to sustainable development? Resources on fair trade, Brussels rural development briefings a series of meetings on ACP-EU development issues

Bull B, McNeill D (2019a) From market multilateralism to governance by goal setting: SDGs and the changing role of partnerships in a new global order. Bus Polit 21:464–486

Bull B, McNeill D (2019b) From market multilateralism to governance by goal setting: SDGs and the changing role of partnerships in a new global order. Cambridge University Press

Calo M, Wise TA (2005) Revaluing peasant coffee production: organic and fair trade markets in Mexico, Global Development and Environment Institute. Tufts University, Medford

Carroll AB (1991) The pyramid of corporate social responsibility: toward the moral management of organizational stakeholders. Bus Horizons 34(4):39–48

Charter M (1992) Greener marketing. Greenleaf Publishing, Sheffield

Charter (2009) A charter of fair trade principles. FLO, WFTO. https://www.wfto.com/sites/default/files/Charterof-Fair-Trade-Principles-Final%20(EN).PDF

Cheng Z, Yang Z, Gao H, Tao H, Xu M (2018) Does PPP matter to sustainable tourism. Sustainability. https://doi.org/10.3390/su10114058

Cremona M, Duran GM (2012) 5 Fair trade in the European Union. The processes and practices of fair trade: trust, ethics and governance, p 122. 13
Detomasi D (2014) The multinational corporation as a political actor: 'Varieties of Capitalism' revisited. J Bus Ethics 128:685–700
El-Jardali F, Ataya N, Fadlallah R (2018) Changing roles of universities in the era of SDGs: rising up to the global challenge through institutionalising partnerships with governments and communities. Health Res Policy Syst 16:38
Energy Sector Management Assistance Program (2022) Mini grids for half a billion people: market outlook and handbook for decision makers. In: License: creative commons attribution CC BY 3.0 IGO. The full ESMAP report is available for download at the following website. World Bank, Washington, DC
FLO (2014) Fairtrade standard for hired labour. Fairtrade International, 15.01.2014_v1.3. https://www.fairtrade.net/fileadmin/user_upload/content/2009/standardsdocuuments/HL_EN.pdf
FLO (2015) Annual report 2014–2016, global change, local leadership. Fairtrade International. https://annualreport14-15.fairtrade.net/en/
Glowka L, Burhenne-Guilmin F, Synge H, McNeely JA, Gündling L (1994) A guide to the convention on biological diversity, Chicago
Gusmao Caiado RG, Leal Filho W, Quelhas OLG, de Mattos L, Nascimento D, Ávila LV (2018) A literature-based review on potentials and constraints in the implementation of the sustainable development goals. J Clean Prod 198:1276–1288
Huybrechts B (2010) Explaining organisational diversity in fair trade social enterprises. Doctoral dissertation, HEC-Ecole de Gestion de l'Université de Liège, Liège, Belgique
Iglesias-Sánchez PP, Jambrino C (2024) Special issue: entrepreneurship and innovation management under sustainable development
IRENA (2020) Innovation landscape brief: pay-as-you-go models. International Renewable Energy Agency, Abu Dhabi
Jacobs M (1991) The green economy. Pluto Press, London
Jaffee D (2007) Brewing justice: fair trade coffee, sustainability and survival. University of California Press, Berkeley
Karamchandini A, Kubzansky M, Frandano P (2009) Emerging markets, emerging models. The Monitor Group, Rockville
Ketola T (2009) Five leaps to corporate sustainability through a corporate responsibility portfolio matrix. Corporate social responsibility and environmental management (published online) in Wiley interscience. Latin America, Colorado State University: Fair Trade Research Group
Leal Filho W, Brandli LL, Lange Salvia A, Rayman-Bacchus L, Platje J (2020) COVID-19 and the UN sustainable development goals: threat to solidarity or an opportunity? Sustainability 12. https://doi.org/10.3390/su12135343
Leal Filho W, Wall T, Barbir J, Alverio GN, Dinis MAP, Ramirez J (2022) Relevance of international partnerships in the implementation of the U.N. Sustainable Development Goals Nature Communications, vol 13
Lokman L, Chahine T (2021) Business models for primary health care delivery in low- and middle-income countries: a scoping study of nine social entrepreneurs. BMC Health Serv Res 21:211
Moyo D (2009) Dead aid. Penguin Books, London
Murray DL, Raynolds LT, Taylor PL (2003) One cup at a time: poverty alleviation and fair trade coffee in Latin America. Fair Trade Research Group, Colorado State University
Pearcy DH, Schrock WA (2014) Fair Trade awareness: exploring its relationship to public policy and sustainable outcomes. Res Bus Econ J
Persson Å, Weitz N, Nilsson M (2016) Follow-up and review of the sustainable development goals: alignment vs. internalization. Rev Eur Comp Int Environ Law 25:59–68
Polack P, Warwick M (2013) The business solution to poverty. Berrett-Koehler Publishers, San Francisco
Prahalad CK (2010) The fortune at the bottom of the pyramid. Pearson Education, Upper Saddle River
Reed D (2009) What do corporations have to do with fair trade? Positive and normative analysis from a value chain perspective. J Bus Ethics 86(1):3–26
Reid R (1993) Establishing and sharing collections of valuable global resource
Renard MC (2003) Fair trade: quality, market and conventions. J Rural Stud 19:87–96
Ronchi L (2002) The impact of fair trade on producers and their organizations: a case study with COOCAFE in Costa Rica, Poverty research unit working paper no 11. University of Sussex, Brighton
Singh P, McDermott-Levy R, Keech E, Mariani B, Klingler J, Moncada MV (2013) Challenges and successes in making health care more accessible to rural communities in Waslala, Nicaragua using low-cost telecommunications. In: Proceedings 2013 IEEE global humanitarian technology conference. IEEE, San Jose/Piscataway
Singh P (2022) Collaboration between Landivar University in Guatemala and Villanova University in the US on Humanitarian Projects for Computer Scientists and Engineers. In: Proceedings American Society of Engineering Education conference, Minneapolis
Stefanaska M, Nestorowicz R (2015) Development of fair trade idea in Europe and the United States. In: Stefańska M, Nestorowicz R (eds) Fair trade in CSR strategy of global retailers. Palgrave Macmillan
Stott L, Murphy DF (2020) An inclusive approach to partnerships for the SDGs: Using a relationship lens to explore the potential for transformational collaboration. Sustainability 12(19):7905
Strong C (1997) The problem of translating fair trade principles into consumer purchase behaviour. Mark Intell Plan 15(1):32–37
Suffian S, Lackey J, Dzombak R, Mehta K (2015) Demonstrating demand for preventive health services in rural Kenya. J Hum Eng 3(1):24–35
United Nations (1987) Brundtland report
Walter EB, Talaat KR, Sabharwal C, Gurtman A, Lockhart S, Paulsen GC, Gruber WC (2022) Evaluation of the BNT162b2 Covid-19 vaccine in children 5 to 11 years of age. N Engl J Med 386(1):35–46
WFTO (2008) Charter of fair trade principles. World Fair Trade Organization
WFTO (2013) 10 principles of fair trade. World Fair Trade Organization. https://www.wfto.com/sites/default/files/10-FAIR-TRADE-PRINCIPLES-2013-(RioAGM-and-EGM-2013-approved-modifications).pdf
Wuebben D, Romero-Luis J, Gertrudix M (2020) Citizen science and citizen energy communities: a systematic review and potential alliances for SDGs. Sustainability. https://doi.org/10.3390/su122310096
Young W, Utting K (2005) Fair trade, business and sustainable development. Sustain Dev 13(3):139–142

Sustainable Development and the Global Economy: Pathways to Achieving Food Security

Bisma Farooq, Madiha Farooq, Shahnaz Anjum, Asma Nazir, Abeer Hashem, and Elsayed Fathi Abd-Allah

Abstract

In recent years, there has been a discernible acknowledgment within the global economic landscape of the intricate nexus between sustainable development and food security. This chapter embarks on an exploration of the nuanced relationship between these pivotal elements, delving into the strategic utilization of sustainable development methodologies to fortify food security for both current and forthcoming generations. At its essence, sustainable development embodies the convergence of economic advancement, environmental stewardship, and societal equity. Through the implementation of sustainable agricultural methodologies, the fortification of resilient food systems, and the empowerment of rural communities, nations stand poised to confront the multifaceted challenges stemming from climate variability, dwindling resources, and demographic shifts, thereby safeguarding food security. The strategic integration of sustainable development principles within the contours of the global economy presents a promising trajectory toward achieving universal food security. By embracing a culture of innovation, fostering synergistic collaborations, and ensuring policy coherence, the global community can architect a food system that is not only resilient but also sustainable, ensuring equitable access to nutritious sustenance for present and future generations.

Keywords

Global economy · Sustainable development · Food security · Sustainable agriculture · Resilient food systems · Climate change · Resource scarcity · Population growth

1 Introduction

The worldwide economy remains at an intersection, fighting with the test of guaranteeing manageable improvement while resolving the major food security problem. As the total populace proceeds to develop and ecological tensions strengthen, the requirement for coordinated work to accomplish food security through supportable means has never been more essential. In this section, the interconnectedness of the worldwide economy, economic turn of events, and food security, featuring the meaning of resolving these issues, by and large, to prepare for an additional fair and versatile future will be discussed.

2 Global Economy

The worldwide economy envelops the complex snare of monetary collaborations, exchange relations, and monetary streams across countries and land masses. Over the last 100 years, the worldwide economy has undergone extraordinary development and change, filled by innovative progressions, exchange advancement, and globalization. Even so,

B. Farooq (✉)
Department of Biomedical Engineering, Sathyabama Institute of Science and Technology, Chennai, India

M. Farooq
Department of Microbiology, Hemwati Nandan Bahuguna Garhwal University, Garhwal, India

S. Anjum
Department of Botany, Lovely Professional University, Phagwara, Punjab, India

A. Nazir
Department of Clinical Biochemistry, University of Kashmir, Srinagar, India

A. Hashem
Botany and Microbiology Department, College of Science, King Saud University, Riyadh, Saudi Arabia
e-mail: habeer@ksu.edu.sa

E. F. Abd-Allah
Plant Production Department, College of Food and Agricultural Sciences, King Saud University, Riyadh, Saudi Arabia
e-mail: eabdallah@ksu.edu.sa

J. A. Parray et al. (eds.), *Environmental Landscape and Sustainable Biodiversity for Healthy Green Growth*, Sustainable Landscape Planning and Natural Resources Management, https://doi.org/10.1007/978-3-031-75405-0_6

this development has included some significant disadvantages, with ecological debasement, asset consumption, and broadening monetary incongruities presenting enormous difficulties to long-haul maintainability (United Nations 1987).

2.1 Understanding Manageable

Development, as defined by the United Nations, entails pursuing economic growth, social inclusion, and environmental protection in a manner that meets present needs without compromising the ability of future generations to address their own challenges. It highlights the interdependence of economic, social, and environmental dimensions, recognizing that development must be pursued holistically to be truly sustainable.

2.2 The Basics of Food Security

Food security is a crucial standard of freedom and a foundation for a feasible turn of events. It occurs when every individual consistently has physical and financial access to sufficient, safe, and nutritious food to fulfill their dietary needs and preferences for an active and healthy life. In any case, accomplishing food security isn't just a question of expanding food creation; it likewise requires resolving hidden issues like poverty, disparity, and natural debasement that add to food uncertainty (FAO 1996).

2.3 The Exchange Between Worldwide Economy, Sustainable Development, and Food Security

The nexus between the worldwide economy, economic development, and food security is complicated and diverse. Monetary approaches, economic accords, and market elements impact horticultural creation, food dissemination, and admittance to food assets. In the meantime, unreasonable practices in horticulture, like serious monoculture and overexploitation of regular assets, worsen ecological debasement and undermine the drawn-out practicality of food frameworks (Sachs 2015).

3 Outline of Global Economy

The worldwide economy alludes to the interconnected organization of financial exercises and exchanges among nations and locales worldwide. It incorporates creating, utilizing, and trading worldwide merchandise, administrations, and assets. Critical parts of the worldwide economy incorporate global exchange, monetary business sectors, global enterprises, and financial strategies set by legislatures and worldwide associations.

The worldwide economy has encountered colossal development and change in the last years, driven by mechanical progressions, exchange advancement, and expanded financial joining. Nonetheless, it faces monetary imbalance, instability, and ecological corruption. Understanding the elements of the worldwide economy is urgent for policymakers, organizations, and people, as it impacts monetary development, business, expectations for everyday comforts, and unique open doors for advancement worldwide (O'Brien 2019).

Feasible improvement is an all-encompassing way to deal with monetary. Social, economic, and environmental progress that meets current needs without jeopardizing the ability of future generations to meet their own challenges. It emphasizes the interdependence of economic, social, and ecological systems and aims to promote long-term well-being while conserving natural resources and ecosystems.

4 Critical Standards of Economic Improvement

4.1 Ecological Stewardship

Safeguarding and saving average assets, diminishing contamination and waste, and relieving the effects of human exercises on the environment.

4.2 Social Value

Guaranteeing that monetary open doors, social administrations, and fundamental privileges are available to all individuals, no matter what their orientation, identity, or financial status.

4.3 Financial Suitability

Advancing monetary development and improvement that is comprehensive, fair, and naturally maintainable, cultivating advancement and business while tending to poverty and disparity. Improving requires joint effort and collaboration among state-run administrations, organizations, common society associations, and people. It includes incorporating manageability contemplations into dynamic cycles across areas and levels of administration, from neighborhood networks to the worldwide field.

5 Idea of Food Security

Food security is present when every individual consistently has physical, social, and financial access to sufficient, safe, and nutritious food that fulfills their dietary requirements and preferences for a healthy and active life. This encompasses four dimensions:

- *Accessibility:* Adequate amounts of food are reliably accessible through creation, dissemination, and trade (Sen 1999).
- *Availability:* Individuals have the monetary means to get sufficient food through buying power or social security nets.
- *Use:* Food is used successfully by the body to address nourishing issues, with admittance to safe water, disinfection, and well-being administrations.
- *Strength:* Admittance to Food is steady after some time, without unexpected disturbances because of variables, such as cost spikes, catastrophic events, or clashes (Sachs 2015).

Guaranteeing food security is fundamental for advancing human well-being, decreasing poverty and imbalance, and cultivating social dependability and monetary turn of events. Notwithstanding, it stays a considerable test, especially in locales impacted by neediness, struggle, environmental change, and natural debasement. Tending to food security requires complete techniques that join horticultural turn of events, destitution decrease, social insurance, and practical average asset the executives. It additionally includes elevating evenhanded admittance to land, water, and other valuable assets, enabling smallholder ranchers and weak networks, and fortifying food frameworks' strength to shocks and stresses. Policymakers, professionals, and partners can cooperate to construct a more food-secure and robust future for all by focusing on practical development and tending to the primary drivers of food instability (FAO 1996).

6 Impact of Globalization on Food Production

The influence of globalization on food production is far-reaching and intricate, impacting various facets of agricultural systems and methods.

1. *Market Accessibility and Expansion:* Globalization has streamlined the integration of agricultural markets across borders, empowering farmers to tap into broader and more diverse markets for their agricultural goods. This expanded market access allows farming producers to widen their production and sales beyond local or national boundaries, thus catalyzing agricultural output and fostering economic growth.
2. *Technological Progress:* Globalization has spurred the transfer and dissemination of agricultural technologies and best practices across nations and regions. Breakthroughs in technology, such as enhanced seeds, mechanization, and precision farming techniques, have bolstered farm efficiency and productivity, resulting in heightened food production and supply.
3. *Access to Inputs and Operational Efficiency:* Globalization has streamlined access to crucial agricultural inputs like seeds, fertilizers, pesticides, and machinery through international trade and supply networks. This accessibility to inputs empowers farmers to embrace contemporary agricultural methodologies, optimize resource utilization, and elevate yields, thereby contributing to heightened food production levels (Evenson and Gollin 2003).
4. *Specialization and Comparative Edge:* Globalization has fostered a climate of agricultural specialization and the centralization of production in regions boasting comparative advantages in specific crops or commodities. This specialization enables countries to concentrate on producing goods wherein they hold a competitive edge, driving efficiency and productivity enhancements in food production (Access to Inputs and Operational Efficiency).
5. *Integration of Supply Chains:* Globalization has facilitated the seamless integration of food supply chains, facilitating the smooth movement of food products across borders from production to consumption. This integration has bolstered logistical operations, distribution mechanisms, and market accessibility, guaranteeing a more dependable and diverse food supply for consumers worldwide.
6. *Challenges of Interdependence:* Nevertheless, globalization also presents challenges for food production, including reliance on international markets for inputs, fluctuations in global commodity prices, and susceptibility to disruptions in global supply chains. Dependency on imported inputs or markets can expose farmers to price volatility, trade impediments, and geopolitical tensions, potentially impacting their livelihoods and food security (Pingali and Traxler 2002).

Globalization has instigated substantial transformations in food production systems, offering prospects for market expansion, technological advancements, and efficiency enhancements. Nonetheless, it also poses challenges concerning interdependence, market volatility, and supply chain vulnerabilities. To capitalize on the benefits of globalization while mitigating its drawbacks, policymakers and stakeholders must champion inclusive and sustainable agricultural practices, invest in resilient food systems, and ensure equitable access to markets and resources for all actors within the agricultural value chain (Chror and Manova 2012).

7 Influence of Economic Policies on Agricultural Development

The impact of economic policies on agricultural development holds significant weight, as it encompasses a range of measures that mold the growth and sustainability of the farm sector:

1. *Subsidies:* Government subsidies directed toward farmers wield considerable influence over agricultural production, input utilization, and market dynamics. These subsidies can spur farmers to adopt specific crops or technologies, altering land use and production patterns. Nonetheless, subsidies often favor large-scale producers, potentially distorting market equilibrium and perpetuating inequalities within the agricultural sphere (Reardon et al. 2012).
2. *Taxation:* Tax policies significantly impact agricultural development by influencing production costs, market prices, and investment choices. Tax incentives or exemptions targeting agricultural inputs, machinery, and equipment can promote the adoption of modern farming techniques and technology. Conversely, high taxes on agricultural inputs or land may impede farmers' capacity to expand production and embrace new methodologies, thereby restricting agrarian progress (Brown 2011).
3. *Regulatory Frameworks:* Government regulations about land usage, environmental conservation, and food safety standards significantly shape agricultural development. Regulations that advocate for sustainable land management practices, environmental preservation, and adherence to quality standards contribute to the long-term sustainability of agriculture. However, excessively stringent regulations or bureaucratic obstacles can hinder innovation, productivity, and competitiveness within the agricultural sector (Brown 2011).
4. *Trade Policies:* Trade policies, encompassing tariffs, quotas, and trade agreements, wield considerable influence over agricultural development by shaping market access and competitiveness. Trade liberalization measures can broaden export opportunities for agricultural producers, thereby stimulating production and investment in export-driven crops. Conversely, protectionist policies or trade barriers may shield domestic farmers from international competition but impede agricultural modernization and restrict market access for agricultural exports (Pretty 2008).
5. *Financial Measures:* Access to affordable credit and financial services is a linchpin for agricultural development, particularly among smallholder farmers and rural communities. Government initiatives promoting agricultural lending, microfinance endeavors, and rural credit programs facilitate investments in farm infrastructure, input acquisition, and technology integration. Nevertheless, inadequate access to credit, steep interest rates, and stringent collateral prerequisites may hamper smallholder farmers' ability to invest in activities enhancing productivity and agricultural modernization (Narayanan and Gulati 2008).
6. *Research and Innovation Initiatives:* Government funding for agricultural research, extension services, and innovation policies profoundly influences technological advancements and productivity enhancements within the agricultural sector. Investments in research institutions, agricultural universities, and extension programs bolster developing and disseminating superior crop varieties, pest management strategies, and sustainable farming practices. Additionally, policies incentivizing private sector involvement in agricultural research and development expedite innovation, augmenting farm productivity and competitiveness. Economic policies are pivotal in shaping agricultural development by steering incentives, market dynamics, and investment decisions within the sector. Well-crafted policies that champion sustainable agriculture facilitate market access, and nurture innovation can foster inclusive growth, mitigate poverty, and bolster food security in rural communities. However, policymakers must diligently weigh policy interventions' trade-offs and unintended consequences to ensure that agricultural development policies align harmoniously with broader economic, social, and environmental objectives (Barrett and Heisey 2002).

8 Income Inequality's Impact on Food Access

The ramifications of income inequality on food accessibility are profound:

1. *Restricted Affordability:* Income inequality curtails the purchasing power of individuals or households with lower incomes, posing challenges in affording a balanced and nutritious diet. Consequently, most of their income is allocated to food expenses, leaving less for other essential needs like healthcare, education, and housing (Spielman and Pandya-Lorch 2009).
2. *Disparities in Healthy Food Access:* Individuals with lower incomes often reside in areas with limited access to grocery stores offering fresh, healthy food choices. These regions, known as food deserts, compel residents to rely on convenience stores or fast-food outlets offering cheaper yet less nutritious options. This disparity contrib-

utes to heightened rates of diet-related health issues such as obesity, diabetes, and cardiovascular diseases among lower-income groups.

3. *Imbalance in Nutrition:* Income inequality exacerbates nutritional imbalances as lower-income individuals may lack access to a well-rounded diet comprising fruits, vegetables, and lean proteins. Instead, they may resort to cheaper, energy-dense foods high in sugar, salt, and unhealthy fats. This dietary pattern contributes to malnutrition, encompassing undernutrition and overnutrition, leading to adverse health outcomes (Popkin et al. 2012).
4. *Constrained Food Security:* Income inequality is closely intertwined with food insecurity, where individuals or households face inconsistent access to adequate and nutritious food. Financial constraints, job instability, and limited access to social safety nets heighten the likelihood of food insecurity among lower-income households. Consequently, they may adopt coping mechanisms such as skipping meals, reducing portions, or relying on low-cost, energy-dense foods to stretch their budgets.
5. *Intersection with Other Inequalities:* Income inequality intersects with other forms of disparity, such as gender, race, ethnicity, and geography, further exacerbating discrepancies in food access. Marginalized communities, including women, racial minorities, indigenous populations, and rural residents, bear a disproportionate burden of food insecurity and encounter additional obstacles in accessing nutritious food (Lobell and Field 2007).

To address income inequality's impact on food accessibility, comprehensive strategies addressing root causes and structural determinants are imperative:

1. *Supportive Income Programs:* Implementation of social welfare initiatives like minimum wage laws, income assistance, and food aid programs can mitigate income inequality's effects on food accessibility. Targeted assistance to low-income individuals and families ensures they have the necessary resources to afford nutritious food.
2. *Enhanced Employment Opportunities:* Creating quality employment opportunities with fair wages and benefits empowers individuals to enhance their financial stability and purchasing power. Investments in education, skills training, and workforce development equip individuals with the tools needed to secure employment and achieve economic self-sufficiency (Wada et al. 2014).
3. *Transformation of the Food System:* Advocating for equitable access to healthy food through initiatives like community gardens, farmers' markets, and urban agriculture projects improves food accessibility in underserved areas. Supporting local food systems, small-scale farmers, and sustainable agricultural practices fosters food sovereignty and resilience, reducing reliance on centralized food supply chains.
4. *Policy Measures:* Enactment of policies addressing structural inequalities and promoting social justice, such as progressive taxation, affordable housing, and healthcare reform, can mitigate income inequality and enhance overall well-being. Implementing regulatory measures ensuring fair labor practices, consumer protections, and environmental sustainability in the food system contributes to equitable food access.
5. *Community Empowerment:* Involvement of communities in decision-making processes and empowerment of grassroots organizations and advocacy groups amplifies marginalized voices and addresses local food access challenges. Encouragement of community-led initiatives prioritizing equity, diversity, and inclusion fosters collaborative solutions to food insecurity and inequality (Lal 2015).

Through multifaceted approaches addressing income inequality and its impact on food accessibility, societies can establish more inclusive and resilient food systems, ensuring equitable access to nutritious and culturally appropriate food, irrespective of socioeconomic status.

9 Climate Change and Agricultural Vulnerability

Climate change presents significant hurdles for agriculture, exacerbating vulnerability through various avenues:

1. *Altered Growing Conditions:* Changes in temperature, precipitation patterns, and extreme weather events disturb conventional growing seasons and impact crop productivity. These alterations may also introduce novel pests and diseases, further endangering agricultural yields.
2. *Water Scarcity:* Variations in precipitation patterns and heightened evaporation rates contribute to water scarcity, affecting irrigation systems and diminishing water accessibility for crops. Droughts have the potential to severely impact agricultural output, leading to crop failures and loss of livelihoods for farmers (Tscharntke et al. 2012).
3. *Soil Degradation:* Climate change contributes to soil erosion, salinization, and degradation, diminishing soil fertility and agricultural productivity. Events like floods and storms exacerbate soil erosion, leading to the loss of arable land and decreased crop yields.
4. *Loss of Biodiversity:* Climate change affects biodiversity crucial for agriculture, including pollinators, beneficial insects, and natural predators for pest control. This loss disrupts ecosystem services vital for maintaining soil health, nutrient cycling, and pest regulation, compromising agricultural resilience (Thornton and Herrero 2015).

5. *Adverse Impact on Livestock:* Rising temperatures and altered precipitation patterns stress livestock, reducing their productivity and increasing susceptibility to diseases. Heat stress causes reduced feed intake, reproductive issues, and higher mortality rates among livestock.
6. *Impact on Food Security:* Climate change-induced disruptions in agricultural production undermine food security, especially in vulnerable regions with limited adaptive capacity. Reduced crop yields, increased food prices, and farmer livelihood loss exacerbate hunger and malnutrition, perpetuating poverty and inequality.

Addressing agricultural vulnerability to climate change necessitates implementing adaptation strategies and building resilience within farming systems:

1. *Crop Diversification:* Encouraging diverse cropping systems enhances resilience to climate variability and lowers the risk of crop failure. Planting various crops with different climate tolerances mitigates the impacts of extreme weather events and pest outbreaks.
2. *Improved Water Management:* Investing in water-efficient irrigation technologies, rainwater harvesting, and soil moisture conservation mitigates the effects of water scarcity on agriculture. Adopting water-saving techniques optimizes water use and sustains agricultural productivity in water-stressed areas.
3. *Conservation Agriculture:* Embracing conservation agriculture practices such as minimal tillage, cover cropping, and crop rotation prevents soil erosion, improves soil health, and enhances resilience to climate change. These practices promote sustainable land management and contribute to long-term agricultural sustainability.
4. *Climate-Resilient Crop Varieties:* Developing and promoting climate-resilient crop varieties through breeding programs and genetic engineering enhances agricultural resilience. These varieties withstand environmental stresses like drought, heat, and pests, ensuring more reliable crop yields under changing climate conditions.
5. *Early Warning Systems:* Implementing early warning systems for extreme weather events helps farmers prepare and mitigate climate-related disasters' impacts. Timely information and support services enable proactive measures to protect crops, livestock, and livelihoods.
6. *Policy Support:* Enacting policies promoting sustainable agricultural practices, supporting smallholder farmers, and incentivizing climate-smart agriculture enhances resilience to climate change. Government support through subsidies, extension services, and insurance schemes aids farmers in adopting adaptation measures and coping with climate-related risks.

Addressing agricultural vulnerability to climate change demands coordinated action across multiple sectors, including agriculture, environment, water management, and disaster risk reduction. Investing in climate-resilient agriculture and adopting adaptive strategies bolster food security, livelihoods, and sustainable development amidst climate change's challenges (Wheeler and von Braun 2013).

References

Barrett CB, Heisey KC (2002) How effectively does multilateral food aid respond to fluctuating needs for food assistance? World Dev 30(9):1491–1510
Brown LR (2011) World on the edge: how to prevent environmental and economic collapse. New York, London: W.W. Norton & Company. pp. 240
Chror D, Manova K (2012) Off the cliff and back? Credit conditions and international trade during the global financial crisis. J Int Econ 87:117–133
Evenson RE, Gollin D (2003) Assessing the impact of the green revolution, 1960 to 2000. Science 300:758–762
FAO (1996) Rome declaration on world food security and world food summit plan of action.World Food Summit, Rome p. 43
Lal R (2015) Soil degradation as a reason for inadequate human nutrition. Food Secur 7(1):45–57
Lobell DB, Field CB (2007) Global scale climate–crop yield relationships and the impacts of recent warming. Environ Res Lett 2(1):014002
Narayanan BG, Gulati A (2008) Globalization and Indian agriculture: opportunities and challenges. Indian J Agric Econ 63(1):1–17
O'Brien R (2019) Global political economy: evolution and dynamics. Bloomsbury Publishing
Popkin BM, Adair LS, Ng SW (2012) Global nutrition transition and the pandemic of obesity in developing countries. Nutr Rev 70(1):3–21
Pretty J (2008) Agricultural sustainability: concepts, principles and evidence. Philos Trans R Soc B: Biol Sci 363(1491):447–465
Reardon T, Timmer CP, Minten B (2012) Supermarket revolution in Asia and emerging development strategies to include small farmers Proc Natl Acad Sci 109(31):12332–12337
Sachs JD (2015) The age of sustainable development. Columbia University Press
Sen A (1999) Development as freedom. Oxford University Press, Oxford
Spielman DJ, Pandya-Lorch R (2009) Millions fed: proven successes in agricultural development. International Food Policy Research Institute (IFPRI)
Thornton PK, Herrero M (2015) Adapting to climate change in the mixed crop and livestock farming systems in sub-Saharan Africa. Natl Clim Chang Response Strateg 25(1):605–616
Tscharntke T, Clough Y, Wanger TC, Jackson L, Motzke I, Perfecto I et al (2012) Global food security, biodiversity conservation and the future of agricultural intensification. Biol Conserv 151(1):53–59
United Nations (1987) Our common future: report of the world commission on environment and development. UN Documents: Gathering a Body of Global Agreements has been compiled by the NGO Committee on Education of the Conference of NGOs from United Nations web sites with the invaluable help of information & communications technology
Wada Y, Wisser D, Bierkens MFP (2014) Global modeling of withdrawal, allocation and consumptive use of surface water and groundwater resources. Earth Syst Dynam 5(1):15–40
Wheeler T, von Braun J (2013) Climate change impacts on global food security. Science 341(6145):508–513

Challenges in Establishing Sustainable Innovations to Conserve Marine Resources

Ajay Bhalerao, Ravindra Ashok Sharma, Narayan D. Totewad, and Ashish Jain

Abstract

Marine assets are defined as the biological and physical components that reside in the oceans and seas and benefit society significantly. Their proper management and wise utilization are key to establishing the conservation and sustainability of marine resources. The oceanic ecosystem is one of the most diversified ecosystems on this planet. Our understanding and conservation strategies are mainly focused on terrestrial ecosystems. Less attention is given to conserving marine ecosystems. Oceans are valuable resources for humanity. They have a significant role in assuring food security, coastal protection, and climate regulation. However, overexploitation of many marine ecosystems leads to permanent destruction, creating a significant threat to the existence of marine as well as terrestrial ecosystems. Many anthropogenic activities are accelerating the destruction of biodiversity and contaminating marine environments. Some preservation strategies exist to protect valuable endangered marine ecosystems. However, there are many obstacles, including lack of information and data, challenges with surveillance and regulation, contamination hazards, and insufficient funding. This chapter highlights the significant barriers and difficulties to conserving marine resources and possible approaches to developing and executing practical marine resource conservation approaches.

Keywords

Ocean conservation · Marine resources · Sustainability · Biodiversity · Green energy

A. Bhalerao
Microbiology and Biotechnology Department, BNN College, Dhamankar Naka, Bhiwandi, Maharashtra, India

R. A. Sharma
Biotech Department, Bharat Serums and Vaccines Limited, Ambernath East, Maharashtra, India

N. D. Totewad
Department of Microbiology, BK. Birla College of Arts, Science and Commerce, Kalyan, Maharashtra, India

A. Jain (✉)
Department of Microbiology, Smt. Chandibai Himathmal Mansukhani College, Ulhasnagar University of Mumbai, Ulhasnagar, Maharashtra, India

1 Introduction

The ocean covers three-fourths of the surface of the earth. Marine ecosystems are extensive and diversified, containing a variety of habitats such as coastal waters, bays and estuaries, shallow and deep-sea ecosystems, sea mounts, island arcs, coral reefs, Kelp Forests, mangrove systems, salt marshes, hydrothermal vents, and gyre centers. The marine ecosystem is a source of biotic and abiotic resources. Oceans support life by providing seafood, climate regulation, and offering an earning resource for the population residing in coastal regions. Anthropogenic activities threaten marine ecosystems, resulting in widespread degradation and biodiversity loss (Ahmed et al. 2022). Overexploitation of marine resources is resulting in a decline in global fish stocks attributed to unsustainable fishing practices (Arthur et al. 2022; Bi et al. 2022). Illegal, Unreported, and Unregulated (IUU) fishing disrupts the viability of fisheries and makes conservation efforts more difficult (Garcia et al. 2021). Marine ecosystems are becoming more vulnerable to global climate change and overexploitation of marine resources. Extreme weather events, acidification of the oceans, melting of ice, and increasing ocean temperatures are changing the dynamics of marine ecosystems and adversely affecting marine biodiversity, which is responsible for disruptions of food webs and the degradation of critical ecosystems such as coral reefs and mangroves (Neelmani et al. 2019). Marine pollution represents another threat to ocean ecosystems; nonbiodegradable pollutants,

J. A. Parray et al. (eds.), *Environmental Landscape and Sustainable Biodiversity for Healthy Green Growth*, Sustainable Landscape Planning and Natural Resources Management, https://doi.org/10.1007/978-3-031-75405-0_7

chemical substances, oil spills, and nutrient runoff are significant concerns for marine life and resources (Häder et al. 2020). Despite increased awareness of the issue, marine debris is expanding quickly, highlighting the critical need for joint efforts of the local population, environment activists, policymakers, and government enforcement to address this global challenge. Given the significant constraints that marine conservation efforts face, fresh approaches to sustainable ocean asset management are urgently needed. These solutions encompass a range of approaches, from technological advancements to policy interventions, aimed at fostering resilience and promoting the long-term health of ocean ecosystems (Kleisner et al. 2021; Pedersen Zari et al. 2019).

2 Environmental Challenges

Marine ecosystems face numerous environmental threats that compromise the balance between biotic and abiotic elements and their sustainability. Human activities and global environmental changes compound these issues, creating considerable barriers to realistic marine conservation efforts.

2.1 Overexploitation of Marine Resources

One of the significant threats to conserving marine resources is their overexploitation. Fishing stoke is reducing at an exponential rate due to unsustainable fishing practices. Overfishing is the primary concern for disturbance and unbalancing marine ecosystems, which also negatively affects biodiversity and upsets marine food chains (Zeller and Pauly 2019). This problem is further compounded by illegal, unnoticed, and unlicensed fishing, which jeopardizes the long-term viability of fisheries and undermines conservation efforts (Sumaila et al. 2019).

2.2 Climate Change

Marine ecosystems are at hazard of climate change, including alterations in oceanic temperatures, ocean acidity, weather patterns, melting of ice, sea-level rise, and marine heat waves. These climatic factors negatively influence marine biodiversity (Hall-Spencer and Harvey 2019; Barua et al. 2020; Rastelli et al. 2020). These environmental stresses increase the vulnerability of marine biodiversity to extinction by upsetting marine environments, including mangroves, coral reefs, and other marine ecosystems (O'Hara et al. 2021). In addition, extreme weather events and changes in the climate have exacerbated and devastated habitats and impaired marine ecosystems.

2.3 Marine Pollution

Marine pollution presents another significant environmental challenge, detrimental to marine ecosystems and biodiversity. Marine pollution is widespread, including plastic litter, chemical pollutants, and nutrient runoff (Hassan and Haq 2019; Tuuri and Leterme 2023; Mincer et al. 2016). These pollutants damage water quality, injure marine organisms, and create dead zones. Despite efforts to address marine pollution, the accumulation of marine litter continues, providing a long-term hazard to marine habitats and human health (Filho et al. 2019).

In conclusion, there are several challenges and obstacles to marine conservation initiatives, including marine pollution, climate change, exploitation of marine resources, and lack of effective regulation and policies. Addressing these challenges requires coordinated action at local, regional, and global levels to ensure the sustainability and resilience of aquatic ecosystems.

3 Technological Innovations

Our understanding of marine ecosystems is improving, and technological advancements make more effective conservation strategies possible. There are still many obstacles in marine conservation, which can be resolved by various cutting-edge technologies, from data analysis and communication to monitoring and surveillance.

3.1 Remote Sensing and Monitoring Systems

Researchers can collect essential data on marine ecosystems at large spatial scales thanks to remote sensing technologies like satellites and uncrewed aerial vehicles (UAVs). Research demonstrates the value of remote sensing in monitoring changes in ocean temperature, sea surface temperature, and chlorophyll content. With their insightful information on biodiversity hotspots, ecosystem dynamics, and habitat changes, these data assist conservation efforts and help guide policy decisions (Umbert et al. 2020; Jiménez López and Mulero-Pázmány 2019; Katlane et al. 2023).

3.2 Remotely Operated Vehicles (ROVs) and Autonomous Underwater Vehicles (AUVs)

AUVs and ROVs are vital tools for a more comprehensive study and investigation of maritime environments. Research shows how these vehicles can monitor marine life, collect samples, and conduct underwater surveys. Due to their extreme weather and difficulty accessing deep-sea environments, they are a valuable source of data for the conservation and management of AUVs and ROVs (Agarwala 2020; McLean et al. 2020).

3.3 DNA Barcoding and Molecular Techniques

Research and conservation of marine biodiversity are being revolutionized by DNA barcoding and molecular techniques. Research shows that species identification, genetic diversity assessment, and the detection of invasive species in marine environments are all made possible by DNA barcoding. In addition to monitoring biodiversity and ecosystem health and identifying illegal trade in endangered species, these techniques provide quick and accurate identification of marine organisms (Nagarajan et al. 2020).

3.4 Geographic Information Systems (GIS) and Spatial Analysis

Researchers can integrate and analyze spatial data on marine ecosystems, human activity, and conservation priorities using GIS and spatial analysis tools. Scientists are using excellent mapping tools to create safe havens for ocean creatures. These tools, called GIS, help them find the best underwater neighborhoods for different species. By studying things like habitat and animal movement patterns, scientists can predict where creatures are likely to live. This information helps them create efficient marine protected areas, like underwater parks, that keep human activities away from essential breeding grounds and feeding areas. It is all about using science to make wise decisions and protect our precious oceans (Urcádiz-Cázares et al. 2021; de la Torriente et al. 2019).

3.5 Citizen Science and Crowdsourcing

The public is involved in conservation efforts through citizen science and crowdsourcing projects, which use their enthusiasm and collective knowledge to collect data and track marine biodiversity. Research studies demonstrate the value of citizen scientists in data collection, species occurrence documentation, and environmental change monitoring in maritime habitats (Chen et al. 2020). These programs increase community involvement, raise public awareness, and produce important data for scientific study and conservation management. New tools like underwater cameras, citizen science programs, and fancy DNA tests give scientists superpowers to understand and protect our amazing sea creatures (Kasten et al. 2021; Earp and Liconti 2020).

4 Policy and Governance Challenges

All people desire healthy oceans, but establishing and implementing appropriate legislation can be difficult. Adequate marine governance is required to ensure the sustainability of the ocean's numerous rising human activities and uses.

4.1 Inadequate Policy Frameworks

One of the primary obstacles to conserving marine ecosystems is inadequate policy frameworks at local, national, and international levels. The fundamental problem is that the laws meant to protect marine life must be more stringent; it is like having a fence with many gaps. Because of this, it is not easy to keep our water's unique marine life healthy and in great numbers. Since they are dispersed, they can be challenging regulations, and safeguarding the ocean is relatively easy. To maintain the health of our undersea ecosystem, we need solid and precise regulations that everybody accepts (Berry 2021; Mondré and Kuhn 2022).

4.2 Weak Institutional Capacity

One further significant barrier to marine conservation attempts is weak institutional capability. Establishing solid organizations capable of successfully carrying out and maintaining conservation legislation is essential. However, government agencies and organizations' ability to solve issues related to marine conservation can be limited by a lack of funding, poor training, and institutional inertia, resulting in less-than-ideal results (Joly et al. 2019; Winther et al. 2020).

4.3 Lack of Stakeholder Engagement

It is crucial to consider the viewpoints of many stakeholders to conserve the ocean properly. This includes officials, citizens along the shore, companies, and environmental groups. However, trust or competing interests might make reaching an agreement among these parties easier. Creating effective ocean preservation plans makes it more challenging (Neely et al. 2021; Vallet et al. 2019).

4.4 Enforcement and Compliance

To ensure that conservation measures are followed and end illicit activity, enforcement of marine conservation legislation is essential. According to research, in order to save the ocean, strict regulations and mechanisms for ensuring compliance are required. However, other obstacles make it difficult to implement the law, including a lack of funding, persons who break the law covertly, and a lack of concern by some politicians. This makes it possible for negative things to happen to the ocean, such as humans fishing inappropriately and destroying the homes of sea life (Cremers et al. 2020).

To safeguard the ocean, we need to cope with a few issues. First and foremost, the policies ought to make feel when taken collectively. Second, the personnel who put in force the regulations require extra resources and schooling (to grow institutional potential). Third, every person who cares about the sea deserves a say. Finally, we need to ensure that individuals admire the policies (higher enforcement systems) (Cremers et al. 2020; Iacarella et al. 2021).

5 Future Directions and Challenges

Although we have been getting to know greater protection, the ocean continues to be challenging. We should deal with some critical issues to ensure that the ocean stays wholesome and gives to anybody. To improve even further, we should focus on some essential areas.

5.1 Integration of Ecosystem-Based Approaches

Thinking of the sea as a whole gadget, in which everything is interrelated, from microscopic organisms to giant whales, is the most excellent technique to guard it in the future. We can take care of all the stunning species that name that location home in this manner. Research suggests how crucial it is to enforce integrated management plans for species interactions, habitat connectivity, and ecological techniques. By using environment-based totally strategies, making plans for marine conservation may additionally be more able to adapt to changes in the surroundings and sell the long-term conservation of marine ecosystems (Renzi et al. 2019; Heinen et al. 2020).

5.2 Enhancing Governance and Enforcement

Stronger guidelines and enforcement are essential to efficiently protecting the ocean. Studies highlight the significance of adaptive systems of governance that facilitate responsive choice-making, promote openness, and engage individuals. Increasing governance and enforcement talents at neighborhood, national, and global tiers is essential to halt unlawful fishing, prevent habitat harm, and promote adherence to conservation measures. Increasing governance and enforcement competencies at local, countrywide, and worldwide ranges is critical to halting illegal fishing, preventing habitat damage, and promoting adherence to conservation measures (Dutra et al. 2019; di Franco et al. 2020).

5.3 Embracing Technological Innovations

Advances in technology are expected to improve marine conservation monitoring and enforcement activities. Studies have also suggested that far-flung sensing technology, uncrewed aerial automobiles, and satellite monitoring structures will be utilized to display the fishing sports and pick out the unlawful fishing vessels to put into effect the marine covered areas. Technological advancements can enhance conservation packages' functionality and efficiency, even producing applicable statistics to help make decisions and move toward evidence-based and adaptive management (Appleby et al. 2021).

5.4 Fostering Community Engagement and Participation

For long-term sustainability, empowering neighborhood communities and promoting energetic participation in marine conservation programs is essential. Research findings have also shown the significance of inclusive and participatory methods containing conventional knowledge, local experience, and network-based management frameworks. Participation in decision-making, capacity building, and promoting social justice will ensure that the players have a say in the conservation programs. Consequently, they will establish proper ownership of the initiatives, resulting in sustainable and enhanced marine biodiversity (Kohsaka and Uchiyama 2021; Geekiyanage et al. 2020; Steenbergen et al. 2022; Adade Williams et al. 2020).

6 Conclusion

In conclusion, marine conservation efforts face complex and multifaceted challenges concerning the environmental, socio-financial, and governance dimensions of the problem. Analysis of these demanding situations based on the supplied research articles confirms the pressing need for collective efforts to enhance the sustainable management of marine resources and guard marine ecosystems' fitness and resilience. It is essential to understand that the latter's health and biodiversity are currently exposed to severe risks due to overexploitation of resources, marine pollution, and climate change.

It is not easy to protect the ocean because so much is at stake, and so large. In other words, so many human jobs depend on it, and so does food, so it is hard or impossible to enact strong protections, and some people may have no legal right to fish or use the ocean. When protections are put in place, they might need to be more effectively enforced. Therefore, we need creative solutions and everybody to sign up for forces to appreciate the fitness of our oceans (Rastelli et al. 2020; Tuuri and Leterme 2023).

Research highlights the capability of these techniques to advance marine conservation dreams and promote sustainability (Heinen et al. 2020; Dutra et al. 2019; Appleby et al. 2021; Adade Williams et al. 2020).

Everyone ought to cooperate to manage marine assets efficiently and resolve problems in the ocean. To create and implement effective strategies, scientists, governments, and other stakeholders must paint collectively.

Thus, it means:

Implement strict laws to stop illegal fishing and protect the environment.

Use today's time to monitor extensive underwater areas and ensure the code is in place.

Community members should be encouraged to participate in marine conservation to ensure the long-term stability and management of marine ecosystems.

Through navigation prevention and highway creation, we can reduce the effects of environmental degradation and the economic burden on coastal communities and ensures marine ecosystems' resilience and long-term viability for future generations.

References

Adade Williams P, Sikutshwa L, Shackleton S (2020) Acknowledging indigenous and local knowledge to facilitate collaboration in landscape approaches—lessons from a systematic review. Land 9(9):331. https://doi.org/10.3390/land9090331

Agarwala N (2020) Monitoring the ocean environment using robotic systems: advancements, trends, and challenges. Mar Technol Soc J 54(5):42–60. https://doi.org/10.4031/MTSJ.54.5.7

Ahmed SF, Kumar PS, Kabir M, Zuhara FT, Mehjabin A, Tasannum N, Hoang AT, Kabir Z, Mofijur M (2022) Threats, challenges and sustainable conservation strategies for freshwater biodiversity. Environ Res 214:113808. https://doi.org/10.1016/j.envres.2022.113808

Appleby T, Studley M, Moorhouse B, Brown J, Staddon C, Bean E (2021) Sea of possibilities: old and new uses of remote sensing data for the enforcement of the Ascension Island marine protected area. Mar Policy 127:103184. https://doi.org/10.1016/j.marpol.2018.06.012

Arthur RI, Skerritt DJ, Schuhbauer A, Ebrahim N, Friend RM, Sumaila UR (2022) Small-scale fisheries and local food systems: transformations, threats and opportunities. Fish Fish 23(1):109–124. https://doi.org/10.1111/faf.12602

Barua P, Rahman SH, Barua S, Rahman IMM (2020) Climate change vulnerability and responses of fisherfolk communities in the South-Eastern coast of Bangladesh. Water Conserv Manag 4(1):20–31. https://doi.org/10.26480/wcm.01.2020.20.31

Berry DS (2021) Unity or fragmentation in the deep blue: choices in institutional design for marine biological diversity in areas beyond national jurisdiction. Front Mar Sci 8. https://doi.org/10.3389/fmars.2021.761552

Bi M, Zhang Z, Guo X, Wan L (2022) Evaluation of sustainable utilization of African marine fishery resources. Aust Fish 8(1):4. https://doi.org/10.3390/fishes8010004

Chen S, de Bruyne C, Bollempalli M (2020) Blue economy: community case studies addressing the poverty-environment nexus in ocean and coastal management. Sustainability 12(11):4654. https://doi.org/10.3390/su12114654

Cremers K, Wright G, Rochette J (2020) Strengthening monitoring, control, and surveillance of human activities in marine areas beyond national jurisdiction: challenges and opportunities for an international legally binding instrument. Mar Policy 122:103976. https://doi.org/10.1016/j.marpol.2020.103976

de la Torriente A, González-Irusta JM, Aguilar R, Fernández-Salas LM, Punzón A, Serrano A (2019) Benthic habitat modeling and mapping as a conservation tool for marine protected areas: a seamount in the western Mediterranean. Aquat Conserv Mar Freshwat Ecosyst 29(5):732–750. https://doi.org/10.1002/aqc.3075

di Franco A, Hogg KE, Calò A, Bennett NJ, Sévin-Allouet M-A, Esparza Alaminos O, Lang M, Koutsoubas D, Prvan M, Santarossa L, Niccolini F, Milazzo M, Guidetti P (2020) Improving marine protected area governance through collaboration and co-production. J Environ Manag 269:110757. https://doi.org/10.1016/j.jenvman.2020.110757

Dutra LXC, Sporne I, Haward M, Aswani S, Cochrane KL, Frusher S, Gasalla MA, Gianesella SMF, Grant T, Hobday AJ, Jennings S, Plagányi É, Pecl G, Salim SS, Sauer W, Taboada MB, van Putten IE (2019) Governance mapping: a framework for assessing the adaptive capacity of marine resource governance to environmental change. Mar Policy 106:103392. https://doi.org/10.1016/j.marpol.2018.12.011

Earp H, Liconti A (2020) Science for the future: using citizen science in marine research and conservation. In: Jungblut S, Liebich V, Bode-Dalby M (eds) YOUMARES 9 – the oceans: our research, our future. Springer International Publishing, pp 1–19. https://doi.org/10.1007/978-3-030-20389-4

Filho WL, Havea PH, Balogun A-L, Boenecke J, Maharaj AA, Ha'apio M, Hemstock SL (2019) Plastic debris on Pacific Islands: ecological and health implications. Sci Total Environ 670:181–187. https://doi.org/10.1016/j.scitotenv.2019.03.181

Garcia S, Barclay K, Nicholls R (2021) Can anti-illegal, unreported, and unregulated (IUU) fishing trade measures spread internationally? Case study of Australia. Ocean Coast Manag 202:105494. https://doi.org/10.1016/j.ocecoaman.2020.105494

Geekiyanage D, Fernando T, Keraminiyage K (2020) Assessing the state of the art in community engagement for participatory decision-making in disaster risk-sensitive urban development. Int J Disaster Risk Reduct 51:101847. https://doi.org/10.1016/j.ijdrr.2020.101847

Häder D-P, Banaszak AT, Villafañe VE, Narvarte MA, González RA, Helbling EW (2020) Anthropogenic pollution of aquatic ecosystems: emerging problems with global implications. Sci Total Environ 713:136586. https://doi.org/10.1016/j.scitotenv.2020.136586

Hall-Spencer JM, Harvey BP (2019) Ocean acidification impacts on coastal ecosystem services due to habitat degradation. Emerg Top Life Sci 3(2):197–206. https://doi.org/10.1042/ETLS20180117

Hassan S, Haq IU (2019) Pervasive pollution problems caused by plastics and its degradation. Int J Online Biomed Eng (IJOE) 15(10):29. https://doi.org/10.3991/ijoe.v15i10.10873

Heinen JH, Rahbek C, Borregaard MK (2020) Conservation of species interactions to achieve self-sustaining ecosystems. Ecography 43(11):1603–1611. https://doi.org/10.1111/ecog.04980

Iacarella JC, Clyde G, Bergseth BJ, Ban NC (2021) A synthesis of the prevalence and drivers of non-compliance in marine protected areas. Biol Conserv 255:108992. https://doi.org/10.1016/j.biocon.2021.108992

Jiménez López J, Mulero-Pázmány M (2019) Drones for conservation in protected areas: present and future. Drones 3(1):10. https://doi.org/10.3390/drones3010010

Joly CA, Scarano FR, Bustamante M, Gadda TMC, Metzger JPW, Seixas CS, Ometto JPH, Pires APF, Boesing AL, Sousa FDR, Quintão JMB, Gonçalves LR, de Padgurschi MCG, dos Santos de Aquino MF, de Castro PFD, dos Santos IL (2019) Brazilian assessment on biodiversity and ecosystem services: summary for policymakers. Biota Neotrop 19(4). https://doi.org/10.1590/1676-0611-bn-2019-0865

Kasten P, Jenkins SR, Christofoletti RA (2021) Participatory monitoring—a citizen science approach for coastal environments. Front Mar Sci 8. https://doi.org/10.3389/fmars.2021.681969

Katlane R, el Kilani B, Dhaoui O, Kateb F, Chehata N (2023) Monitoring sea surface temperature, chlorophyll, and turbidity in Tunisian waters from 2005 to 2020 using MODIS imagery and the Google earth engine. Reg Stud Mar Sci 66:103143. https://doi.org/10.1016/j.rsma.2023.103143

Kleisner KM, Ojea E, Battista W, Burden M, Cunningham E, Fujita R, Karr K, Amorós S, Mason J, Rader D, Rovegno N, Thomas-Smyth A (2021) Identifying policy approaches to build social-ecological resilience in marine fisheries with differing capacities and contexts. ICES J Mar Sci 79(2):552–572. https://doi.org/10.1093/icesjms/fsab080

Kohsaka R, Uchiyama Y (2021) Geographical indications and regional trade agreements: facilitating international partnerships for sustainable development. In: Leal Filho W, Marisa Azul A, Brandli L, Lange Salvia A, Wall T (eds) Partnerships for the goals. Springer International Publishing, pp 477–489. https://doi.org/10.1007/978-3-319-95963-4

McLean DL, Parsons MJG, Gates AR, Benfield MC, Bond T, Booth DJ, Bunce M, Fowler AM, Harvey ES, Macreadie PI, Pattiaratchi CB, Rouse S, Partridge JC, Thomson PG, Todd VLG, Jones DOB (2020) Enhancing the scientific value of industry remotely operated vehicles (ROVs) in our oceans. Front Mar Sci 7. https://doi.org/10.3389/fmars.2020.00220

Mincer TJ, Zettler ER, Amaral-Zettler LA (2016) Biofilms on plastic debris and their influence on marine nutrient cycling, productivity, and hazardous chemical mobility. In: Handbook of environmental chemistry, vol 78. Springer, pp 221–233. https://doi.org/10.1007/698_2016_12

Mondré A, Kuhn A (2022) Authority in ocean governance architecture. Polit Gov 10(3):5–13. https://doi.org/10.17645/pag.v10i3.5334

Nagarajan M, Parambath AN, Prabhu VR (2020) DNA barcoding: a potential tool for invasive species identification. In: Trivedi S, Rehman H, Saggu S, Panneerselvam C, Ghosh SK (eds) DNA barcoding and molecular phylogeny. Springer International Publishing, pp 31–43. https://doi.org/10.1007/978-3-030-50075-7

Neelmani RC, Pal M, Sarman V, Vyas UD, Muniya TN (2019) Impacts of climate change on marine biodiversity. J Entomol Zool Stud 7(2):425–430

Neely CL, Bourne M, Chesterman S, Vågen T-G, Lekaram V, Winowiecki LA, Prabhu R (2021) Inclusive, cross-sectoral, and evidence-based decision-making for resilience planning and decision-making in a devolved context. Eur J Dev Res 33(4):1115–1140. https://doi.org/10.1057/s41287-021-00410-3

O'Hara CC, Frazier M, Halpern BS (2021) At-risk marine biodiversity faces extensive, expanding, and intensifying human impacts. Science 372(6537):84–87. https://doi.org/10.1126/science.abe6731

Pedersen Zari M, Kiddle GL, Blaschke P, Gawler S, Loubser D (2019) Utilizing nature-based solutions to increase resilience in Pacific Ocean Cities. Ecosyst Serv 38:100968. https://doi.org/10.1016/j.ecoser.2019.100968

Rastelli E, Petani B, Corinaldesi C, Dell'Anno A, Lo Martire M, Cerrano C, Danovaro R (2020) A high biodiversity mitigates the impact of ocean acidification on hard-bottom ecosystems. Sci Rep 10(1):2948. https://doi.org/10.1038/s41598-020-59886-4

Renzi JJ, He Q, Silliman BR (2019) Harnessing positive species interactions to enhance coastal wetland restoration. Front Ecol Evol 7(APR). https://doi.org/10.3389/fevo.2019.00131

Steenbergen DJ, Song AM, Andrew N (2022) A theory of scaling for community-based fisheries management. Ambio 51(3):666–677. https://doi.org/10.1007/s13280-021-01563-5

Sumaila UR, Ebrahim N, Schuhbauer A, Skerritt D, Li Y, Kim HS, Mallory TG, Lam VWL, Pauly D (2019) Updated estimates and analysis of global fisheries subsidies. Mar Policy 109:103695. https://doi.org/10.1016/j.marpol.2019.103695

Tuuri EM, Leterme SC (2023) How plastic debris and associated chemicals impact the marine food web: a review. Environ Pollut 321:121156. https://doi.org/10.1016/j.envpol.2023.121156

Umbert M, Guimbard S, Ballabrera Poy J, Turiel A (2020) The synergy between ocean variables: remotely sensed surface temperature and chlorophyll concentration coherence. Remote Sens 12(7):1153. https://doi.org/10.3390/rs12071153

Urcádiz-Cázares FJ, Cruz-Escalona VH, Peterson MS, Aguilar-Medrano R, Marín-Enríquez E, González-Peláez SS, del Pino-Machado A, Enríquez-García AB, Borges-Souza JM, Ortega-Rubio A (2021) Linking habitat and associated abiotic conditions to predict fish hotspots distribution areas within La Paz Bay: evaluating marine conservation areas. Diversity 13(5):212. https://doi.org/10.3390/d13050212

Vallet A, Locatelli B, Levrel H, Dendoncker N, Barnaud C, Quispe Conde Y (2019) Linking equity, power, and stakeholders' roles in relation to ecosystem services. Ecol Soc 24(2):art14. https://doi.org/10.5751/ES-10904-240214

Winther J-G, Dai M, Rist T, Hoel AH, Li Y, Trice A, Morrissey K, Juinio-Meñez MA, Fernandes L, Unger S, Scarano FR, Halpin P, Whitehouse S (2020) Integrated ocean management for a sustainable ocean economy. Nat Ecol Evol 4(11):1451–1458. https://doi.org/10.1038/s41559-020-1259-6

Zeller D, Pauly D (2019) Viewpoint: back to the future for fisheries, where will we choose to go? Glob Sustain 2:e11. https://doi.org/10.1017/sus.2019.8

Interdisciplinary Approaches to Environmental Sustainability and End Poverty

Garima Bartariya and Mukta Sharma

Abstract

Sustainable development may be jointly pursued more effectively if key stakeholders are identified and brought together and their capacities, resources, and motivations are linked. This can be done through collaborative action. Utilizing an integrated rural economy can lead to sustainable development and poverty reduction in rural areas. The dedication and participation of all stakeholders from all sectors are crucial for the sustainable development of a place. Interdisciplinary approaches to environmental sustainability and poverty alleviation are critical for addressing complex global challenges. By integrating insights and methodologies from various fields, such as environmental science, economics, sociology, public policy, and technology, we can develop more comprehensive and practical strategies to tackle these interconnected issues. This chapter focuses on the integration of stakeholders that can foster cooperation and confidence to build environmental sustainability.

Keywords

Poverty · Sustainability · Economy · Integrating approach · Environment

1 Introduction

Interdisciplinary approaches to environmental sustainability and poverty alleviation are critical for addressing complex global challenges. By integrating insights and methodologies from various fields, such as environmental science, economics, sociology, public policy, and technology, we can develop more comprehensive and practical strategies to tackle these interconnected issues.

Poverty is an essential component pervasive throughout most of the world and one of humanity's most significant problems in the twenty-first century. These include rapid population growth, urbanization, and rising poverty rates, but they also relate to climate change, environmental degradation, and increased natural disasters. It is possible to define poverty as the state where impoverished individuals or groups live together; in this sense, entire nations are occasionally considered needy. Depending on the circumstances, people experience poverty in many different ways. Furthermore, locals frequently have stronger connections to realities related to the environment, gender, and very many conflicts than they create to the economic issues that typically dominate assistance and government policy (OECD 2019). However, mostly the meaning of poverty includes low income or lack of required income. In this regard, it is essential to remember Kofi Annan's (UN Secretary-General) statement, which reads, "Wherever we lift one person's soul out of a life of poverty, we are defending human rights. And whenever we fail in this mission, we fail human rights" (UNHCR 2004).

Most people agree that poverty is a social injustice that must be addressed with determined actions organized through political action. By understanding poverty-related policy-motivated action, it is critical to ascertain the type and limit of absolute as well as relative poverty between countries or within a country. However, the question of how to assess poverty is surprisingly contentious. Measuring efforts' intricate intricacies is merely one aspect of the situation. While the other term "poverty" really has normative connotations: judgments made about examples of poverty throughout the globe include both descriptive and evaluative elements, and they are frequently paired with recommendations on how to reduce poverty. First, there needs to be more clarity on how to define poverty. Monetary metrics define poverty as a state in which an individual's income, expenditures, and consumption are below a given threshold. This threshold may be a constant amount or ratio considering the

G. Bartariya (✉) · M. Sharma
School of Life Science and Technology, IIMT University, Meerut, UP, India

J. A. Parray et al. (eds.), *Environmental Landscape and Sustainable Biodiversity for Healthy Green Growth*, Sustainable Landscape Planning and Natural Resources Management, https://doi.org/10.1007/978-3-031-75405-0_8

degree of economic success within a given society. While monetary metrics are frequently used in international and national reference, the most well-known examples in modern times are the global poverty measures of the World Bank, sometimes known as "international poverty lines" and briefly noted above (Beck et al. 2020).

2 Environmental Sustainability, Poverty, and Inequality Interrelationship

Ecological sustainability, poverty, and inequality are closely interconnected and can exacerbate each other in various ways.

2.1 Access to Resources

In many cases, impoverished communities lack access to clean water, adequate sanitation, and reliable energy sources. This lack of access to essential environmental resources can perpetuate poverty by hindering health, education, and economic opportunities.

2.2 Climate Change Impacts

Poorer communities are often more vulnerable to the climate change, like variations in weather events, rise in sea level, and fluctuating precipitation patterns. These events can lead to loss of livelihoods, displacement, and increased food insecurity, further deepening poverty.

2.3 Environmental Degradation

Environmental degradation, deforestation, soil erosion, and pollution disproportionately affect marginalized and impoverished communities. This degradation can reduce agricultural productivity, degrade ecosystems that provide essential services, and exacerbate health problems, all contributing to poverty.

2.4 Inequitable Distribution of Environmental Benefits and Harms

Wealthier individuals and nations often have greater access to and control over environmental resources, while poorer populations bear an unjustified burden of environmental disturbance. For example, industrial pollution and waste disposal sites are generally found near low-income areas, exposing residents to health risks.

2.5 Limited Participation in Decision-Making

Poorer communities may have limited voice and representation in environmental decision-making processes, exacerbating environmental injustices and perpetuating inequalities in resource allocation and access.

The term "environmental poverty," defined by Liu (2012), refers to the state where human activity-induced environmental degradation results in a lack of a healthy environment essential for society's existence and development. Adopting the "grow first, clean up later" approach, commonly in the first half of the EKC, can lead to economic advantages for a nation, state, or city but also leads to environmental sacrifice, excessive social unfairness, and disparities in income and environment. This is illustrated by a decomposed curve called EKC (Environmental Kuznets Curves). There could be communities with differently shaped curves in the exact location. Specific communities might thrive while causing environmental poverty, irrevocable environmental damage, and economic collapse in other areas. Locations that adhere to alternatives or "flat EKCs" could take a while to exit (Fig. 1).

Addressing environmental poverty and inequality requires integrated approaches considering ecological sustainability and social justice. This includes policies and initiatives that prioritize equitable access to ecological resources, empower marginalized communities, mitigate the effect of climate change, promote resource management, and ensure inclusive decision-making processes. By addressing environmental and social dimensions, efforts to reduce poverty and inequality can become more effective and sustainable.

3 An Integrated Approach to Poverty and Environment

Integration has been more challenging to achieve in reality. However, there are lessons to be learned from the slow but steady growth of progressive environmental and development institutions. The implications of climate-related changes demonstrate that low-income people are the first to suffer from environmental deterioration and be the victims. The impoverished are more susceptible than others to environmental risks and conflicts about the environment and are least equipped to handle problems when they arise. They

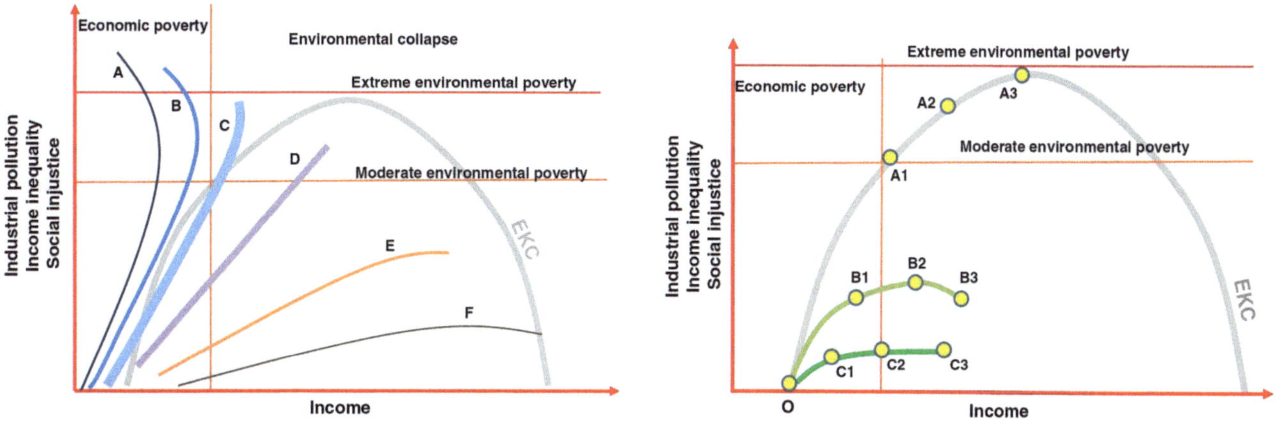

Fig. 1 A analysis of poverty economics (Liu 2012)

also frequently rely more on the nature and the direct application of their resources, making them more vulnerable to the adverse effects of environmental deterioration and restricted access to natural resources. From isolated approaches to protections to utilizing synergies when feasible, these institutions today strive for fully integrated sustainable development. Pollution and land degradation are dramatically increasing, biodiversity is decreasing, and the climate is changing. These issues affect others, poverty and inequality in particular. Both men and women end up stuck in poverty in severely damaged environments, where they are unable to adjust to climatic shocks. Global civilization faces enormous, intricate, and critical issues concerning poverty, the environment, pandemic catastrophes, and climate-related change. These issues affect local lives, making it impossible for certain people to make a living as farmers and fishermen.

Poverty and environmental issues frequently have similar root causes indicative of structural flaws. These include the unbridled authority that certain groups hold over others, the diminution of the rights and capacities of marginalized groups, and flawed economic signals that incentivize the conversion of environmental assets such as deforestation into new uses at the expense of conservation. Environmental measures to address poverty and ecological issues concurrently and coherently changes in management, policy, investment, and governance practices must be integrated (Asian Development Bank and United Nations Environment Programme 2019). The only sensible method for handling related issues is to take an integrated and well-rounded strategy. It integrates forces and assets and promotes systemic change. In this approach, we seek to transition from specific "magic bullets" to systemic institutional reform, from a vicious downward spiral of cascade crises to a constructive cycle of supporting solutions. Governments realized when they adopted the SDGs that a cross-government comprehensive strategy integrated approach for execution and evaluation was necessary because of the objectives' interconnected character. In actuality, 93 indicators from the 17 SDGs concern the environment (Bann 2019).

4 Integration Types Needed to Achieve Sustainable Development for Linked Poverty-Environment Challenges

More effective institutions are required to address associated poverty and environmental challenges.

4.1 Vertical Type of Integration

Vertical integration can forge a strong connection between societal capabilities, demand from below, and top-down policy vision. Multilevel strategies that connect pertinent interests at various levels, i.e., national as well as international enable the development of coherent policies and plans, particularly for public goods, and maximize the advantages of poverty-environment coordination across these hierarchies.

4.2 Horizontal Type of Integration

Horizontal integration is the ability to connect disciplines and topics. Working in multi-sector, multi-theme, and multidisciplinary approaches may break through sectoral barriers, support cogent policies and initiatives, and maximize the advantages of the poverty-environment for all sectors, populations, and resources.

4.3 Temporal Type of Integration

The ability to integrate time can facilitate ongoing development; gradually integrating across the cycle of decision-making, learning, and developing will enable the evolution of the required longer-term institutional reforms.

4.4 Stakeholder Type of Integration

Integration of stakeholders can foster cooperation and confidence. Sustainable development may be jointly pursued more effectively if key stakeholders are identified and brought together and their capacities, resources, and motivations are linked. This can be done through collaborative action.

Utilizing an Integrated Rural Economy the dedication and participation of all stakeholders from all sectors is crucial for the sustainable development of a place. As demonstrated in Fig. 2, it is also clear that an integrated system approach can be used to promote food security and higher living standards. The programs for rural infrastructure development, agriculture, and cooperatives rely on one another and are only feasible if not in isolation. This is because these integrated approaches interrelate many sectors of the national economy together (Francis and David 2012).

Essential vows under the 2030 Agenda to enhance development cooperation are outlined in the Development Assistance Committee (OECD 2021) of the Organization for Economic Co-operation and Development, including (i) systematic integration of climate and environmental objectives, (ii) coordinated approaches on the environment and climate, and (iii) assistance to make sustainability transitions for developing countries.

4.5 Education for Environmental Sustainable Development and Poverty

Education for Sustainable Development (ESD) is a critical approach based on education that equips individuals with the knowledge, skills, and values required to address the interconnected challenges of the twenty-first century, including poverty, climate change, and inequality. It emphasizes the importance of understanding and acting upon these significant challenges so that they meet their needs of present as well as future generations. Encouraging future environmental scientists and engineers to identify the factors causing environmental issues is one of the main benefits of receiving a high-quality, multidisciplinary environmental education (Fig. 2). Furthermore, environmental education is recommended to prioritize critical and integrative thinking, foster

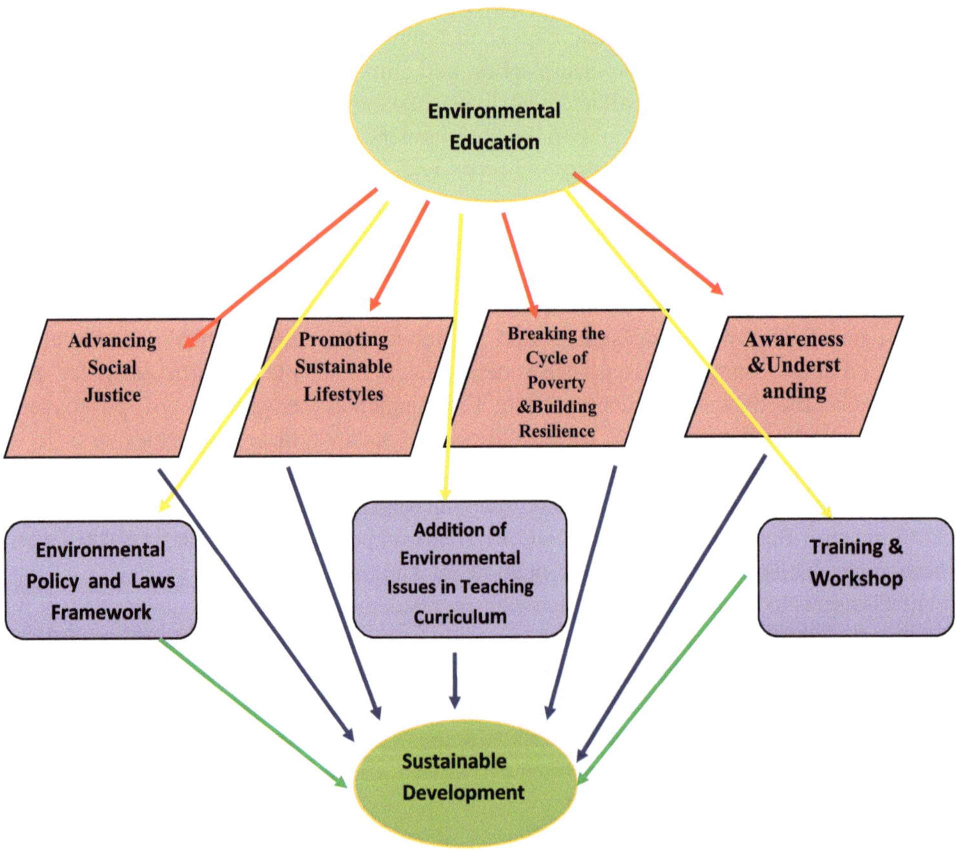

Fig. 2 Education for environmental sustainable development

the development of communication and problem-solving abilities, and underscore the significance of attitudes, beliefs, and commitments in addressing environmental concerns (Simmons 2000). Organize workshops, seminars, and community events focused on environmental issues and poverty alleviation. Encourage participation from diverse stakeholders, including local communities, NGOs, government agencies, and businesses, to facilitate knowledge sharing and collaborative problem-solving. Education for Sustainable Development (ESD) addresses poverty and promotes sustainable development.

4.6 Awareness and Understanding

Education for sustainable development (ESD) fosters awareness and understanding of the interconnectedness between environmental, social, and economic issues. ESD encourages informed decision-making and responsible actions by educating individuals about sustainable development goals and the impacts of poverty on communities and ecosystems. Education empowers individuals, particularly those living in poverty, with the knowledge and skills needed to improve their livelihoods sustainably. This includes practical skills in agriculture, resource management, renewable energy, and entrepreneurship.

De Alba (2020) reports that during confinement, a significant number of children, youngsters, and females were neglected in their own homes due to education-based influence. This resulted in a fall of "opportunity to learn" (Reimers and Schleicher 2020), a decrease in the acquisition of essential life survival skills, and an emotional detachment from education among the most susceptible students. According to the discussion given by McKinney 2021 this disengagement will result in a rise in school dropout rates because students won't have access to engaging teachers and more socio-cultural activities. It will also bring about the implementation of digital education and unequal educational achievement because of social and financial factors that affect low-income families. Recently, there has been a push toward a concurrent understanding of lifetime learning, including vocational education, technical education, and complementary provisions for regular schooling and various adult education programs. In addition, the poorest developing nations, particularly those in South Asia and Sub-Saharan Africa, continue to implement inadequate elementary and secondary school reforms, and the need for lifelong learning needs to be given more priority (Ahmed 2010).

Entering the field of axiology, the importance of the natural environment as a value must be recognized. Any normative culture needs to carefully analyze this issue because they are constructing their hierarchies of values. To achieve this, they are forced to use a variety of assessment standards, such as material, ethical, spiritual, and practical. All the ideas and opinions about nature and how humans (or society) fit into it over the relevant historical period constitute a standard of awareness common to the culture in question (Górka et al. 2001). Western societies have primarily relied on anthropocentric content to shape their understanding of the natural world. Within this paradigm, man is the apex of creation, the fount of all morality, and the standard by which everything else is measured.

As a result, men and their wants should come first in the creation of norms and in how the world is perceived (Seed 1994). This method has roots in medieval and ancient ontology, the Early Modern era's concept of science, and the Enlightenment era's progress goal. Appreciating the technocratic paradigm of progress centered on the pursuit of continuous increases in the production and consumption of tangible goods is correlated with advancement. Real-life stories and case studies that illustrate the impacts of poverty on the environment and highlight successful strategies for addressing these challenges can spread awareness among ordinary people. Personal narratives can resonate with audiences and inspire empathy, action, and advocacy.

4.7 Breaking the Cycle of Poverty and Building Resilience

Quality education is the only way to break the cycle of poverty by equipping individuals with the skills and tools to achieve better employment opportunities, increase their income, and improve their living standards. Environmental sustainable development emphasizes inclusive and equitable education, ensuring marginalized groups have equal learning opportunities. There is no historical precedence for the discrimination of social and economic status that divides the wealthiest and the poorest on the planet. This divide is not only significant but getting more significant. The wealthiest billion get richer yearly, while the poorest remain stuck at starvation. The differences in life expectancy, disease patterns, family size, education, and nutrition indicate an economic divide (Brown 2008).

Sustainable development helps communities build resilience to environmental and economic challenges like climate change, natural disasters, and market fluctuations. By promoting sustainable practices and innovative solutions, education equips individuals with the capacity to adapt to changing circumstances and mitigate risks.

4.8 Promoting Sustainable Lifestyles

Environmental sustainable development encourages the adoption of sustainable lifestyles that minimize resource consumption, reduce waste, and promote social equity. Individuals learn about sustainable consumption and production principles through education, empowering them to make environmentally and socially responsible choices.

People's daily behaviors that support sustainability define their role as individuals and as a community in promoting a sustainable lifestyle. As individuals, we must help one another. Disposal of unneeded furniture, food, books, toys, and other items is essential. If possible, create jobs. Many people don't support businesses that pay cheap salaries because they don't get paid well for their work. This is the digital age's reality. Education for all is the goal of this objective. They are upgrading themselves by enrolling in universities or free online courses. Giving away books is helpful to those who cannot purchase them. Stay up to date on workers' rights. Don't support businesses that offer cheap wages. Encourage your community's farmers and fair-trade companies. Foster collaboration among various sectors such as agriculture, environmental resources, health, and education to address interconnected challenges related to poverty and the environment. Integrated approaches can promote synergies, optimize resource use, and maximize social and environmental co-benefits.

4.9 Advancing Social Justice

Education is a powerful tool for advancing social justice and promoting inclusive development. Addressing inequalities in access to education and resources based on gender, caste, and other criteria. Environmentally sustainable development contributes to poverty reduction efforts. It promotes a more just and equitable society.

Analysis of this fact showed that among world's poorest people two-thirds are women only, connecting to the fact that more females are unemployed, malnourished, and uneducated than males. In addition, females often have less access to jobs, less user of healthcare, no involvement in property distribution, and significantly less involvement in politics. More research is required to determine the connections between gender differentiation, poverty, the environment, and macroeconomic policies. Wee and Heyzer (1995) try to demonstrate how national and international economic policies that worsen the environment also encourage the feminization of poverty using case studies from various Asian nations. Women are primarily responsible for raising their children and providing enough food, healthcare, and education resources. Women's lives in ancient civilizations have always been intricately connected to how natural resources are used. This symbiotic relationship has frequently changed due to economic, technical, and opportunity expansion, putting additional strain on women's lives and environment also (Chant 2006).

Persons with disabilities and special needs, persons of linguistic minorities, slum dwellers, and those with special needs are hard to reach. They are still not included in most education programs. For these groups, specialized and more narrowly focused initiatives are needed to ensure inclusive, responsive, and suitable programs in their instructional strategies. Ahmed (2010) provides training and capacity-building opportunities for local communities, grassroots organizations, and vulnerable populations to enhance their knowledge, skills, and resources for sustainable livelihoods and environmental stewardship.

Education for Sustainable Development fosters a sense of global citizenship and collective responsibility for addressing poverty and sustainability challenges. Through education, individuals develop a deeper understanding of global interdependence and their role in creating a more sustainable and equitable world. Education for Sustainable Development is essential for tackling poverty and promoting holistic development that balances social, economic, and environmental priorities. Integrating sustainability principles into education systems at all levels can empower individuals and communities to build a resilient future.

5 Environmental Sustainability with Technology to Reduce Poverty

To improve environmental quality overall, ecological engineering and environmental biotechnology have recently developed along individual subspecialties engineering methods with the physical, chemical, and mathematical components of various environmental issues (Wiesner and Theis 1996). This curriculum reform aims to create intricately woven interdisciplinary programs that inspire students to participate in integrated learning workshops and apply systemic holistic approaches to comprehend and resolve intricately complicated environmental problems (Bishop 2000; Hargreave 2000).

According to Gartner, digital solutions with sustainable technology are a set that can help an organization achieve environmental, social, and governance (ESG) goals. Internal IT, corporate IT, and customer operations are the three key business domains where opportunities are created by sustainable technology. Transparency and honesty in sourcing and trade processes, increased material and energy efficiency, lower emissions, and ethical labor practices are all part of enterprise operations. It comprises cloud computing, advanced analytics, automation, and artificial intelligence. Additionally, customer operations using sustainable technology serve as a prism to deliver goods and services that support individuals' sustainability objectives.

In the end, cutting-edge and proven technologies will be combined with sustainable technology. Cloud services boost the use of shared resources, decreasing environmental effects and the adaptability of cloud service models. Artificial Intelligence for Sustainability can enhance company operations and streamline challenging procedures to lower an organization's carbon and environmental impact and minimize significant hazards. Additionally, software for sustainability and ESG includes tools for data discovery, collection, analysis, insight, and reporting relevant to sustainability. These tools might be in many different formats (Jackie Wiles 2023).

The idea of technical determinism is closely associated with the technocratic culture. Today, the technological determinism in digital transformation and its implications for global society, especially in the context of the environment is one of the highlighted applications. In the discussion of sustainable development, the experience of being placed under lockdown during the early months of the 2020 coronavirus pandemic, the move to remote work and other forms of work, and the potential persistence of this tendency could all be significant points of reference. The remaining sectors focus on improving digital competence and implementing new technologies like artificial intelligence. They also aim to reduce economic inequalities between regions of the community.

Specific measures outlined in the adaptive strategy can include investments in environmentally friendly technological measures (based on the Best Available Techniques (BAT)), encouragement of industrial innovation, and introduction of more affordable, healthier, and cleaner modes of transportation (private and public), decreased emissions from the energy-based sectors, higher building energy efficiency, and collaboration with foreign mates to raise GES (global environmental standards) (EC 2019). These technologies offer a fresh perspective on topics like preserving biodiversity, conducting studies to guarantee a more beneficial food chain, establishing favorable conditions for sustainable cropping, looking for eco-friendly and alternative sources of energy, creating more eco-friendly production cycles, constructing energy-efficient constructions and retrofitting older ones to reduce energy costs, encouraging more ecologically friendly modes of transportation (like self-driving cars and alternative fuel vehicles), getting rid of pollution, and reaching net-zero greenhouse gas emissions.

6 Economic Dimension for Sustainable Development of Environment and Poverty Eradication

The economic dimension of sustainable development focuses on economic growth and prosperity to meet the needs of future generations. This means finding ways to grow economics, protecting the environment, and ensuring social equity. These dimensions can help to reduce poverty in several ways:

- Promoting economic growth creates jobs and raises incomes.
- Investing in education and healthcare can help people escape poverty and improve their livelihoods.
- Providing social safety nets, such as cash transfers, can help protect people from falling into poverty.
- Promoting gender equality can empower women to participate more fully in the economy.

Poverty is a significant barrier to sustainable development. Education, healthcare, and other essential services are frequently inaccessible to those who are impoverished. They are also more vulnerable to environmental shocks, such as climate change. Economic policies can help to reduce poverty in developing Asian countries.

6.1 Microfinance

Microfinance provides small loans to impoverished communities so they can start or grow their businesses. Microfinance is a valuable strategy for addressing poverty and vulnerability because it can lead to entrepreneurship, which is especially significant in areas with low economic growth. Enhancing the capabilities of vulnerable groups, particularly women, is another critical objective of microcredit (Gatto et al. 2016; Gatto 2020). Grameen Bank is the industry standard for microfinance as it exists today. As a practical substitute for charity and usury, the Grameen Bank has emerged as the first "village bank" to offer financial services to the underprivileged (Yunus and Yusus 2007).

In India, there are about 260 million people who live in poverty. Major obstacles to the Indian economy's rapid growth have included low per capita income, high population pressure, widespread massive underemployment, youngster unemployment and low capital formation rate, unequal wealth and asset distribution, and low technology. Many of India's poorest communities depend directly on natural resources such as land, forests, and water for survival. However, unsustainable exploitation of these resources can lead to environmental degradation, threatening the livelihoods of these populations in the long term. Therefore, promoting environmental sustainability is essential for poverty reduction efforts to be effective and sustainable. The microfinance market connects in India through the Banks Linkage Programme, which aims to create financial relationships between unofficial groupings of people and formal organizations like banks for the poor, particularly women, who need financial services. In India, the SHG-Bank (Self Help Group) connection model has become a vital tactic for the banking sector to reach the poorest of the poor (Upadhyaya 2018).

Microfinance institutions can integrate environmental considerations into their lending practices by offering loans for eco-friendly technologies, sustainable agriculture practices, and renewable energy initiatives. By supporting sustainable livelihoods, microfinance can help communities adapt to environmental challenges while reducing their vulnerability to poverty.

6.2 Conditional Cash Transfers

The conditional cash transfer (CCT) program provides cash to low-income families to send their children to school, keep them up-to-date with modern education and trends, and differentiate between right and wrong. Poor quality of life in low-income households significantly impacts children susceptible to growth challenges due to inadequate access to clean water, good sanitation, proper nutrition, and primary essential education (Arriagada et al. 2020). Large-scale, government-run cash transfer initiatives have effectively lowered poverty and enhanced recipients' economic independence, attendance in school, child nutrition, women's independence, and understanding of health care usage. Cash transfers have been shown to alleviate poverty in several ways. These include the removal of psychological and economic barriers brought about by the transfer and positive effects on nonbeneficiaries (Richterman et al. 2023).

These programs often involve community-based approaches, which can be leveraged to promote environmental awareness and conservation efforts at the grassroots level. Empowering communities to manage their natural resources sustainably can contribute to poverty alleviation and ecological preservation. CCTs are a crucial policy tool for climate change adaptation because, instead of lessening the effects of any one hazard, the goal should be to reduce susceptibility by addressing its socioeconomic causes. Given their capacity to provide the resources required to lessen risk, CCTs may be a valuable treatment for vulnerability (Davies et al. 2009).

Adding conditionality, like participation in activities that improve adaptive capacity and educate people about the implications of climate change, would be one way to "climate-proof" CCTs. These projects could include training courses, focus groups or assessments of community susceptibility, and workshops on adaptation and climate change vulnerability. Examples of community preparation initiatives taken from the literature on disaster risks, such as "participatory capacity and vulnerability assessments," may provide some thought-provoking insights (Oxfam 2001).

6.3 Public Employment Programs

Public employment programs can provide jobs for impoverished people and help stimulate the economy. Among the most popular anti-poverty initiatives in developing nations are public employment-related programs, in which the government offers jobless people opportunities to earn. The National Rural Employment Guarantee Scheme (NREGS) in India, which has six hundred million rural inhabitants eligible to receive, funding equivalent to 0.5% of India's GDP, is the most extensive program of its kind in the world. Complicating matters, the program's direct income benefits may be amplified or attenuated by the labor-market general equilibrium (GE) effects on market wages and employment. For example, program earnings may overestimate or underestimate net income increases for low-income people to the extent that such type of public-related employment programs enhances market salaries (Muralidharan et al. 2020).

At the local, national, and regional levels, the Asian and Pacific Region has launched some programs to combat poverty and related concerns. However, because these have been implemented piecemeal and in isolation, their overall influence still needs to be improved. There aren't many instances of nations tying environmental preservation to social investments like better health care, education, and creating jobs for the underprivileged, particularly women (UNEP 1997). The countries in the region have also tried to address the issue of villagers' landlessness by implementing two main techniques to lessen poverty. Government initiatives have included relocating the landless to accessible arable land in a few of these nations (including Indonesia, Sri Lanka, and Malaysia); one such initiative is the transmigration program in Indonesia. On the other hand, land and tenancy reforms partially meet the need for cultivable land in several other nations in the region (including India and Pakistan) with an uneven land distribution.

Apart from land allocation, several official programs have focused on agricultural policy (credit availability, pricing, infrastructure, inputs, and productivity) to reduce poverty (e.g., India, Pakistan, the Philippines, Indonesia, and Sri Lanka). Rural residents' fundamental needs, particularly those for safe drinking water and housing, have been the focus of social policy. Education and training have been given top importance in developing human resources.

Social policies have prioritized meeting the basic needs of rural residents, particularly access to clean drinking water and housing. Priority has also been given to the development of human resources, emphasizing training and education. An attempt has been made recently to address the issue of natural disasters methodically. To lessen the impact of disasters, steps have been taken to establish organizations, create strategies, pass laws, and support initiatives. Significant advancements have been made in forecasting, early warning, and environmental restoration, especially about volcanic and climatic dangers. Operations for providing aid after a disaster have been greatly improved. The joint WHO program is still centered on operating and maintaining water supply systems, along with observing drinking water quality. The Jakarta Declaration provides a clear vision for developing and promoting health, which was the outcome of the Fourth International Conference on Health Promotion. A sanitation strategy for high-risk populations has also been introduced (ESCAP).

7 Environmental Approach for Vulnerability and Poverty

Vulnerability can be defined as the susceptibility to harm, shock, or stress due to a lack of resources, capabilities, or social protection. It is important to understand that vulnerability is not synonymous with poverty. While all poor individuals are inherently more vulnerable due to their limited resources, not everyone in poverty will be equally vulnerable, and individuals who are not necessarily poor can also be vulnerable due to other factors. Unlike poverty, vulnerability shows a society's current state and has a predictive component that indicates the future if a specific danger materializes. It is said that vulnerability is a means of imagining what might occur to a particular population under specific risk and hazard conditions. Therefore, vulnerability should have a predictive aspect. Cannon and his co-workers (2003) highlight its forward-looking aspect. According to Thywisson (2006), vulnerability is "an intrinsic characteristic of a community" that exists constantly and only becomes apparent when a hazard materializes.

Some authors considered vulnerability as one factor that might contribute to poverty or make it more difficult for individuals to escape it. For example, according to a few studies cited by Prowse (2003), vulnerability is defined as "part of the multiple dimensions of poverty." Since poverty is no longer solely defined by income, vulnerability should be considered in poverty analyses. Poverty is now measured through a broader framework of "well-being," which aims to consider a holistic view of human livelihoods. Figure 3 indicates how the development cooperation community views the relationship between vulnerability and poverty. They also consider vulnerability as a major factor that affects poverty.

Most research on the connections between poverty and vulnerability conveys the opinion that these two methods are clearly distinct from one another but also strongly related and have a profound impact on one another. Vulnerability "may be seen as a cause of poverty, as a reason why the poor remain poor, or as an effect of poverty," according to German Technical Cooperation (GTZ) in particular (GTZ 2005).

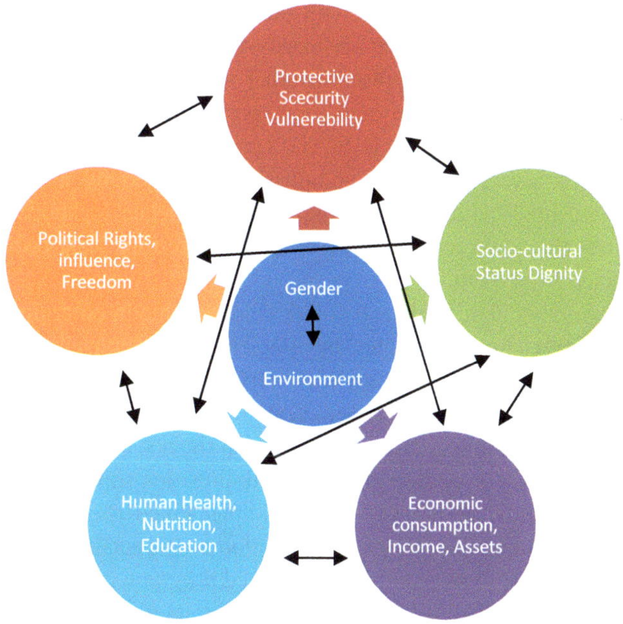

Fig. 3 Poverty dimensions (GTZ 2005)

When he discusses the "mutually reinforcing nature of poverty and vulnerability," Prowse (2003) makes a similar argument.

Vulnerability to environmental hazards is not solely determined by physical exposure but also by social factors such as income, education, access to resources, and governance structures. Marginalized and impoverished communities often bear the brunt of environmental disasters due to inadequate infrastructure, limited access to resources, and social exclusion, which hinders their ability to recover from ecological shocks.

7.1 Poverty and Vulnerable Groups

Due to factors like rapid urbanization, population growth, and environmental degradation, the vulnerability can no longer be disregarded. The international community is now emphasizing the necessity of addressing vulnerability with a comprehensive approach. An approach that considers its social, economic, and cultural components to effectively address the catastrophic effects of these risks and how they push people below the poverty line or destitute households deeper into poverty.

Vulnerable populations make up sizable portions of developing countries. A sizable fraction of vulnerable groups includes households with HIV/AIDS-affected members in many developing nations overall. Such households impacted by HIV/AIDS are more susceptible to falling into poverty during a shock, be it a natural disaster like a flood or an economic shock like rising prices. The fact that they typically lack the resources to deal with the shock when it happens exacerbates the problem. Primarily due to the HIV/AIDS pandemic, orphans represent another vulnerable population that is becoming more and more prevalent in developing nations. Sometimes, the HIV/AIDS pandemic causes parents to pass away, leaving their households headed by a kid. These households typically have low resilience to shocks since they own relatively few assets and are thus least protected from them.

Another significant and at-risk demographic in emerging nations is households led by women. These households' asset base is poor, making them very susceptible to shock. Since they don't have many coping mechanisms, they are highly vulnerable to the impacts of such shocks. Measures to lessen these populations' susceptibility to shocks and increase their resilience must be implemented to improve their vulnerability to poverty. Children, especially girls, make up the third category. Specialized programs are implemented in most developing nations to guarantee that children receive adequate nutrition, access healthcare, and education resources and lead healthy lives with a lower likelihood of experiencing severe poverty. On the other hand, these programs might not be widely dispersed nationwide, leaving some youngsters susceptible.

Another vulnerable group that exists in both developed and developing nations is the disabled community. The majority of the time, people with disabilities are unable to support themselves and require extra care to make sure they do not descend into or stay in poverty. Immigrants and members of minority groups fall under the third category. In the absence of intentional measures aimed at integrating immigrants and minorities into the broader society of industrialized nations, these groups of individuals could be particularly vulnerable to poverty and be a significant obstacle for the government (Makoka and Kaplan 2005). Various environmental sustainability organizations have established Variations in vulnerability measures to further their work.

7.2 The Commonwealth Vulnerability Index

The heads of government approved the three years of intense research that went into developing the CVI, which the Commonwealth finance ministers mandated. Two guiding ideas guided the development of the index: first, the impact of external shocks and second, a country's ability to endure from such types of shocks. The CVI is a country-level index that rates developing countries according to how resilient and susceptible they are to external shocks based on quantitative criteria. The indicator is based on the observation that the most evident evidence of vulnerability is the volatility of income growth (Commonwealth Secretariat 2000). The diversification index of the United Nations Conference on Trade and Development assesses the degree of export dependency, the absence of diversification, and the impact of natural catastrophes, reflecting the total frequency and effects of these events across time (Makoka and Kaplan 2005).

7.3 The Environmental Vulnerability Index (EVI)

The United Nations Environmental Programme and the South Pacific Applied Geoscience Commission recently developed it. UNEP and SOPAC (2005) state that the EVI was created to offer a quick and uniform way to assess vulnerability and pinpoint concerns that must be addressed to support a country's development. The EVI encourages global cooperation on the natural life support ecosystems and sustainable development. The Environment Vulnerability Index (EVI) uses 50 factors to assess how vulnerable a country's environment is to future shocks. These comprise six indicators related to climate, four indications about geology, six indicators related to geography, 28 indicators related to ecosystem resources and services, and six indicators related to ecological processes and human interactions (Makoka and Kaplan 2005).

7.4 The Prevalent Vulnerability Index (PVI)

The Inter-American Development Bank was responsible for developing the Prevalent Vulnerability Index (PVI). The PVI assesses viability based on social resilience, socioeconomic fragility, and exposure to risk areas (Makoka and Kaplan 2005).

8 Poverty as a Social Relation for Environmental Sustainability

"Poverty as a social relation" refers to the understanding that poverty is not simply a result of individual characteristics or lack of resources but is fundamentally shaped by social structures, relationships, and power dynamics within societies. This perspective emphasizes that poverty is not just about the absence of material wealth but also unequal access to resources, opportunities, and social inclusion. The branch of normative theory holds that achieving "social equality," or "relational equality," entails more than equally distributing rights, freedoms, and tangible and immaterial goods within a particular setting. Instead, it is believed that the desire for equal treatment necessitates the more basic realization of egalitarian social interactions, which generates specific dis-

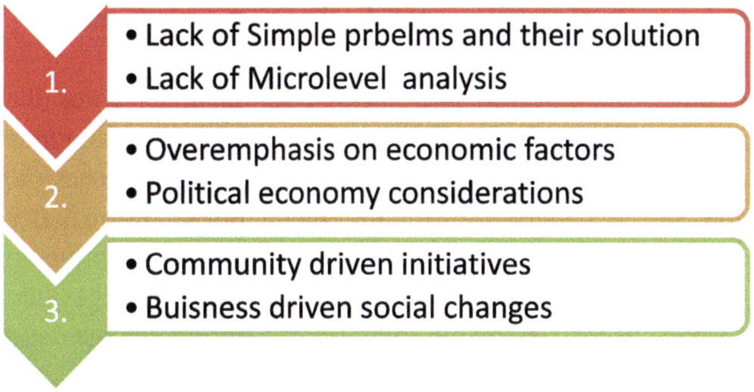

Fig. 4 Barriers to poverty in social relations

tributional needs. A more profound and convincing explanation of poverty may also be obtained from the social egalitarian viewpoint, which holds that poverty is more complex than a lack of resources. Rather than ignoring monetary indicators entirely, we need to comprehensively grasp the multiple detrimental effects of economic hardship on social interactions (Oelofsen 2018).

According to Daniel Putnam, to comprehend poverty as a social connection, one must concentrate on the risks that pose to meaningful interpersonal relationships. He also considers the variety of relational disadvantages that impoverished people face lead to two distinct duties for academics: broader justice-related obligations to engage in social capital and relational goods and second obligations connected to poverty (Fig. 4) (Beck et al. 2020).

8.1 Structural Causes

Poverty is often rooted in systemic inequalities and structural barriers that limit individuals' capabilities and opportunities. These structural factors can include unequal distribution of wealth, limited access to education and healthcare, discrimination based on race, gender, or ethnicity, and economic policies that favor the wealthy. From this angle, the incapacity to lead a respectable or "ordinary" social life may weaken social networks, relationships, and social engagement, thereby igniting a chain reaction of bad luck that exacerbates disadvantages in several spheres of life (Paugam 1995).

Numerous studies, such as those by Etongo et al. (2016) and Moseley (2004), provide a wealth of information demonstrating the connection between farm size and land growth. It was noted that the wealthy, who cleared additional land for agricultural expansion, were the main forces behind the rise in farm sizes. Most participants, who stated that unfarmed lands covered in woods and grasses needed to be cleared for farming, frequently highlighted this finding. These results corroborate studies by Ravnborg (2003), who found that nonpoor farmers—rather than the poorest people—were the direct cause of environmental degradation. Reardon and Vosti (1995) also highlight the financial hardships that impoverished people experience, which restricts their capacity to purchase tools and employ extra workers to remove trees and bushes to start new farms.

8.2 Social Exclusion

Poverty is about material deprivation, social exclusion, and marginalization. People living in poverty may face stigmatization, discrimination, and lack of access to social networks and institutions, which further perpetuate their disadvantaged position in society.

Using the economic deprivation measure of poverty, Mood and Jonsson (2016) begin this reference by presenting descriptive findings about the relationship between poverty and outcome factors. Figure 1 demonstrates that impoverished people participate less in organizations and politics and have weaker social interactions. Thus, there is a connection between poverty and both primary and secondary social interactions.

People who are socially excluded are often more likely to be living in poverty, and poverty is a significant driver of environmental degradation. When people struggle to meet and resort to practices detrimental to society and the environment, poor urban planning and inadequate infrastructure can result in disproportionately exposure to environmental hazards in marginalized communities. For example, low-income neighborhoods are often located in flood-prone areas or near industrial sites, increasing their vulnerability to natural disasters and pollution and causing health issues among marginalized populations who cannot afford alternative sources. Such people also face barriers to accessing environmentally sustainable job opportunities. The transition to green energy may lead to job losses in traditional industries such as coal mining, disproportionately affecting communities reliant on these sectors for employment.

Addressing social exclusion related to poverty and environmental sustainability requires holistic approaches prioritizing equity, community participation, and environmental justice. This involves implementing policies and initiatives that promote sustainable development, ensure equitable access to resources, and empower marginalized communities to participate in decision-making processes.

8.3 Power Dynamics

While environmental conservation is crucial, it can sometimes displace poor communities traditionally living in protected areas. Without proper resettlement and livelihood support, these communities can become worse off. Poverty is shaped by power dynamics within societies, including political, economic, and social power structures. Those with power and privilege often have more significant influence over resource allocation, policymaking, and decision-making processes, which can exacerbate inequalities and perpetuate poverty among marginalized groups. Power dynamics usually underlie systems of inequality, oppression, and discrimination based on factors such as race, gender, class, ethnicity, sexuality, and ability. Dominant groups may use their power to maintain privilege and control over marginalized groups, perpetuating social hierarchies and injustices.

The narrative that links poverty and environmental degradation persists because of the effect of power, wealth, and socio-political forces on ecological deterioration. Regardless of poverty levels, some influential groups or individuals contribute disproportionately to environmental degradation (Paulson et al. 2003; Walker 2005; Robbins 2019; Loftus 2020). This clearly indicates that how power disparities and limited access to the process of decision-making may make the studied area's environmental deterioration worse. One's capacity to contribute to ecological degradation might be gauged by the amount of chemicals used in farming. Farmers who could afford to buy and use pesticides and herbicides were usually not impoverished. They could forgo tedious manual weeding in favor of more convenient chemical alternatives since these herbicides were more affordable and acknowledged knowledge of the possible harm to the quality of the water, the health of the land, and the balance of the ecosystem as a whole. One significant barrier to purchasing vast amounts of chemicals is the financial capacity of the impoverished. As a result, they contributed more to water contamination than the needy farmers, who could only buy in lesser amounts or not at all (Nachibi and Morgan 2023).

The political ecology paradigm strengthens our understanding of the subject by highlighting the impact of social structures, political-economic processes, and power dynamics on environmental degradation. As a result of their need for the environment for their livelihoods, impoverished farmers adopt more sustainable methods. In contrast, wealthy farmers who are getting maximum advantages may prioritize more and more profit over environment-friendly considerations. Regardless of poverty levels, this highlights the disproportionate contributions made by powerful groups to environmental deterioration (Loftus 2020; Paulson et al. 2003; Robbins 2019; Walker 2005). It emphasizes achieving sustainable ecological management by addressing power disparities and unequal or partiality-based access to resources and other decision-making processes.

9 Community and Collective Solutions

Recognizing poverty as a social relation underscores the importance of community-based and collective solutions that empower marginalized groups, promote social cohesion, and address underlying structural inequalities. This includes initiatives that strengthen social support networks, promote grassroots organizing, and advocate for policy changes that advance social justice and equity. Numerous changes and initiatives at all levels, community, local, national, and regional are required to combat urban poverty. Suppose community-based planning is achieved by collective action. The literature on social movements and common-pool resources offers some of the most complex and comprehensive theoretical and empirical insights, which can help reduce the tangible signs of poverty, such as lack of access to essential services and infrastructure (Boonyabancha and Mitlin 2012). Communities can work together to tackle poverty and environmental sustainability interconnected challenges by making efforts like the following:

(i) Community-based organizations can implement customized solutions that empower underprivileged communities and promote sustainable development since they frequently have a thorough awareness of local requirements.
(ii) Cooperatives can support ecologically benign activities like sustainable agriculture or the generation of renewable energy while giving underprivileged people access to the financial system.
(iii) Communities can lessen their dependency on agricultural techniques that harm the environment and strengthen their resistance to poverty by working together to manage and cultivate local green spaces.
(iv) Environmental conservation and alleviating poverty simultaneously can encourage people by involving local populations in managing natural resources like fisheries, forests, and water sources.
(v) Communities may foster environmental stewardship and build routes out of poverty by providing individuals with the information and resources they need to adopt eco-friendly activities.
(vi) Community-led waste management programs can reduce environmental pollution and generate income by reusing and recycling materials. Examples of these programs include recycling campaigns and composting schemes.
(vii) By working together, government agencies, academic institutions, civil society organizations, and the private sector can use various resources and experiences to create comprehensive solutions that address underlying issues and foster long-term resilience.

(viii) Locals profit financially from community-owned renewable energy projects, such as wind farms or solar cooperatives, which offer inexpensive, clean energy.

10 Approach for Poverty Reduction with Environment in India

India's battle against poverty is intricately linked to its environment. More than a singular approach will be required. Here's how a multidisciplinary approach can be the key. Economists and environmentalists designed projects and programs that incentivize sustainable practices. Here, an example of such a program is provided.

The program is an urban risk reduction effort in two of New Delhi's poorer neighborhoods. The project's primary goal is to incorporate risk mitigation into urban planning. The goal is to improve the safety of low-income, vulnerable urban communities' lives and means of subsistence. Recurrent disaster risk is a significant contributor to urban poverty; hence, steps must be taken to lower the risk to reduce poverty. The project was carried out in two communities in Delhi between April 1997 and March 2000: a low-income neighborhood in the old city district of Delhi and a slum-squatter neighborhood inside the Yamuna River's riverbed. As the project's primary stakeholders, local community leaders supported the holding of community workshops. Public organizations that functioned as a link between the community and government authorities included the Indian Institute of Public Administration. SEEDS made the procedure easier.

Department for International Development of The United Kingdom funded the most famous project that targeted about 1000 households in each neighborhood. In India, the Indian Institute of Public Administration, the Center for Development and Emergency Practice, the National Centre for Disaster Management, Oxford Brookes University, and the Oxford Centre for Disaster Studies are some famous organizations that raise funds for these tasks. Through the promotion of school-level education, a country can improve the income generation capabilities of people and provide shelters with hard-won assets in upcoming disasters. These disasters can successfully advocate risk reduction at the national policy level, the three-year project helped to reduce poverty (GTZ 2005).

Another initiative in India targets the most vulnerable people through interventions tailored to their age, gender, and livelihood, including calamity risk reduction, into development in Malda District, West Bengal State, northeastern India. The project began in February 2006, and training sessions were conducted in 2007–2008. When the project is anticipated to be finished, its primary goals are to perform the following:

- Encourage community members to plan alternate income sources to defend their vulnerability to disasters.
- Encourage them to become trained in disaster management-related activities and preparedness attitudes so that people can deal with disasters.
- Enhance community members' skills to get and use of suitable shelter, water, and sanitation facilities.

The methodology employed for this is as follows:

- Conducting vulnerability assessments to quantify environmental and economic elements and pinpoint areas with potential for catastrophe risk reduction.
- Performing activities in Participatory Learning and Action (PLA), particularly in trend analysis, seasonality, wealth mapping, and village mapping.
- Teaching people about poverty reduction strategies and disaster readiness, as determined by community needs and stakeholder recommendations (GTZ 2005).

References

Ahmed M (2010) Economic dimensions of sustainable development, the fight against poverty and educational responses. Int Rev Educ 56:235–253

Arriagada A, Rawlings LB, Zumaeta-Aurazo M (2020) Combining cash transfers and parenting programs: operational design considerations. Glob Soc Policy 20(1):15–20

Asian Development Bank and United Nations Environment Programme (2019) We are strengthening the environmental dimensions of the sustainable development goals in Asia and the Pacific: tool compendium

Bann C (2019) Final programme evaluation of Joint UNDP–UN Environment Poverty-Environment Initiative (PEI) – 2013–2018 final report

Beck V, Hahn H, Lepenies R (2020) Interdisciplinary perspectives on poverty measurement, epistemic injustices and social activism. In: Beck V, Hahn H, Lepenies R (eds) Dimensions of poverty. Philosophy and poverty, vol 2. Springer, Cham

Bishop P (2000) Environmental engineering education in North America. Water Sci Technol 41(2):9–16

Boonyabancha S, Mitlin D (2012) Urban poverty reduction: learning by doing in Asia. Environ Urban 24(2):403–421

Brown L (2008) Plan B.3, mobilizing to save civilization. W.W. Norton and Company, London and New York

Cannon T, Twigg J, Rowell J (2003) Social vulnerability, sustainable livelihoods and disasters. Report to DFID. Conflict and Humanitarian Assistance Department (CHAD) and Sustainable Livelihoods Office London

Chant S (2006) Re-visiting the 'feminization of poverty' and the UNDP gender indices: the case for a gendered poverty index? UNDP, New York

Commonwealth Secretariat (2000) A commonwealth Vulnerability Index for Developing Countries. Commonwealth Publication

Davies M, Guenther B, Leavy J, Mitchell T, Tanner T (2009) Climate change adaptation, disaster risk reduction and social protection: complementary roles in agriculture and rural growth? IDS Working Papers 320:001–037

De Alba A (2020) Curriculum and pedagogical operation during COVID-19. Uncertain future. In: Garcia J (ed) Education and pandemic. Una Visión Academica. IISUE, Mexico, pp 289–294

EC (2019) Directive on liability for the environment. Obtained (17.07.2020) from https://ec.europa.eu/environment/legal/liability/pdf/factsheet/ELDfactsheet_PL

Etongo D, Djenontin INS, Kanninen M (2016) Poverty and environmental degradation in Southern Burkina Faso: an assessment based on participatory methods. Land 5(3):20

Francis F, David K (2012) The challenges of agriculture and rural development in Africa: the case of Nigeria. Int J Acad Res Progress Educ Dev 1(3):45–61

Gatto A (2020) A pluralistic approach to economic and business sustainability: a critical meta-synthesis of foundations, metrics, and evidence of human and local development corp. Soc Responsib Environ Manag 27(4):1525–1539

Gatto A, Polselli N, Bloom G (2016) Bloom empowering gender equality through rural development: rural markets and micro-finance in Kyrgyzstan. L'Europa e la Comunità Internazionale Difronte alle Sfide dello Sviluppo, pp 65–89

Górka K, Poskrobko B, Radecki W (2001) Ochrona środowiska. Problemy społeczne, ekonomiczne i prawne. PWE, Warsaw

GTZ (2005) Linking poverty reduction and disaster risk management. German Technical Cooperation, Eschborn

Hargreave D(2000) What constitutes an interdisciplinary environmental education? Environmental writings at Western Michigan University. http://www.wmic.edu/environmental-studies/Writings/hargreave2.Html. 2 July 2002

Liu L (2012) Environmental poverty, a decomposed environmental Kuznets curve, and alternatives: sustainability lessons from China. Ecol Econ 73:86–92

Loftus A (2020) Political ecology II: whither the state? Prog Hum Geogr 44(1):139–149

Makoka D, Kaplan M (2005) Poverty and vulnerability – an interdisciplinary approach. MPRA Paper, MPRA_paper_6964

McKinney S (2021) Poverty and education during the Covid-19 pandemic: editorial. Res Educ Bull 10:4–8

Mood C, Jonsson JO (2016) The social consequences of poverty: an empirical test on longitudinal data. Soc Indic Res 127:633–652

Moseley WG (2004) Environmental degradation and "poor" smallholders in the West African Sudano-Sahel: global discourses and local realities. In: African environment and development rhetoric, programs, realities. Ashgate, pp 41–62

Muralidharan K, Niehaus P, Sukhtankar S (2020) General equilibrium effects of (Improving) public employment programs: experimental evidence from India. Econom: J Econom Soc 91(4):1261–1295

Nachibi SU, Morgan AK (2023) Understanding the relationship between poverty, environmental degradation, and power dynamics: a qualitative study in Northern Ghana. Environ Dev Sustain, pp 1–29

OECD (Organisation for Economic Co-operation and Development) (2019) Greening development co-operation: lessons from the OECD Development Assistance Committee. OECD Publishing, Paris

OECD (Organisation for Economic Co-operation and Development) (2021) Integrating environmental and climate action into development co-operation: reporting on DAC members' high-level meeting commitments. OECD Publishing, Paris

Oelofsen R (2018) Women and ubuntu: Does ubuntu condone the subordination of women? In: African philosophy and the epistemic marginalization of women. Routledge, pp 42–56

Oxfam (2001) Participatory capacities and vulnerabilities assessment: finding the link between disasters and development. Oxfam Great Britain, Philippines Programme

Paugam S (1995) The spiral of precariousness: A multidimensional approach to the process of social disqualification in France. In: Room G (ed) Beyond the threshold: the measurement and analysis of social exclusion. Policy Press, Bristol, pp 47–79

Paulson S, Gezon LL, Watts M (2003) Locating the political in political ecology: an introduction. Hum Organ 62(3):205–217

Prowse M (2003) Towards a clearer understanding of poverty. CPRC Working paper no 24 Manchester

Ravnborg HM (2003) Poverty and environmental degradation in the Nicaraguan hillsides. World Dev 31(11):1933–1946

Reardon T, Vosti SA (1995) Links between rural poverty and the environment in developing countries: asset categories and investment poverty. World Dev 23(9):1495–1506

Reimers FM, Schleicher A (2020) Guidance to guide the educational response to the COVID-19 pandemic. OECD, Paris

Richterman A, Millien C, Bair EF (2023) The effects of cash transfers on adult and child mortality in low- and middle-income countries. Nature 618:575–582

Robbins P (2019) Political ecology: a critical introduction. Wiley

Seed J (1994) Antropocentryzm. In: Devall B, Sessions G (eds) Ekologia głęboka. Pusty Obłok, Warsaw, pp 316–327

Simmons B (2000) Towards excellence in environmental education: A view from the United States. Water Air Soil Pollut 123:517–524

Thywisson K (2006) Glossary of core terms of disaster risk management. In: Birkmann J (ed) Measuring vulnerability to hazards of natural origin, Tokyo

UNEP (1997) Nairobi Declaration on the Role and Mandate of the United Nations Environment Programme February, UNEP/GC.10/34, pp 52–55

UNEP and SOPAC (2005) Building resilience in SIDS: the environmental vulnerability index. SOPAC Secretariat, Fiji Islands; UNEP, Nairobi

United Nations High Commissioner for Human Rights (2004) Human rights dimension of poverty. Available at http://www.ohchr.org/english/issues/poverty/. Last Assessed 21/11/2005

Upadhyaya AS (2018) Micro finance in sustainable development: a descriptive study on Indian experience. J Econ Dev 13(3):77–88

Walker PA (2005) Political ecology: where is the ecology? Prog Hum Geogr 29(1):73–82

Wee V, Heyzer N (1995) Gender, poverty and sustainable development: towards a holistic framework of understanding and action. Engender and UNDP, New York

Wiesner M, Theis T (1996) Environmental engineering education: application area and discipline. J Environ Eng 122(2):89–90

Wiles J (2023) Are you thinking too small about sustainable technology? Article in Gartner

Yunus M, Yusus AJM (2007) Banker to the poor. Penguin Books India

Sustainability in Environmental Engineering: Professional Perspectives on Combating Climate Change through Science

Farah Naaz, Mohd. Zeeshan, and Mohd. Farhan

Abstract

The chapter aims to elucidate the significance of environmental science and engineering professionals in promoting sustainability amid severe climate change. Here, we focus on how the field of environmental science and engineering has long recognized the importance of sustainability in combating climate change and ensuring environmental sustainability. Climate change resulting from human actions is one of the biggest concerns of this era. A number of studies have suggested that emissions are increasing global temperatures, which in turn are posing challenges to the environment, economies, and global society as a whole. To prevent even-worse effects from occurring, we must work to limit the volume of greenhouse gas emissions, all the while making sure that human civilizations and ecosystems adjust to the changes brought on by high greenhouse gas emissions. This climate issue can be addressed through policy, conservation, and technological innovations. This chapter outlines the viewpoints of professionals on the topic of sustainability and their possible approaches to tackling the issue; it then elaborates on these approaches and outlines the gaps where more research and discussion are needed. By overcoming cross-discipline barriers and building practical solutions, environmental science and engineering professionals become key experts in shaping sustainability.

Keywords

climate change · sustainability · environment · stewardship · technology

1 Introduction

One of the most critical issues of our day is climate change, which profoundly impacts ecosystems, economics, and communities worldwide. Due to the growing number of environmental disasters, sustainability plays a crucial part in environmental engineering and research. The global low-income population extensively uses goods and services provided by a shared pool of terrestrial and aquatic natural resources. Natural resources are also known to support an essential social safety net during times of economic downturn and constrained food supplies (WRI 2005). Sustainability, which aims to meet people's current needs without compromising the ability of future generations to meet their own, is fundamental to environmental engineering and research. Many environmental engineers have focused their efforts in developing regions on the provision and study of water, sanitation, and hygiene (WASH) services, which include (1) identifying environmental and governance barriers to achieving sanitation coverage (Fry et al. 2008); (2) determining the costs required to meet global coverage for safe water and sanitation (Hutton and Bartram 2008); (3) assessing the factors that lead to changes in handwashing behavior (Naughton et al. 2015a); (4) assessing water quality during the movement of water from the source (e.g., wells, boreholes, or communal taps) to households (Harris et al. 2013), household water storage (Schafer and Mihelcic 2012), and self-supply ground water systems (Akers et al. 2015); (5) improving our understanding of the design and operation of conventional and resource recovery–based sanitation systems (Mehl et al. 2011; Verbyla et al. 2013; Exley et al. 2015); and (6) improving the design and management of hand pumps and point-of-use treatment technologies (Clasen

F. Naaz (✉)
National JALMA Institute for Leprosy and Other Mycobacterial Diseases, Tajganj, Agra, India

M. Zeeshan
Dr. Bhim Rao Ambedkar University, Agra, India

M. Farhan
Jamia Millia Islamia University, New Delhi, India

J. A. Parray et al. (eds.), *Environmental Landscape and Sustainable Biodiversity for Healthy Green Growth*, Sustainable Landscape Planning and Natural Resources Management, https://doi.org/10.1007/978-3-031-75405-0_9

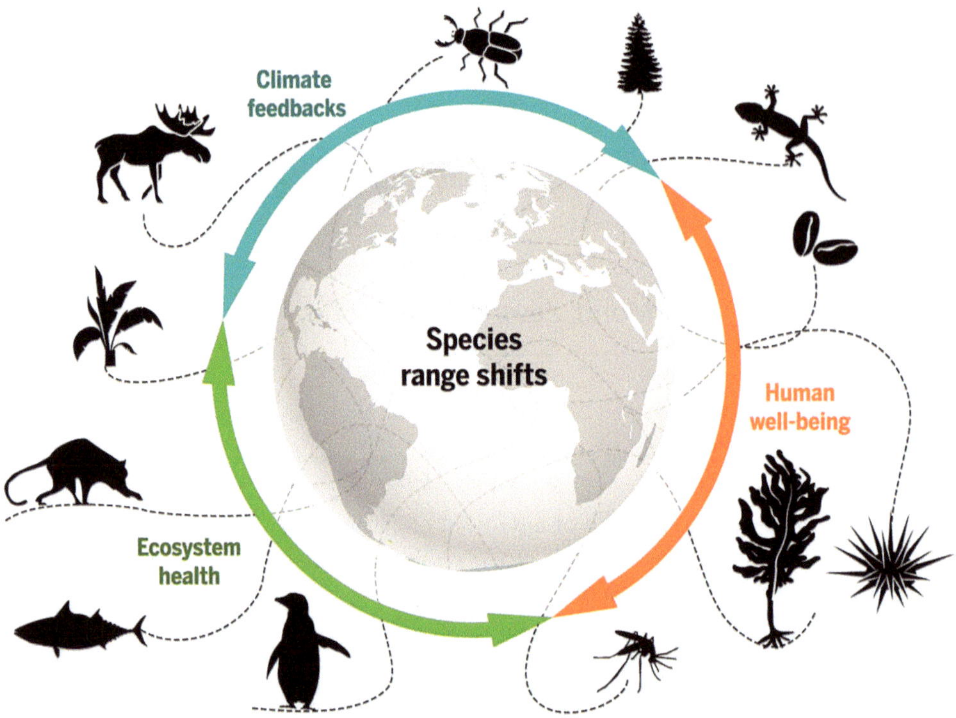

Fig. 1 The altering geographic distribution of life is having an increasing impact on human well-being, ecosystem function, and even climate itself as the world's climate changes

et al. 2004; Schweitzer et al. 2013; Marks et al. 2014) (Fig. 1). Sustainability is an integrated strategy for economic growth, social justice, and environmental stewardship that balances ecological integrity and human development. Sustainability in the context of climate change is not only a desirable goal but also a necessity for survival. Everyone is subject to climate change, and environmentalists can hold us accountable for systemic choices that undermine ecological sustainability.

1.1 Recognizing Sustainability in Environmental Science and Engineering

Engineering the environment could be a way to harmonize human health with efforts in making progress toward sustainable development. Researchers have carried out experiments to remediate the environment, by purifying the air, removing water pollutants, preventing the harmful impacts of environmental contaminants from affecting humans, and isolating the environment from human activities (Mihelcic et al. 2003; AAEE 2009). The meaning and principles of sustainability can be found in this chapter and in the United Nations' Sustainable Development Goals (SDGs), which play roles in putting countries on track toward sustainable business practices. Environmental engineering has come close to determining the global funds needed to improve the lives of those living in impoverished regions. For instance, over the past 15 years, international development efforts have been guided by the eight Millennium Development Goals (MDGs), with environmental engineering most closely aligning with MDG 7, which pertains to environmental sustainability, particularly regarding water and health issues.

Environmental sustainability has long been a significant responsibility of environmental engineering. However, current environmental problems are more nuanced and intricate, and global climate change must now be considered while developing solutions (Anastas 2012). Sustainability is the most challenging domain in environmental science and engineering because it requires analyzing real-world applications. Current and recent generations have emphasized the satisfaction of their current sets of wants without considering or attending to the survival of other species or future generations of humans. However, the main problem is ensuring that the present generation even endures despite our having already consumed shortages of the world's limited resources.

1.2 Environmental Stewardship

Responsible treatment for and taking care of natural resources by ensuring their health and sustainability for the future generally entails practices that conserve, protect, and restore ecosystems, biodiversity, and natural habitats. These practices aim at reducing human disturbances to the environment. Protecting natural resources from overexploitation and preserving them entail managing them responsibly. To do so, the erosion of ecological systems must be halted, wildlife must remain diverse, pollution from industry must be remediated, the development of cities must adopt sustainable infrastructure, and noxious emissions must be reduced.

1.3 Social Equity

Sustainability is no small task. It entails, among other factors, a fair system that makes sure that everyone in society receives equal opportunities and earns thriving outcomes. Social equity is the principle that should bring about equal access to resources, opportunities, and rights for everyone, irrespective of ethnicity or social status, and that considers how best to meet the needs of the disadvantaged groups in societies. It focuses on demolishing systemic inequalities by guaranteeing education, healthcare, jobs, and legal rights, to reduce the discrepancies in outcomes within society. Individuals' experiences of sustainability and social equality differ depending on their ethnicity, gender, socioeconomic status, and other criteria. In short, environmental justice aims to ensure that low-income people are able to benefit from their society's resources and that communities breathe unpolluted air and drink and use clean water. Moreover, environmental justice advocates for decision-making processes that include everyone and everything affected.

1.4 Economic Viability

Sustainable economic viability entails that all projects, businesses, and initiatives financially constrain themselves such that they fulfill both environmental and social goals. It ensures that economic activities indeed create sufficient profit or value to meet costs and continue over time without depleting natural resources or harming human well-being. Individuals, businesses, and technologies can balance profit with environmental protection when trying to find sustainable options.

1.5 Interconnectedness

Sustainability considers the following three elements: society, economies, and the environment. It is an integrative approach that brings these elements together and triers to ensure that they thrive and complement one another. It also reflects that environmental systems are interdependent. And international initiatives such as reducing global hunger cannot be solved by taking an approach that considers only one of the above elements. Instead, various approaches must be incorporated and implemented together. Interconnectedness stimulates taking a systemic viewpoint, one that should ensure that policies and actions include and address root causes and ensuring that unintended consequences stemming from one element do not undermine another.

1.6 Long-Term View

Taking a long-term view fosters decisions and actions that respond to needs in the present without compromising the ability of future generations to meet their needs. It helps to foresee, prevent, and offset long-term environmental, social, and economic consequences. Hence, it underlines the need to balance immediate benefits and future outcomes, rather than taking actions that would solve a short-term problem at the expense of long-term well-being. Such prioritization in intergenerational equity generates resilience, adaptability, and commitment toward conserving resources, ecosystems, and opportunities for the future prosperity of humanity and the Earth. This prioritization encourages leaders to go a step further, and it defines their strategies. Conversely, leaders' strategies must be guided by long-term views; otherwise, they will ever achieve short-term success. Therefore, developing new technologies and sourcing new materials are not enough. We must in the present also properly utilize the ones that we already have.

2 The Importance of Climate Change

Climate change impacts various areas differently. Sea levels have already pushed some communities to move out. Others live in fear of hunger because of droughts. For the world to maintain a climate that is still habitable for humans, the increase in average global temperature must not exceed 1.5 degrees Celsius (34.7 degrees Fahrenheit). Some solutions can lead to this achievement, but their deployment has to be prioritized. Cutting emissions, adapting to climate impacts, and financing needed adjustments are three broad categories of action. In reducing emissions, divesting from fossil fuels in favor of renewable energy is often one of the first steps considered. Popular choices are solar and wind power. Using solar or wind energy can, however, be expensive and impractical at times, which leaves room for innovation in this field. Engineers and technologists can develop climate change engineering solutions and tools to counter environmental degradation, which would be major contributions toward a greener planet. Munasinghe and Swart (2005) and Swart et al. (2003) asserted that sustainable development measures can strengthen adaptation in climate change policies and

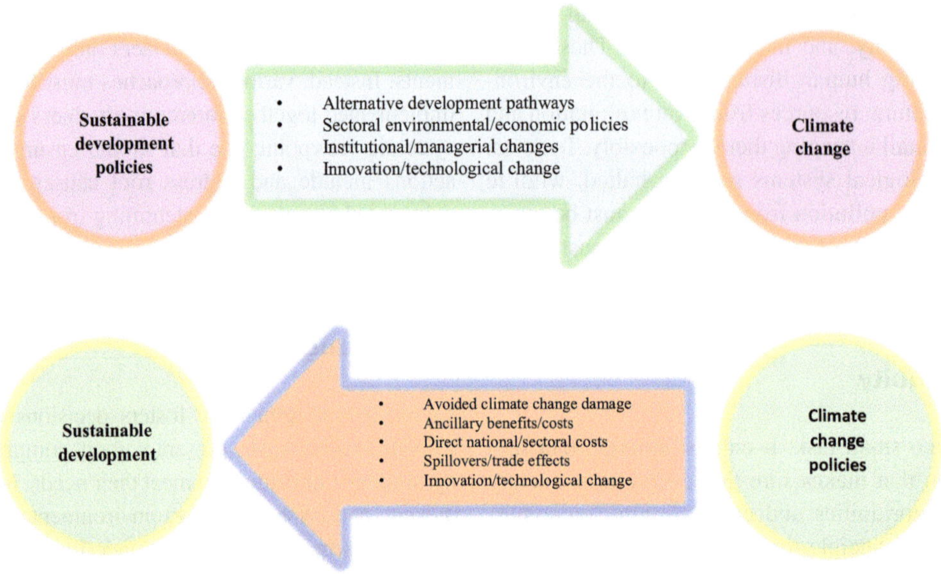

Fig. 2 Two-way correlations between climate and sustainable development. (Source: Swart et al. (2003))

vice versa: Fig. 2 reflects some of the context of the interaction that they envisioned.

3 The Roles of Professionals in Climate Change: Mitigation and Adaptation

- Climate change is a boundary-crossing issue that increasingly affects advice and decision-making in many professions. Some examples, provided as an illustrative, not exhaustive list, include the following:
  - Architects and engineers advising an industrial company that wants to build a new factory on matters related to energy efficiency (reducing greenhouse gas emissions) and susceptibility to future climate change impacts (climate change adaptation)
  - Professional foresters modeling the future growth patterns of forests for forestry companies on the basis of a changing climate to be used as the basis for decisions about logging and replanting
  - Biology professionals advising governments at different levels on how to manage salmon runs in light of climate change
  - Insurance professionals looking at the risks related to climate change impacts and providing clients with insurance products that are designed to protect them against identified risks
  - Accountants and other professionals designing systems for measuring the greenhouse gas emissions of companies; in some cases, providing third-party validation of those measurements; and counseling on how to comply with applicable reporting requirements
  - City planners giving advice on how cities and other communities can change development patterns to address and prepare for the impacts of climate change, such as extreme weather events, sea level rise, and drought conditions.
- Professionals should be providing advice on these and other climate-related matters on the basis of the best science and information about climate change because many policymakers have never formally received training related to climate change.
- But many professionals are proactively involved in exploring the impacts of climate change on their work, learning by doing and contributing to the knowledge of the profession. In addition, concerning climate change, professionals conduct much of the grassroots work on identifying how to adopt strategies for reducing greenhouse gas emissions (mitigation) and for planning and managing responses to the effects of increasing temperatures (adaptation). Their work can make them able innovators in developing solutions and responses and in taking an active, ongoing role in adaptive management as the full implications of climate change become clearer.
- Professionals may also play roles as trusted commentators or spokespeople on the issue as it relates to their fields of expertise, including as communicators of information to the public while encouraging action on climate change.
- Professionals therefore have both an incentive and a responsibility to be aware of climate change in their work.

4 Outlooks on Professional Perspectives

Outlooks on the professional roles in sustainability and its application in various fields in overcoming the environmental, social, and economic challenges of the world include a professional's ethical responsibility for devising new ideas and cross-disciplinary methods to reduce environmental adversity and enhance sustainability gains. Moreover, resource efficiency, equity, and resilience should be priorities in professional practice and in work on designing programs, policymaking, and management. Sustainability is engaged in building solutions aligned with the larger objectives of sustainable development and intergenerational equity.

Engineers can share their firsthand experiences with climate issues and solutions to them in a search for technological advances. Engineers draw on expert opinions when producing new concepts and solutions for industry. These opinions cover a broad range of subjects, which are covered in the following subsections.

4.1 Obstacles to Implementation

However, we face some barriers from the activities outlined in environmentally sustainable solutions. Engineers and technical specialists solve difficult tasks. Although some difficulties result from the limitations of our current technologies and although financial priorities continue to stand in the way, laws could ensure that we meet our sustainable development goals.

4.2 Success Stories

Professionals have a social responsibility to help guide decisions on what is best for humankind and identify the most effective strategies for sustainability. Such guidance should stem from professional expertise that is informed by stakeholders' real-life experiences. This approach must not only address the challenges of adopting sustainable practices but also highlight their benefits, such as by detailing the strategies that have produced significant beneficial environmental effects.

4.3 Regulation and Policy

Instead of relegating the responsibilities of safety and health officials to basic science or engineering, these safety and health officials seek to coordinate efforts and interact with the public more regularly. We can assess the efficacy of present directives by putting into practice animal species conservation initiatives and creating environmental laws and regulations that solve the core of complex challenges. Using scientific knowledge to strengthen and support environmental management is one more approach in which scientific data can be used. In addition to these difficulties in environmental regulation are shortcomings in the development of new legislation.

4.4 Technological Innovation

Many aspects of science and engineering interact with environmental issues, but in this chapter, we focus on education, research, policies, collaboration, and economy in particular. Environmentalists and renewable energy experts will see positive change when ecological engineering leads the way in the scientific community. The public life A crucial component of public relations is technology, which differs in terms of manufacturing quality, the quantity of scientific evidence supporting it, and its conformity to contemporary trends. It generally plays a crucial role in public life.

4.5 Interdisciplinary Collaboration

Interdisciplinary degree programs require learning from many other fields—the same information used by engineers and doctors. This interdisciplinary approach establishes uniform terminology, harmonizing academic disciplines to better overcome environmental challenges. The partnership between ecological engineers and scientists should be highlighted because the priorities of policymakers, economists, and social scientists, along with other experts, comprise the most essential elements of research. Professionals are best suited to lead the way in offering direct approaches to solving environmental issues and achieving sustainable development goals.

5 Implementing Sustainable Practices

Professionals' roles in sustainability and its applications in various fields aim to overcome the environmental, social, and economic challenges of the world, and this includes professionals' ethical responsibility to devise new ideas and cross-disciplinary methods to reduce adversity and improve progress toward sustainability. Moreover, resource efficiency, social equity, and resilience should be priorities within the professional practices of policy design, policymaking, and management. Sustainability in every professional practice should mean building solutions aligned with the larger objectives of sustainable development and intergenerational equity. These factors enable cities and their inhabitants to better confront challenges in the future. Although the major share of the environmental engineering industry is spent in developing countries, the people working in this industry have been in a tug-of-war as their efforts try to solve environmental problems. These efforts have included (1) identifying the governance and environmental barriers to achieving sanitation coverage (Fry et al. 2008); (2) calculating the costs needed to meet global coverage of safe water and sanitation (Hutton and Bartram 2008); and (3) evaluating the factors that lead to changes in handwashing. This chapter demonstrates the effects of sustainable practices in mitigating climate change and promoting resilience by exemplifying accomplished projects and initiatives. Environmental engineering's role in controlling pandemic-related diseases that may be brought on by viruses that infect agricultural and urban water cycles should also be considered (Table 1) (Wigginton et al. 2015).

5.1 Renewable Energy

To lower greenhouse gas emissions and mitigate climate change, sustainable energy sources, including solar, wind, hydroelectric power, and biomass, are essential. Initiatives that use renewable energy show how these technologies can be used to reduce dependency on fossil fuels and move toward a low-carbon-energy future at different scales, from individual homes to massive power plants.

5.2 Waste Management and Recycling

Recycled materials, composting, and waste-to-energy conversion are examples of sustainable waste management techniques that help reduce emissions from landfills, protect the environment, and conserve resources. Successful waste management projects include creative trash-reduction methods, recycling infrastructure, and public education campaigns encouraging recycling and waste-reduction behaviors.

5.3 Sustainable Urban Planning

This approach incorporates social, economic, and environmental factors into the planning, development, and construction of cities and neighborhoods. Compact, mixed-use development, green infrastructure, public transit, and energy-efficient buildings are all prioritized in these projects, which show how sustainable urban planning can improve people's quality of life, lower carbon emissions, and increase overall resiliency to the effects of climate change.

5.4 Water Resource Management

Sustainable techniques for water resource management aim to guarantee water availability and quality for current and future generations. Initiatives for sustainable water management include reusing and treating wastewater, conserving water, collecting rainwater, and restoring aquatic ecosystems to preserve and promote biodiversity.

Table 1 Energy-management approaches: the water and wastewater sector

Approach	Contribution to Energy Efficiency
Energy Preservation	Directly decreases energy needs by lowering the energy required for particular tasks
Recovering Heat	Lowers energy consumption in homes and businesses by recovering heat directly from a water stream
Anaerobic Medicine	Creates energy sources and stabilizes biodegradable organic materials by converting them to biogas
Heat Generation	Produces heat that can be used directly or to produce energy, when organic matter burns
Hydropower from Effluent	Produces energy from wastewater by using turbines or other inductively mounted equipment
On-Site Solar and Wind Energy	Generates electricity by using wastewater treatment plants' enormous land areas
Systems of Bioelectrochemistry	Directly produces power from the organic materials in wastewater

Source: Lazarova et al. (2012), Mo and Zhang (2013)

5.5 Biodiversity Conservation

Given the vital role that biodiversity plays in sustaining ecosystem services and human well-being, biodiversity conservation initiatives work to preserve and restore ecosystems, habitats, and species diversity. Biodiversity protection can be achieved via sustainable methods such as managing protected areas, restoring habitats, reintroducing species, and including local communities in conservation efforts through community-based conservation initiatives.

5.6 Sustainable Forestry and Agriculture

Sustainable agricultural and forestry methods save the environment by conserving water, improving soil quality, lowering chemical inputs, and preserving biodiversity. Agroforestry, organic farming, sustainable forest management, and reforestation initiatives that will enhance ecosystem resilience and trap carbon are a few examples of sustainable agricultural and forestry methods.

6 Possibilities and Difficulties

Environmental engineers and scientists encounter numerous obstacles to implementing sustainable solutions, though they have made advancements in sustainability. Understanding how people interact with waste-reclamation technologies and products can also be aided by the social sciences (Po et al. 2005; Whiteford and Whiteford 2005; Nancarrow et al. 2008; Rodriguez et al. 2009; Russell and Lux 2009; Mankad 2012). Applying marketing knowledge and techniques for nonprofit goals, like altering attitudes and behaviors to improve health, is known as social marketing. This means that social scientists implementing ethnographic research techniques (such as focus groups and interviews) and program evaluations in the field of environmental engineering can affect the target population's voluntary behavior for the public's benefit (Kotler and Zaltman 1971; Brown 1997; Lee and Kotler 2011; Guang and Borges 2012). Several educational programs exist to develop the global competency of early-career engineers (Mihelcic et al. 2006; Mihelcic et al. 2008; Trotz et al. 2009; Mihelcic 2010; Jesiek et al. 2014). Locating resources from wastewater has proven to be a marriage made in heaven between this method of approach and this novel contribution (Johnson 2003; Doria et al. 2009; Dolnicar and Hurlimann 2011; Dolnicar et al. 2010; Kemp et al. 2012). This chapter discusses how the implementation of sustainability in various practical contexts is complicated by various circumstances, such as a lack of resources, friction, and uncertainties. Moreover, these barriers can evolve into such things as creativity, teamwork, and systemic change, which are vital in fast-tracking sustainability objectives. The following subsection reveals possible pathways for progress and innovation. It details the prevailing obstacles and dilemmas that experts encounter when ensuring the successful implementation of sustainability ideas.

6.1 Economic Challenges

One of the main obstacles is the cost incurred by engaging in sustainable practices. In contrast to traditional yet unsustainable alternatives, allocating preliminary capital to green technologies, sustainable construction and transportation, and sustainable water and power triggers economic growth. Projects, renewable-power-generation facilities, and renewable energy infrastructure are unusually seen as burdensome due to their costs. To counteract this perception, fossil fuel subsidies should be abolished.

6.2 Technological Hurdles

Sustainability in technology must overcome such challenges as high-cost renewable energy technologies and inefficient energy-storage systems. These challenges act as barriers to clean energy adoption. Manufacturing is another obstacle because it is hampered by its reliance on nonrenewable materials, accompanied by their inefficient recycling processes—thus hindering sustainable production. Waste management is grappling with a growing pile of e-waste alongside poor infrastructure, hampering circular economy practices. Desalination and wastewater-recycling technologies are prohibitively expensive and consume a lot of energy; these characteristics of such technologies act as hurdles in the implementation of water sustainability. Urban development has to deal with the renovation costs incurred while refurbishing old buildings and the expenditure required for green infrastructure, whereas agriculture has to deal with a lack of will to transition to resource-efficient practices. Carbon emission reduction initiatives can leverage costly capture technologies, but they measuring their effectiveness is complicated. Public and private resistance, policy gaps, and ethical issues, such as equitable access and job displacement, further hamper sustainability efforts. Innovative research, political will, collaborative global action, and widespread education on sustainability will help us to overcome the challenges mentioned above.

6.3 Social and Behavioral Factors

Sustainability should be rooted in the consumer psyche to position sustainable solutions as the most persuasive means for conservation efforts and changing societal behaviors. Social and behavioral factors thus seem to play significant roles in the success of sustainability, and indeed, many times, the public's attitudes, habits, and cultural norms are responsible for the failure of a sustainability initiative. Consumer resistance to change, customers' lack of awareness on the environmental impacts of their behaviors, and convenience-seeking approaches that dissuade green choices are hindrances to the success of sustainable practices. Environmental justice also gives rise to challenges for sustainability in that disadvantaged neighborhoods tend to have less access to green technologies or suffer more-dangerous environmental exposures. Behavioral inertia also exists in companies that are typically driven primarily by profit and short-term views on their goals. Moreover, cultural and lifestyle expectations, such as those perpetuating overconsumption and dependence on fossil fuels, often do not comport with sustainability objectives. Countering behavioral inertia involves targeted education, community mobilization, participatory policies, and positive motivation toward changing behavior at the individual level and equal participation in forming environmental solutions. The sustainability of a given community will also be compromised if its development doesn't address inequality or evenly distribute access to resources and technologies.

6.4 Global Issues and Interconnectedness

Global interrelations thus form the crux of sustainability because the environmental, social, and economic problems have borderless interconnections. These problems include climate change, the depletion of resources, and the loss of biodiversity, which are faced by all countries more than ever; however, developing regions are subjected to higher burdens from climate change, and their populations are more vulnerable to such burdens. Global supply chains lead to environmental degradation, caused by unsustainable consumption and production in one part of the planet affecting ecosystems or community life somewhere else. The interdependence of nations implies that cooperation encourages solutions for sustainability, such as their international cooperation in policies, technology sharing, or resource management. More often, however, global economic systems favor short-term advantages over long-term healthy ecosystems—that is, favoring economic development over environmental protection. We recommend that environmental concerns take priority in decision-making to combat pollution, the loss of biodiversity, and climate change. All these problems should be attended to in a global, inclusive way, one that considers all their interdependences and holds us collectively responsible for bringing about a sustainable future.

7 Multidisciplinary Cooperation for Sustainability

For these issues of sustainability to be addressed effectively, multidisciplinary cooperation is a requisite. Such issues extend their consideration beyond environmental concerns and into the fields of economics, engineering, the social sciences, and policymaking. Any comprehensive solution to these concerns will need input from experts in those fields. An example is engineers dealing with renewable energy technologies, where they should work with economists in financial viability assessments and work with social scientists who investigate access issues. Policymakers help in devising frameworks that foster sustainable practices by industry. Good communication and joint efforts between sectors encourage innovation and optimize the use of resources toward encouraging sustainable practices at the local and global levels. All those detailed approaches will help to tackle the complicated nature of sustainability challenges. In this way, a holistic approach that brings in actors from all sectors is the start to a solution that benefits the most stakeholders. A community-based sustainability approach is being proposed, where academics, other experts, and locals come together to brainstorm the best possible solutions. The key toward realizing this broadly beneficial solution for sustainability is developing multidisciplinary dialogue among governments, industry, and nongovernmental organizations (NGOs).

7.1 Integration of Diverse Perspectives

The integration of diverse perspectives is a prerequisite for achieving sustainability because it ensures inclusive, equitable, and context-specific solutions. Voices from different cultures, societies, and geographical locales can help us to understand the unique challenges facing diverse communities and ecosystems. Sustainable practices respect local traditions while recognizing global environmental and economic considerations by hearing from multiple voices. For example, Indigenous involvement in conservation efforts yields holistic approaches that include both traditional knowledge and modern science, while having policymakers, businesses, and local communities on board helps to ensure that sustainability solutions are practical, and broadly accepted, and ultimately successful. Cross-disciplinary teamwork's dynamics allows different kinds of competencies and cultures to cooperate on finding holistic solutions to preserving the environment.

7.2 Systems Thinking

Systems thinking is an important means for finding and considering the interlinkages between the environment, societies, and economics. Systems thinking evaluates a situation beyond a few isolated concerns and toward including a wider context; this approach recognizes that changes made in one part of a system can have far-reaching effects on other parts and brings to fore the root causes behind sustainability problems, such as high consumption, waste generation, and resource depletion. Moreover, it promotes developmental solutions addressing multiple facets of an issue at the same time. In this approach, for example, sustainable farming methods must consider not only the health of the farm's ecosystem but also economic viability and the health of human communities. Designing systems-thinking solutions would create positive feedback loops whereby progress in one area reproduces benefits in another, all toward achieving more-sustainable and more-resilient outcomes.

7.3 Collaboration Across Sectors

To make sustainability a realistic endeavor, collaboration is essential. In order to address environmental, social, and economic challenges all at once, it combines a variety of perspectives, resources, and areas of expertise into a multidisciplinary approach. Businesses, organizations, and communities must collaborate to create efficient laws, tools, and plans in order to accomplish sustainable development. For instance, NGOs can increase awareness and involve local communities, while governments can regulate companies and provide incentives to encourage creative solutions. Maintaining transparency in a multi-agency approach to sustainability encourages a comprehensive viewpoint and enables all stakeholders to appropriately assess progress. We can successfully address some of our most urgent issues, such as resource scarcity, inequality, and climate change, with workable and fair solutions if we all align with the priorities set forth in this cooperative strategy.

7.4 Transdisciplinary Research and Education

Sustainability has always been praised in transdisciplinary research and education, which not only introduce knowledge from various disciplines to stakeholders outside of universities, such as those in communities, businesses, and policymaking organizations, but also foster deeper understandings of the problem by combining science with the problem's social and practical aspects. Transdisciplinary research claims that new, innovative solutions emerge beyond theoretical knowledge when education meets real-world experiences and responds to people's expressed needs. By transforming education, young students will start acquiring a culture of holistic thought and transdisciplinary collaboration, which should help to better prepare them to face a highly diverse set of questions around sustainability in the future. Thus, actions that cross the divide between disciplines in academia and sectors of society can produce well-rounded knowledge to induce action and can help to assign people with appropriate skills to suitable tasks, which should accelerate change toward sustainability. In transdisciplinary research, communities, policymakers, and professionals have to be included because they will be partners in increasing knowledge, devising research agendas, fundraising, and implementing social justice–focused and context-specific policy solutions. Through interdisciplinary collaboration, initiatives are better able to include all stakeholders and other considerations and are more adaptable to widening the scope of initiatives to include different perspectives and forms of knowledge. Multidisciplinary teams, which bring together various professionals with various backgrounds, can provide highly sophisticated ideas, techniques, and technologies that are much more effective than would be feasible if scientific disciplines were left siloed.

7.5 Knowledge Exchange and Capacity Building

People from a wide range of fields—such as medicine and social work—can contribute to knowledge exchange, clarify their capacities, and thereby engage in interdisciplinary collaboration. Many people with different backgrounds in one team can exchange success stories, methods, and experiences by engaging in joint research projects, workshops, and training programs to support their overall joint research, enabling them to build cultures of learning and to continue their development of sustainability in science and in practice.

7.6 Governance and Policy

Governance and policy are critical for promoting sustainability in that they establish the necessary frameworks, regulations, and incentives for both the public and private sectors to act. In practice, effective governance ensures the prioritization of sustainability across all levels of decision-making—from local governments to international institutions. Standards can be set for environmental protection, social equity, or economic development through such policies, and innovations in green technology and sustainable practices can be incentivized. Carbon-pricing mechanisms, renewable energy subsidies, and waste-reduction regulations, for example, can encourage industries to develop better-adapted sustainable models. Such governance becomes more transparent and inclusive when it involves more stakeholders across more sectors, which is crucial to making policies fair and flexible for present and future generations. Detailed and well-articulated policy frameworks also provide stability and direction to align global, national, and local efforts toward a common end—sustainability. A team of interprofessional health experts can help to make the system more inclusive, just, and fair and can promote policies that address and rectify gendered differences in outcomes.

8 Future Prospective and Suggestions

We have an opportunity to address and prepare for an environmentally disrupted planet, which we need to seize to work toward long-term sustainability. Geographic information systems and satellite technology evaluate indicators around the world, which can be incorporated into development projects in the developing world to measure outcomes (Pettorelli et al. 2014; Naughton et al. 2015b). Recent advances in technologies and environmental objectives aimed at mitigating climate change were described herein. Additionally, this chapter laid out how to emphasize sustainable development goals and build the capacities of institutions to bring about revolutionary changes. These goals help experts, policymakers, teachers, and researchers to focus their efforts. This chapter serves as a strategy for motivating groups to continue collaborating on sustainable development goals while being guided by consultations with experts.

8.1 Future Directions in Sustainability

The future of sustainability must include driving a transition toward a more collaborative, more equitable, and greener world. Key areas of advancement include replacing polluting technologies with clean ones, such as renewable energy, electric vehicles, and carbon capture; storage facilities and energy-efficient systems that support a reliable transition toward green energy should also be advocated. The emerging circular economy, wherein products are designed for reuse and recycling with minimal waste, is gaining momentum. Digital technologies like artificial intelligence (AI) and big data are expected to play pivotal roles in optimizing resource consumption, measuring environmental effects, and developing more-intelligent and more-efficient systems. Sustainability dimensions will more and more be integrated into social ones to ensure that green development benefits marginalized populations and prioritizes equity. Global collective action and its backing by robust policies are necessary to address complicated and interlinked problems such as climate change, biodiversity loss, and social inequality. Experts can in advance pinpoint and list viable solutions that would address such problems.

8.2 Priorities for Future Study

Further study on sustainability should above all take account of today's most pressing global issues while promoting long-term solutions. Such study should include the development of new clean energy technologies, such as energy storage and integrating it into energy grids to achieve a reliable transition to renewable sources. Important fields of study include research into adaptation to climate change and mitigation strategies, such as carbon capture and nature-based solutions, which may help to keep environmental effects to a minimum. In agriculture, much work needs to be done on sustainable practices to improve food security while reducing resource use, especially given the ever-growing global population. The social and behavioral aspects of sustainability, including consumer behavior, equity, and policy effectiveness, must also be considered, because without them, solutions will not be accepted by the public at large. Last but not least, interdisciplinary research needs to be conducted into the roles that digital technologies, such as AI and data analytics, can play in optimizing sustainability and resource management. Including people's real-world experiences into the development of scientific solutions and taking interdisciplinary approaches yield more-practical solutions that benefit stakeholders and the environment because they incorporate as many scientific, social, economic, and technological factors as possible.

8.3 Technological Innovation

Surprisingly, environmental sustainable concerns are mounting along with the evolution of technology, though also significant technological innovation can facilitate sustainability by creating new solutions that address environmental, social, and/or economic problems. Clean energy innovations like new types of solar panels, wind turbines, and energy-storage systems facilitate the transition from our dependency on fossil fuels to mitigating climate change. Technologies such as precision farming, vertical farming, and genetically modified crops can improve resource efficiency and alleviate food security. Digital innovations such as artificial intelligence and big data allow for more smart resource management, leading to less waste and better-optimized supply chains. Innovations in a circular economy include recyclable materials and ecofriendly manufacturing processes, which are transitioning industries toward reducing their carbon footprints. However, to make the most of technological innovation, regulation must be accompanied by thoughtful collaboration across sectors. Such innovation has many consequences, including increasing our responsiveness to challenges that would otherwise result in pollution and other environmental crises. Moreover, technologies need to be accessible and beneficial to the communities that they are introduced into. Technology should also be used to reduce the costs of sustainable solutions.

8.4 Policy Suggestions

Well-functioning democracies can consult with representatives and constituents to continue to improve legislation on sustainability while driving economic growth. Policy suggestions to promote sustainability should emphasize establishing frameworks that motivate innovation in green technologies, encourage social equity, and promote responsible consumption. Governments could set up carbon-pricing mechanisms for industries as incentives for them to reduce their emissions; for instance, carbon taxes and cap-and-trade systems encourage industries to fund clean energy investments. Regulations and subsidies can fund renewable energy development, energy efficiency improvements, and a shift to circular economies. Policies promoting the sustainable development of agriculture, water conservation, and biodiversity protection are all key building blocks for the future of natural resource conservation. In addition, environmental justice should include vulnerable communities' access to green technologies and reduce negative impacts on these populations due to climate change. Finally, global cooperation and stronger agreements on climate action are needed to mitigate climate change and align efforts toward an inclusive sustainability approach.

8.5 Education and Capacity Building

Furthermore, current generations need to be prepared for and educated in the technologies and skills that they will need to improve their communities today and in the future. We should widen our already-ecofriendly approach to include not only plastics all single-use consumer products. The first step toward sustainable development requires us to focus our attention on benefiting people. Given the latest developments in environmental science, sustainability should be included in education systems, first by raising awareness and second by educating teachers on how to teach this subject. Teachers will need new educational materials to help students study sustainability, and interdepartmental cooperation among teachers will need to be organized around finding holistic solutions. Education will also foster future leaders who will work toward finding solutions to global challenges. Systemic changes across industries will require skilled professionals trained in green technologies, sustainable practices, and sound policy development built through capacity-building initiatives. Public awareness campaigns and community-based programs may offer responsible ways of keeping people involved in sustainable behavior and decision-making. With continued investment in education and capacity building, societies can develop workforces that are both informed and equipped to contribute toward a sustainable future.

8.6 Partnerships and Collaboration

In short, sustainability, which should be the main goal of safeguarding the future, may be considered as a form of rigorous thinking that works toward strategic consolidation and the involvement of all stakeholders in the development of plans that feature partnerships and collaborations between many professionals, disciplines, and modalities across borders. By coming together and sharing resources, knowledge, and expertise, the various parties can develop innovative, scalable, inclusive solutions. Most of all, cross-sector partnerships will help in the development and implementation of sustainable technologies, policies, and practices while fostering a sense of shared responsibility. In addition, transnational partnerships are as much a part of this agenda as national partnerships are because many issues, such as climate change and biodiversity loss, are global challenges requiring various countries to work together. Thus, partnerships mobilize various strengths from different sectors and communities to stimulate progress toward sustainability goals and finding fair, adaptable, and impactful solutions. The actions to implement such solutions from partnerships will be taken mostly by government agencies and NGOs, though.

9 Conclusions

Sustainability is an integrated strategy for economic growth, social justice, and environmental stewardship that balances ecological integrity and human development. Sustainability in the context of climate change is not only a desirable goal but also a necessity for survival. Because of the scope and diversity of professional activity in areas of practice relevant to climate change, the recognition by professional associations of their responsibility to mitigate and adapt to climate change could make a significant difference to both mitigation and adaptation. Professional associations should add these requirements and introduce new ones that explicitly address climate change. This can be brought about by professional associations' acting on their responsibilities to their clients and especially their responsibility to the public, which goes together with their expertise and should be reflected in their self-regulation. Ideally, action on the important subject of climate change mitigation should come from the associations themselves; in the absence of that, governments could step in to help ensure that a whole range of professionals consider climate change in their work and take action to address it.

References

Akers DB, MacCarthy MF, Cunningham JA, Annis J, Mihelcic JR (2015) Lead (Pb) contamination of self-supply groundwater systems in coastal Madagascar and pre-dictions of blood lead levels in exposed children. Environ Sci Technol 49:2685

American Academy of Environmental Engineers (AAEE) (2009) Environmental engineering body of knowledge. AAEE, the Environmental Engineering Body of Knowledge Task Force. Available at: www.aaees.org/publications-eebodyofknowledge.php. Accessed June 8, 2016

Anastas PT (2012) Fundamental changes to EPA's research enterprise: the path forward. Environ Sci Technol 46:580

Andreasen AR (1995) Marketing social change: changing behavior to promote health, social development and the environment. Jossey-Bass, San Francisco

Brown C (1997) Anthropology and social marketing: a powerful combination. Pract Anthropol 19:27

Clasen TF, Brown J, Collin S, Suntura O, Cairncross S (2004) Reducing diarrhea through the use of household-based ceramic water filters: a randomized, controlled trial inrural Bolivia. Am J Trop Med Hyg 70:651

Dolnicar S, Hurlimann A (2011) Water alternatives: who and what influences public acceptance? J Public Affairs 11:49

Dolnicar S, Hurlimann A, Nghiem LD (2010) The effect of information on public acceptance—the case of water from alternative sources. J Environ Manag 91:1288

Doria MF, Pidgeon N, Hunter PR (2009) Perceptions of drinking water quality and risk and its effect on behavior: a cross-national study. Sci Total Environ 407:5455

Exley JLR, Liseka B, Cumming O, Ensink JHJ (2015) The sanitation ladder, what constitutes an improved form of sanitation? Environ Sci Technol 49:1086

Fry LM, Mihelcic JR, Watkins DW (2008) Water and non watery-related challenges of achieving global sanitation coverage. Environ Sci Technol 42:4298

Guang T, Borges L (2012) The effectiveness of social marketing mix strategy: towards an anthropological approach. Int J Bus Anthropol 3:102

Harris AR, Davis J, Boehm AB (2013) Mechanism of post-supply contamination of drinking water in Bagamoyo, Tanzania. J Water Health 11:543

Hutton G, Bartram J (2008) Global costs of attaining the Millennium Development Goal for water supply and sanitation. Bull World Health Organ 86:1

Jesiek BK, Haller Y, Thompson J (2014) Developing globally competent engineering researchers: outcomes-based instructional and assessment strategies from the IREE 2010China research abroad program. Adv Eng Educ 4:1

Johnson BB (2003) Do reports on drinking water quality affect customers' concerns? Experiments in report content. Risk Anal 23:985

Kemp B, Randle M, Hurlimann A, Dolnicar S (2012) Community acceptance of recycled water: can we inoculate the public against scare campaigns? J Public Aff 12:337

Kotler P, Zaltman G (1971) Social marketing: an approach to planned social change. J Mark 35:3

Lazarova V, Choo KH, Cornel P (2012) Water-energy interactions in water reuse, vol 11. IWA Publishing, London

Lee NR, Kotler P (2011) Social marketing: influencing behaviors for good. Sage Publications, Inc., Los Angeles

Mankad A (2012) Decentralised water systems: emotional influences on resource decision making. Environ Int 44:128

Marks SJ, Komives K, Davis J (2014) Community participation and water supply sustainability: evidence from handpump projects in rural Ghana. J Plann Educ Res 34:276

Mehl J, Kaiser J, Hurtado D, Gibson DA, Izurieta R, Mihelcic JR (2011) Pathogen destruction and solids decomposition in composting latrines: study of fundamental mechanisms and user operation in Rural Panama. J Water Health 9:187

Mihelcic JR (2010) The right thing to do: graduate education and research in a global and human context. In: Downey GL, Beddoes K (eds) What is global engineering education for? The making of international and global engineering educators. Morgan & Claypool Publishers, San Francisco, pp 235–250

Mihelcic JR, Crittenden JC, Small MJ, Shonnard DR, Hokanson DR, Zhang Q, Chen H, Sorby SA, James VU, Sutherland JW, Schnoor JL (2003) Sustainability science and engineering: the emergence of a new meta discipline. Environ Sci Technol 37:5314

Mihelcic JR, Phillips LD, Watkins DW (2006) Integrating a global perspective into engineering education and research: engineering international sustainable development. Environ Eng Sci 23:426

Mihelcic JR, Paterson KG, Phillips LD, Zhang Q, Wat-kins DW, Barkdoll B, Fuchs VJ, Fry LM, Hokanson DR (2008) We are educating engineers in the sustainable futures model with a global perspective. Civil Eng Environ Syst 25:255

Mo W, Zhang Q (2013) Energy-nutrients-water Nexus: integrated resource recovery in municipal wastewater treatment plants. J Environ Manag 127:255

Munasinghe M, Swart R (2005) Primer on climate change and sustainable development facts, policy analysis and applications. Cambridge University Press, Cambridge, p 445

Nancarrow BE, Leviston Z, Po M, Porter NB, Tucker DI (2008) What drives communities' decisions and behaviors in the reuse of wastewater. Water Sci Technol 57:485

Naughton CC, Sissoko HT, Mihelcic JR (2015a) Assessing factors that lead to usage of appropriate technology hand washing stations in Mali, West Africa. J Water Sanit Hyg Dev 5:279

Naughton CC, Lovett PN, Mihelcic JR (2015b) Land suitability of shea (Vitellaria paradoxa) distribution across Sub-Saharan Africa. Appl Geogr 58:217

Pettorelli N, Safi K, Turner W (2014) Satellite remote sensing, biodiversity research and conservation of the future. Philos Trans R Soc Lond Ser B Biol Sci 369:20130190

Po M, Nancarrow BE, Leviston Z, Porter NB, Syme GJ, Kaercher JD (2005) Predicting community behaviour in relation to wastewater reuse: what drives decisions to accept or reject. CSIRO Land and Water, Perth, Australia

Rodriguez C, Van Buynder P, Lugg R, Blair P, Devine B, Cook A, Weinstein P (2009) Indirect potable reuse: a sustainable water supply alternative. Int J Environ Res Public Health 6:1174

Russell S, Lux C (2009) Getting over yuck: moving from psychological to cultural and sociotechnical analyses of responses to water recycling. Water Policy 11:21

Schafer CA, Mihelcic JR (2012) Effect of storage tank material and maintenance on household water quality. J Am Water Works Assoc 104:E521

Schweitzer RW, Cunningham JA, Mihelcic JR (2013) Hydraulic modeling of clay ceramic water filters for point-of-use water treatment. Environ Sci Technol 47:429

Swart R, Robinson J, Cohen S (2003) Climate change and sustainable development expanding the options. Clim Pol 3:S19–S40

Trotz MA, Muga HE, Phillips LD, Yeh D, Stuart A, Mihelcic JR (2009) Non-traditional university research partners that facilitate service learning and graduate research for sustainable development. Proceedings 2009 World Environmental & Water Resources Congress, Kansas City, MO, pp 17–21

Verbyla MW, Oakley SM, Lizima LA, Zhang J, Iriarte M, Tejada-Martinez AE, Mihelcic JR (2013) Taenia eggs in a stabilization pond system with poor hydraulics: concern for human cysticercosis? Water Sci Technol 68:2698

Whiteford LM, Whiteford S (2005) Globalization, water and health: resources in times of scarcity. School of American Research Press, Santa Fe

Wigginton KR, Ye Y, Ellenberg RM (2015) Emerging investigators series: the source and fate of pandemic viruses in the urban water cycle. Environ Sci Water Res Technol 1:735

World Resources Institute (WRI) in Collaboration with United Nations Development Programme, United Nations Environment Programme, and World Bank (2005) World Resources2005: the wealth of the poor—managing ecosystems to fight poverty. WRI, Washington, DC

Integrating Sustainability into Equal Education Opportunities

Savita Jangde and Md. Afjal Ahmad

Abstract

Integrating sustainability into equitable education opportunities is an intricate and diverse undertaking that encompasses legislation, social justice, and instructional methods. The United Nations Decade on Education for Sustainable Development (UNDESD) has driven the global agenda toward incorporating sustainability across educational sectors. Sustainability education, alternatively referred to as Education for Sustainable Development (ESD), endeavors to cultivate in individuals the requisite knowledge, competencies, principles, and dispositions that enable them to actively participate in the construction of a society that is both sustainable and equitable. Core principles of sustainability education include an interdisciplinary approach, systems thinking, holistic understanding, action and engagement, ethical responsibility, and lifelong learning. Infrastructure and resource allocation are crucial factors in integrating sustainability into higher education. Government policies and resource allocation play a significant role in providing necessary infrastructure for ICT integration. Community engagement and partnerships are essential components in promoting sustainability and equal education opportunities. Technology and innovation play a transformative role in promoting sustainability and enhancing equal education opportunities. Integrating sustainability into equal education opportunities must be evaluated by measuring impact and outcomes. Policy frameworks and advocacy efforts are crucial for integrating sustainability into equal education opportunities at both institutional and governmental levels. Student involvement and empowerment are essential components of integrating sustainability into equal education opportunities. In India, the integration of sustainability into equal education opportunities is gaining momentum as a vital component of national development strategies. These endeavors are additionally propelled by partnerships among corporate entities, government organizations, and nongovernmental organizations; they foster inclusive learning environments that enable students of various backgrounds to effectuate constructive transformations within their communities.

Keywords

Sustainability · Higher education · Policy framework · Sustainable development · ICT

1 Introduction

The integration of sustainability into equal education opportunities is a multifaceted endeavor that involves various dimensions such as policy, social justice, and educational practices. The United Nations Decade on Education for Sustainable Development (UNDESD) has been pivotal in driving the global agenda toward incorporating sustainability across educational sectors (Cebrián and Junyent 2015). Achieving equal opportunities in education is not only a matter of fairness but also a crucial determinant of societal equality and intergenerational mobility (Schütz et al. 2008). Sustainable education systems aim to provide equal opportunities for all individuals, ensuring inclusivity and diversity (Medina-García et al. 2020). When considering the intersection of sustainability and education, the emphasis goes beyond conventional academic prowess to include principles of cultural sustainability, respect, and tolerance (Zygmunt 2016). Teachers play a vital role in shaping sustainable futures by integrating sustainability into the curriculum and emphasizing its importance (Janhonen-Abruquah et al. 2018). Moreover, the shift toward sustainability in higher education, particularly in fields like engineering, highlights the impor-

S. Jangde (✉) · M. A. Ahmad
Department of Plant Physiology, Institute of Agricultural Sciences, Banaras Hindu University, Varanasi, India
e-mail: savi.bhu31@gmai.com

J. A. Parray et al. (eds.), *Environmental Landscape and Sustainable Biodiversity for Healthy Green Growth*, Sustainable Landscape Planning and Natural Resources Management, https://doi.org/10.1007/978-3-031-75405-0_10

tance of incorporating sustainability principles across disciplines (Breznik et al. 2021). Ensuring equal educational opportunities involves addressing issues of social justice, multicultural education, and empowerment of marginalized groups (Choi and Lee 2020). The quest for sustainability in education aligns with broader societal goals of achieving gender equality, cultural integration, and effective management of educational systems (Abdullahi et al. 2021; Ennaji 2018). By promoting equal opportunities in education, societies can work toward enhancing social cohesion, democratic values, and fair social structures (Holden 2011; Musi et al. 2022). This chapter delves into various approaches to incorporating sustainability principles into the field of education in order to guarantee equitable opportunities and access for all learners.

2 Foundations of Sustainability Education

Sustainability education, also referred to as Education for Sustainable Development (ESD), is a pedagogical approach that strives to cultivate the values, attitudes, knowledge, and abilities required for people to contribute to a more equitable and sustainable society (UNESCO 2014). Sustainability education fundamentally acknowledges the interdependence of environmental, social, and economic systems and aims to equip learners with the knowledge and skills to comprehend and tackle intricate sustainability issues.

The historical background and evolution of sustainability education have been shaped by significant global initiatives and shifts in educational paradigms. The beginning of the United Nations' "Decade of Education for Sustainable Development" in 2004 was a watershed moment that was intended to bring the concepts, values, and practices of sustainable development into both formal and informal education, according to Hallinger and Chatpinyakoop (2019). This study emphasized the increasing recognition of the significance of education in advancing sustainable development goals on a global level (Hallinger and Nguyen 2020). Over the years, there has been a growing emphasis on management education for sustainable development, reflecting a broader societal shift toward sustainability considerations in various fields (Adomßent et al. 2014). Universities have emerged as key players in driving transformational sustainability efforts, aligning with the agenda of delivering the Sustainable Development Goals (Purcell et al. 2019). The evolution of sustainability education has seen a surge in research output and literature focusing on diverse aspects such as sustainable consumption, transformative learning, and mainstreaming sustainability (Grosseck et al. 2019). The integration of sustainability principles into curricula of higher education has progressed gradually since the introduction of environmental engineering into academic discourse in the mid-1990s (Thürer et al. 2018). Changing industrialization, technological advancements, and the shifting industrial landscape have all contributed to the evolution of sustainability education by requiring a reassessment of educational objectives and results (Dan et al. 2019). Moreover, sustainability education has expanded beyond traditional academic boundaries to encompass interdisciplinary approaches, such as the study of sustainable architecture and heritage conservation (Ryoo and Youn 2019). The incorporation of sustainability principles into engineering curricula and the exploration of sustainable consumption patterns further highlight the diverse trajectories that sustainability education has taken over the years.

3 Core Principles of Sustainability Education

3.1 Interdisciplinary Approach

Sustainability education incorporates knowledge from several academic fields such as environmental science, social sciences, economics, and humanities to offer a holistic comprehension of sustainability concerns (Sterling 2001).

3.2 Systems Thinking

The core of sustainability education is around the principle of systems thinking, which entails comprehending the dynamic relationships and interconnections among ecological, social, and economic systems (Meadows 2008). By embracing a systems viewpoint, learners are able to examine the fundamental origins of sustainability concerns and investigate comprehensive solutions.

3.3 Holistic Understanding

Sustainability education fosters a comprehensive comprehension of sustainability, which includes the environmental, social, cultural, and economic aspects (Wals and Jickling 2002). This comprehensive approach motivates learners to acknowledge the interwoven nature of different sustainability concerns and to contemplate their interdependent effects.

3.4 Action and Engagement

An essential tenet of sustainability education is the prioritization of hands-on learning, analytical reasoning, and practical implementation in real-life scenarios (Sterling 2001). Through advocacy, service-learning initiatives, and practical experiences, students are motivated to address sustainability challenges in their communities both individually and collectively.

3.5 Ethical Responsibility

Sustainability education promotes the development of moral values including stewardship, justice, and accountability to the environment and future generations (Orr 1992). In regard to sustainability, students are encouraged to contemplate the ethical ramifications of their choices and to evaluate their own values and behaviors.

3.6 Lifelong Learning

Sustainability education is an ongoing and continuous activity that goes beyond traditional schooling (UNESCO 2014). The text highlights the significance of continuous learning, adjustment, and innovation in addressing the ever-changing sustainability concerns. Individuals who engage in continuous learning possess the necessary abilities and adaptability to successfully traverse intricate situations and make valuable contributions toward beneficial transformations.

4 Addressing Socioeconomic Disparities

To address socioeconomic disparities in sustainability education, it is critical that educational systems consider the interdependence of environmental, economic, and social variables. Integrating inclusive practices and sustainable development concepts can facilitate the reduction of disparities and the advancement of fairness in education. Multiple researchers have contributed significant perspectives on how to effectively tackle socioeconomic gaps. A few examples are outlined below:

Lozano et al. (2013) highlight the significance of focusing on the university system to improve sustainability in higher education. Universities may have a significant impact on promoting sustainability and eliminating social inequities by prioritizing leadership and implementing systemic changes. In their 2020 study, Nguyen et al. explore the application of pedagogical strategies in STEM education for imparting knowledge on sustainability principles. Through the incorporation of sustainable development ideas into STEM education, students can cultivate a more profound comprehension of sustainability concerns and actively contribute to the resolution of socioeconomic inequalities.

In their study, Naukkarinen and Jouhkimo (2021) emphasize the need of collaboration between schools and universities in advancing integrated and inclusive education for sustainability. Partnerships between educational institutions can bolster sustainability education and successfully tackle socioeconomic inequalities. Williams and Zeman (2021) investigate the influence of internship programs on the attitudes of interns regarding environmental health equity. Internships and other practical experiences can enhance learning and promote a more profound comprehension of sustainability and social equity concerns. By utilizing these sources and their knowledge, educational establishments can create all-encompassing plans to incorporate sustainability education, foster inclusivity, and successfully tackle socioeconomic inequalities in their courses and methods.

5 The Role of Curriculum Development

Curriculum development plays a vital role in higher education by including sustainability, promoting environmental consciousness, and equipping students to address sustainability issues. Cotton et al. (2009) emphasize the need of designing educational programs that enable students to acquire the values, skills, and information necessary for sustainable development. Furthermore, Chankseliani and Mccowan (2020) emphasize the importance of higher education in offering lifelong learning opportunities and ensuring equal access to tertiary education, in accordance with the Sustainable Development Goals. Sibbel (2009) explores the objective of curriculum creation in higher education, which is to synchronize educational programs with global sustainability concerns by employing inventive methods. According to Stough et al. (2018), colleges may effectively include students in environmental matters and encourage the adoption of sustainable practices by incorporating sustainability principles into various academic areas.

6 Teacher Training and Professional Development

Professional development and teacher training are essential for the incorporation of sustainability principles into higher education. Bertschy et al. (2013) emphasize the importance of educators delving into the notion of sustainability and auditing their personal values in order to implement educational initiatives in sustainability that are effective. Albareda-Tiana et al. (2019) noted that by incorporating active teaching-learning strategies into the curriculum, prospective educators can acquire sustainability competencies. Furthermore, Dahl (2019) discovered that although student teachers believe they possess sufficient knowledge and skills in various aspects of professionalism, they still necessitate additional growth to educate for sustainability in an effective manner. In an effort to ascertain the professional competencies required of educators to advocate for "learning to be" and provide pragmatic resolutions toward sustainable development, Ghorbani et al. (2018) undertook an investigation.

7 Infrastructure and Resource Allocation

Integrating sustainability into higher education relies heavily on infrastructure and resource allocation. Garba et al. (2015) highlight the importance of government policies and financial allocation in establishing the essential infrastructure for integrating ICT in the education sector. Lambrechts et al. (2017) underscore the challenges associated with the integration of sustainability principles into higher education systems and stress the importance of professional development in addressing these complexities. In addition, Cha et al. (2020) emphasized the significance of a comprehensive strategy for incorporating ICT into education, with a specific emphasis on establishing a solid educational infrastructure to guarantee high-quality access, particularly in developing nations. Rahat et al. (2023) analyzed the impacts of infrastructure initiatives on environmental, social, and economic spheres, emphasizing the criticality of higher education institutions adopting sustainable infrastructure development practices. By taking into account these viewpoints, higher education institutions can strategically distribute resources to improve infrastructure, advocate for sustainability, and create a favorable atmosphere for incorporating sustainability ideas into their teaching programs. In their study, Ghorbani et al. (2018) aimed to identify the specific professional skills that instructors need in order to effectively promote the concept of "learning to be" and provide practical strategies for achieving sustainable development.

8 Community Engagement and Partnerships

Collaborations with the community and active participation are critical elements in the advancement of equal educational opportunities and sustainability. Through active community engagement, academic establishments can acquire significant knowledge regarding the distinct obstacles and requirements of varied demographics. This ensures that endeavors related to sustainability are both culturally appropriate and all-encompassing (Schultz et al. 2017). Collaborations with government agencies, businesses, and non-profit organizations can offer valuable support, resources, and expertise to bolster educational initiatives and broaden the reach of sustainable learning prospects (Cameron et al. 2019). Through the implementation of community-based initiatives and projects, students are afforded the chance to apply theoretical knowledge gained in the classroom to real-world scenarios. This process fosters a deeper understanding of sustainability issues and empowers students to take proactive measures to bring about positive changes in their immediate surroundings (Hart and Nolan 2018). Furthermore, community engagement initiatives have the potential to foster discourse and cooperation among relevant parties, thereby encouraging the harmonization of efforts and a unified front in the pursuit of sustainable development objectives (Dyer et al. 2019). Through the establishment of robust collaborations and active involvement in nearby communities, academic establishments have the capacity to foster inclusive and equitable learning environments that empower every student to flourish.

9 Technology and Innovation

Innovation and technology have a profound impact on the advancement of sustainability and the improvement of educational opportunity equality. Educational institutions have the ability to expand their reach and ensure that all individuals, irrespective of socioeconomic status or geographic location, have access to high-quality educational resources through the utilization of digital platforms (Belfield et al. 2015). Educational applications, virtual classrooms, and online learning platforms provide adaptable learning opportunities that cater to a wide range of learning styles and preferences, thereby ensuring equal opportunities for all students (Bates and Sangrà 2011). Moreover, experiential learning is facilitated and student engagement is increased through the use of technology-enabled tools, including virtual reality simulations, augmented reality experiences, and interactive multimedia resources (Dalgarno and Lee 2010). These tools enable students to investigate intricate sustainability concepts in immersive and interactive environments. In addition, the integration of state-of-the-art technologies, such as open educational resources (OER) and massive open online courses (MOOCs), enables educators to collaborate and share information globally, thus fostering an environment that values ongoing education and the interchange of ideas (UNESCO 2019). Through the utilization of technology and innovation, academic establishments have the capacity to propel sustainability education forward and advocate for equitable access to educational opportunities. This, in turn, will enable students to acquire knowledge and engage actively in the construction of a future that is both sustainable and equitable.

10 Measuring Impact and Outcomes

Assessing the efficacy of incorporating sustainability principles into equal education opportunities requires the measurement of outcomes and impact. By employing rigorous evaluation techniques, policymakers and educators can determine the degree to which sustainability principles are integrated into academic methodologies and the subsequent effects on students and communities. Tilbury (2011) asserts that quantifiable indicators, including graduation rates, academic performance metrics, and enrollment in sustainability-related courses, offer valuable insights into the concrete results that sustainability education initiatives attain. Furthermore, qualitative evaluation methods, such as focus groups, interviews, and surveys, enable stakeholders to collect input from community members, educators, and students regarding their attitudes, perceptions, and actions concerning sustainability (UNESCO 2014). Educational institutions can exhibit accountability to stakeholders, monitor progress over time, and identify areas requiring improvement through the triangulation of data from multiple sources (Wals and Jickling 2002). In addition, by employing longitudinal studies and comparative analyses, scholars are able to investigate the enduring effects of sustainability education on the academic performance, career trajectories, and civic involvement of students. This underscores the revolutionary capacity that can be unleashed by incorporating sustainability principles into equitable educational opportunities (Orr 1992). Overall, promoting equitable access to high-quality learning experiences for all and furthering sustainability in education are facilitated by the insights gained from measuring impact and outcomes.

11 Policy Frameworks and Advocacy

Policy frameworks and advocacy initiatives play a vital role in promoting the incorporation of sustainability into equitable education opportunities at both institutional and governmental levels. Strong policy frameworks create a favorable climate for educational institutions to prioritize sustainability and advance equity in education. As an illustration, national education policies have the authority to require the integration of sustainability education into curriculum standards, teacher training programs, and school accreditation processes (UNESCO 2005). Likewise, local government policies can encourage schools to adopt sustainability projects by providing financial support, establishing programs to acknowledge their efforts, and fostering collaborations with community organizations (Cameron et al. 2019). In addition, the active promotion and support from nongovernmental organizations, professional groups, and local community initiatives are crucial in increasing public consciousness, rallying public backing, and exerting influence on policy choices concerning sustainability in education (Leal Filho et al. 2019). Advocates can actively participate in the establishment of inclusive and equitable education systems that prioritize sustainability and enable all learners to succeed by involving stakeholders, conducting research, and lobbying for policy change (Hart and Nolan 2018). In summary, legislative frameworks and advocacy initiatives play a crucial role in promoting sustainability in education and guaranteeing equitable access to high-quality learning opportunities for everyone.

12 Student Involvement and Empowerment

Student empowerment and engagement are fundamental elements that must be incorporated into the framework of equitable educational opportunities. Facilitating student engagement in sustainability initiatives not only enriches their educational journey but also cultivates a sense of proprietorship and accountability toward effecting constructive transformations within their localities. Through the active engagement of students in decision-making processes, curriculum development, and the execution of sustainability projects, academic institutions can effectively solicit and incorporate their viewpoints and preferences (Sterling 2001). Furthermore, student-led endeavors, including research projects, green teams, and sustainability societies, offer avenues for pupils to collaborate, generate novel ideas, and champion sustainability not only within their academic institutions but also on a global scale (UNESCO 2014). In addition, extracurricular activities with a sustainability focus, internships, service-learning projects, and service-learning projects are examples of experiential learning opportunities that allow students to develop practical skills for addressing sustainability challenges and applying classroom knowledge to real-world scenarios (Hart and Nolan 2018). Educational institutions can advance equitable access to high-quality educational opportunities for all students and cultivate a culture of sustainability by enabling students to take on proactive roles as agents of change.

13 Indian Scenario

Sustainability is increasingly being incorporated into equitable education opportunities in India, since it is recognized as a crucial element of national development initiatives. The National Education Policy (NEP) 2020 places a strong emphasis on integrating environmental literacy and sustainability throughout all levels of education. This underscores the significance of ecological awareness and accountability. The Environmental Studies (EVS) curriculum for elementary and secondary education, mandated by the Central Board of Secondary Education (CBSE), plays a crucial role in introducing sustainability principles at an early stage. In addition, the Centre for Environment Education (CEE) and TERI School of Advanced Studies play an active role in teacher training programs, ensuring that educators are well prepared to conduct education focused on sustainability. The Sustainable Development Goals (SDGs), namely, Goal 4 (Quality Education) and Goal 13 (Climate Action), offer a worldwide structure for integrating educational goals with the need for sustainability. These initiatives are strengthened through collaborations with business entities, government institutions, and nongovernmental organizations (NGOs). As a result, inclusive educational environments are created, empowering students from diverse backgrounds to act as agents of positive change within their communities. India aims to build a generation of ecologically aware and socially responsible individuals who are capable of tackling the intricate difficulties of sustainable development in the twenty-first century through these collective efforts.

14 Conclusion

In conclusion, the journey toward integrating sustainability into equal education opportunities is not only imperative but also transformative. By embedding sustainability principles across all levels of education, we lay the foundation for a more equitable and environmentally conscious society. As we navigate the complexities of the twenty-first century, this holistic approach to education not only empowers individuals to address pressing environmental and social challenges but also fosters a collective commitment to sustainable development. Through collaborative efforts, policy support, and innovative educational practices, we can ensure that every learner has access to quality education that equips students with the understanding, abilities, and principles that are necessary to construct a future that is more sustainable for individuals and future generations.

References

Abdullahi NJK, Akanmu IB, Adetiba AA, Ovigueraye EPA (2021) National consciousness and management of secondary education for sustainable development in Kwara state, Nigeria. Int J Innov Teach Learn 7(2):1–18

Adomßent M, Fischer D, Godemann J, Herzig C, Otte I, Rieckmann M et al (2014) Emerging areas in research on higher education for sustainable development—management education, sustainable consumption and perspectives from central and eastern Europe. J Clean Prod 62:1–7

Albareda-Tiana S, García-González E, Jiménez-Fontana R, Solís-Espallargas C (2019) Implementing pedagogical approaches for esd in initial teacher training at spanish universities. Sustain For 11(18):4927

Bates AW, Sangrà A (2011) Managing technology in higher education: strategies for transforming teaching and learning. Wiley

Belfield C, Levin HM, Gao Q (2015) The price we pay: economic and social consequences of inadequate education. Brookings Institution Press

Bertschy F, Künzli C, Lehmann M (2013) Teachers' competencies for the implementation of educational offers in the field of education for sustainable development. Sustain For 5(12):5067–5080

Breznik K, Law KMY, Zeme J (2021) Mission in higher education in Slovenia: sustainability in engineering versus others sector. Sustain For 13(14):7947

Cameron M, Lundy L, Murray A (2019) Partnerships and sustainability in higher education: a review and reflection. J High Educ Policy Manag 41(1):104–121

Cebrián G, Junyent M (2015) Competencies in education for sustainable development: exploring the student teachers' views. Sustain For 7(3):2768–2786

Cha H, Park T, Seo J (2020) What should be considered when developing ict-integrated classroom models for a developing country? Sustain For 12(7):2967

Chankseliani M, McCowan T (2020) Higher education and the sustainable development goals. High Educ 81(1):1–8

Choi S, Lee SW (2020) Enhancing teacher self-efficacy in multicultural classrooms and school climate: the role of professional development in multicultural education in the United States and South Korea. AERA Open 6(4):233285842097357

Cotton D, Bailey I, Warren M, Bissell S (2009) Revolutions and second-best solutions: education for sustainable development in higher education. Stud High Educ 34(7):719–733

Dahl T (2019) Prepared to teach for sustainable development? Student teachers' beliefs in their ability to teach for sustainable development. Sustain For 11(7):1993

Dalgarno B, Lee MJW (2010) What are the learning affordances of 3-D virtual environments? Br J Educ Technol 41(1):10–32

Dan H, Zheng M, Cheng W, Lau Y, Yin Q (2019) Interaction between higher education outputs and industrial structure evolution: evidence from Hubei province, China. Sustainability 11(10):2923

Dyer K, Gregg L, Huxtable M (2019) Strengthening community engagement: building trust and promoting inclusive dialogue for sustainable regional development. J Environ Plan Manag 62(7):1135–1153

Ennaji M (2018) Morocco's experience with gender gap reduction in education. Gender Women Stud 2(1):5

Garba SA, Yusuf B, Busthami AH (2015) Toward the use of technology and 21st century teaching-learning approaches: the trend of development in malaysian schools within the context of Asia pacific. Int J Emerg Technol Learn 10(4):72

Ghorbani S, Jafari SEM, Sharifian F (2018) Learning to be: teachers' competences and practical solutions: a step towards sustainable development. J Teach Educ Sustain 20(1):20–45

Grosseck G, Ţîru LG, Bran R (2019) Education for sustainable development: evolution and perspectives: a bibliometric review of research, 1992–2018. Sustain For 11(21):6136

Hallinger P, Chatpinyakoop C (2019) A bibliometric review of research on higher education for sustainable development, 1998–2018. Sustain For 11(8):2401

Hallinger P, Nguyen V (2020) Mapping the landscape and structure of research on education for sustainable development: a bibliometric review. Sustain For 12(5):1947

Hart PM, Nolan M (2018) Embedding community engagement into the curriculum: lessons from an Australian university. High Educ Res Dev 37(5):1011–1024
Holden M (2011) Urban policy engagement with social sustainability in metro Vancouver. Urban Stud 49(3):527–542
Janhonen-Abruquah H, Topp J, Posti-Ahokas H (2018) Educating professionals for sustainable futures. Sustain For 10(3):592
Lambrechts W, Verhulst E, Rymenams S (2017) Professional development of sustainability competences in higher education. Int J Sustain High Educ 18(5):697–714
Leal Filho W et al (2019) Handbook of theory and practice of sustainable development in higher education. Springer
Lozano R, Lukman RK, Lozano FJ, Huisingh D, Lambrechts W (2013) Declarations for sustainability in higher education: becoming better leaders, through addressing the university system. J Clean Prod 48:10–19
Meadows DH (2008) Thinking in systems: a primer. Chelsea Green Publishing
Medina-García M, Toledo LD, Higueras-Rodríguez L (2020) Equal opportunities in an inclusive and sustainable education system: an explanatory model. Sustain For 12(11):4626
Ministry of Education, Government of India (2020) National Education Policy (NEP) 2020 [PDF] https://www.education.gov.in/sites/upload_files/NEP_Final_English_0.pdf
Musi MA, Bachtiar MY, Herlina et al (2022) Local wisdom values of the bugis community in early childhood multicultural learning. Jurnal Pendidikan Anak Usia Dini Undiksha 10(2):255–264
Naukkarinen J, Jouhkimo L (2021) Toward integrated and inclusive education for sustainability with school–university cooperation. Sustain For 13(22):12486
Nguyen TPL, Nguyen TH, Tran TK (2020) Stem education in secondary schools: teachers' perspective towards sustainable development. Sustain For 12(21):8865
Orr DW (1992) Ecological literacy: education and the transition to a postmodern world. State University of New York Press
Purcell WM, Henriksen H, Spengler JD (2019) Universities as the engine of transformational sustainability toward delivering the sustainable development goals. Int J Sustain High Educ 20(8):1343–1357
Rahat R, Pradhananga P, ElZomor M (2023) A step toward nurturing infrastructure sustainability and rating systems through construction management curricula. Int J Sustain High Educ 24(8):1877–1896
Ryoo SL, Youn HC (2019) The evolutionary use of curved wood in Korean traditional architecture. Sustain For 11(23):6557
Schultz BD, Lofton S, Strachan E (2017) Community engagement in sustainability: a framework for developing professional skills and competencies. J Sustain Educ 13(1):1–20
Schütz G, Ursprung HW, Wößmann L (2008) Education policy and equality of opportunity. Kyklos 61(2):279–308
Sibbel A (2009) Pathways towards sustainability through higher education. Int J Sustain High Educ 10(1):68–82
Sterling S (2001) Sustainable education: re-visioning learning and change. Green Books
Stough T, Ceulemans K, Lambrechts W, Cappuyns V (2018) Assessing sustainability in higher education curricula: a critical reflection on validity issues. J Clean Prod 172:4456–4466
Thürer M, Tomašević I, Stevenson M, Qu T, Huisingh D (2018) A systematic review of the literature on integrating sustainability into engineering curricula. J Clean Prod 181:608–617
UNESCO (2005) United Nations decade of education for sustainable development (2005–2014): international implementation scheme. UNESCO
UNESCO (2014) Shaping the future we want: UN decade of education for sustainable development (2005–2014) final report. UNESCO
UNESCO (2019) Open Educational Resources (OER): global report 2017. UNESCO Publishing
Wals AE, Jickling B (2002) "Sustainability" in higher education: from doublethink and newspeak to critical thinking and meaningful learning. Int J Sustain High Educ 3(3):221–232
Williams D, Zeman C (2021) Case study evaluation: pieer internship program and its impact on interns' self-reported attitudes and behaviors regarding environmental health equity. Int J Sustain High Educ 22(7):1646–1663
Zygmunt T (2016) Language education for sustainable development. Discour Commun Sustain Educ 7(1):112–124

Gender Equality in the Sustainable Development Framework

Joyeeta Bhattacharjee, Niraj Singh, and Dilfuza Egamberdieva

Abstract

The concept of sustainable development, as articulated by the Commission on Environment and Development, entails meeting present needs while safeguarding the ability of future generations to fulfill their own. The importance of empowering women, ensuring their full participation, and promoting equality as core elements for peace and sustainable development was underscored during the Fourth World Conference on Women in Beijing in September 1995. Central to achieving gender equity is empowering women and girls, which entails enabling them to take charge of their lives. This empowerment is intricately linked to sustainable development, which relies on an equitable distribution of resources.

Gender equity as a concept thus becomes essential to sustainable development. It involves fair allocation of resources, programs, and decision-making without discrimination, addressing imbalances in benefits for all genders.

The UN Population Fund stresses that empowered women are catalysts for social change within and beyond the household. Achieving and maintaining gender equality necessitates closing the gender gap and establishing a level playing field between women and men, with women's empowerment as a fundamental requirement.

At the 2015 Sustainable Development Summit, member states of the United Nations embraced the 2030 Agenda with 17 SDGs, acknowledging women's contributions to sustainable development, especially in education and child-rearing, as essential.

To harness the potential of women as drivers of sustainable development, it is essential to emancipate their roles within the family, community, and society from sociocultural and religious limitations. This liberation can only be realized through the principle of equity. This chapter aims to construct a conceptual framework that incorporates gender equality as a fundamental metric for attaining global sustainability.

Keywords

Gender equality · Framework · MDGs · Empowerment · Sustainable development

1 Introduction

The concept and process of "development" are highly contested in social sciences. One reason is that development affects different groups of people in myriad ways. These effects are based on their diverse social identities and positions in the social structure. Men and women are two such groups affected in manifold ways by development. This was mainly realized during the formulation of the Millennium Development Goals (MDGs) to combat poverty and bring sustainable development (Gurung 2003).To grasp the context of this chapter, one must first understand the concept of MDGs.

The United Nations Millennium Declaration, ratified in September 2000, mandates global leaders to address poverty, hunger, disease, illiteracy, environmental degradation, and

J. Bhattacharjee
Department of Social Work, The Assam Royal Global University, Guwahati, Assam, India

N. Singh (✉)
Department of Microbiology, The Assam Royal Global University, Guwahati, Assam, India

D. Egamberdieva
Centre of Agricultural Landscape Research, Müncheberg, Germany

Uzbek-China Key Lab of Ecobiomes of Arid Lands, National University of Uzbekistan, Tashkent, Uzbekistan

J. A. Parray et al. (eds.), *Environmental Landscape and Sustainable Biodiversity for Healthy Green Growth*, Sustainable Landscape Planning and Natural Resources Management, https://doi.org/10.1007/978-3-031-75405-0_11

discrimination against women. Derived from this Declaration are the eight Millennium Development Goals (MDGs), unanimously adopted by 189 nations in 2000. These goals were designed to improve the lives of socioeconomically marginalized communities.

The Millennium Development Goals (MDGs) achieved significant successes, including a substantial reduction in the number of individuals living below the poverty line of US$1.25 per day, one of the established targets (Griggs et al. 2014).

However, as the MDGs were slated to expire in 2015, it was acknowledged that a successor framework was needed. This led to the formulation of the Sustainable Development Goals, or SDGs, at the United Nations Conference on Sustainable Development, "Rio+20" (Low and Sengupta 1996). Building upon and modifying the MDGs to establish a more comprehensive approach to development developed the SDGs. The aim was to create a set of universally relevant goals and address the social, environmental, and economic aspects of "sustainable development" (Pathania 2017).

The Commission on Environment and Development coined the term "sustainable development," defining it as the ability of present generations to fulfill their needs without compromising the ability of future generations to do the same. Sustainable development endeavors to eliminate poverty, generate more significant opportunities for everyone, diminish inequalities, elevate basic living standards, and promote fair social development and inclusivity. The essential components for sustainable development are harmonizing economic growth, social inclusion, and environmental protection (Griggs et al. 2014).

As the SDGs were being formulated, it was recognized that gender inclusion is essential for sustainable development. Achieving gender equality is crucial in pursuing a fair and just future for the planet. Sustainable development is built upon democratic principles, the rule of law, and the recognition of fundamental rights, such as freedom and equal opportunities for all individuals. Consequently, ensuring gender equity in distribution becomes essential for achieving sustainability (Pathania 2017).

This chapter will, therefore, emphasize how empowering women and promoting gender equality serve as a foundation for achieving sustainability. It seeks to state that sustainable development requires fair treatment of women, and once they are included, "development" as a concept becomes attainable and, as a process, holistic.

So, this chapter is divided into four sections. It begins with accentuating the present-day state of women in the country across several spheres. Next, it describes the issue of women's empowerment in society and its interconnectedness with equity and development. From there, it moves ahead to throw light on the contribution of women in fields of importance, such as leadership, and how this leads to a striking enhancement in said spheres. Finally, primary data from India reflecting grassroots women's contribution to sustainability and representation is provided. These serve as a framework to comprehend development as a holistic and sustainable process.

2 Gender Equity: A Necessity for Sustainable Development

To grasp the significance of gender equality in the pursuit of sustainable development, it is crucial to understand the prevailing circumstances in which genders function within society and globally. Collectively, women constitute the most marginalized group worldwide. While acknowledging the diverse experiences among women—such as variances between those in the Global North and South and disparities based on race, motherhood status, and socioeconomic status—it is essential to recognize the common challenges that shape most women's lives. These challenges significantly influence their life trajectories and experiences (Wanjiru 2012).

To assist in understanding those above, here are some instances from different domains of life:

2.1 The Global Gender Gap Report: Economy, Polity, and Society

In 2016, The Global Gender Gap Report by the World Economic Forum was formulated, which includes the 11th edition of the Global Gender Gap Index. It enumerates the extent of gender-based discrepancies and measures the comparative breaches between women and men across four key areas: health, education, economy, and politics. This report discusses the "triple burden of women," emphasizing their multiple responsibilities, including unpaid care work for children, the sick, and the elderly, as well as tasks like fetching water and preparing food. In addition to the above, in developing nations, women play a crucial role in production, especially in small-scale farming. They also take on animal care, agriculture, and collecting fuel. Socioeconomic factors, societal perspectives, and socioeconomic obligations shape and affect gender roles, responsibilities, and decision-making processes, which lead to the same (Sexsmith 2017).

Prevalent sociocultural traditions and principles limit women's movement, social networks, accessibility to resources, and occupations or ventures that are open for their pursuit. Institutional arrangements—whether official or unofficial—can produce gender-based restrictions or reduce them in a conducive environment (Sexsmith 2017).

In various work sectors, resources and opportunities are divided unequally between men and women for the above-mentioned reasons. This inequality reflects on women's lives, productivity, and societal participation. We shall now explore some such instances across different countries based on occupations and avenues as per the data from the Global Gender Gap Report of 2016.

3 Farming, Agriculture, and Related Sectors

Both play essential but distinct roles in developing countries where men and women may be involved in smallholder farming. However, in many countries (from developing to developed), males and females have differential access to similar human or natural resources. Women comprise 43% of the labor force in Asian countries and over 60% in African countries, and their role in agriculture continues to expand. However, because of their restricted access to productive resources, opportunities, and healthcare, they continue to produce less than their male counterparts (Black et al. 2019).

In the Global South, women are more in touch with the environment through their daily necessities, which begin with housework and expand to agricultural labor in overlapping ways. Acts such as fetching water and raw food ingredients from fields and supporting their male counterparts with winnowing, sowing, watering crops on agricultural land, feeding and clothing the family, cleaning and maintaining family health using naturally derived resources, and so on are accomplished by women. In most African countries, women comprise 70 percent of the agriculture-based labor force, manage 90 percent of all household needs or chores, and take on 60 percent of all harvesting and market work. Despite this, their workload is mainly undocumented, neglected, and underappreciated (Leach et al. 2016).

4 Politics and Leadership

Regarding political representation, women's right to vote in countries like Somalia and Cambodia was only recognized in the late 1950s. Several noteworthy instances of this inequality exist in developed countries like the USA and Japan. Although recognized as a top-ranked legal attorney, Hillary Clinton had her credentials mocked as her two presidential campaigns proved inefficacious consecutively. They were publicly and overtly described as her futile attempts to succeed her "husband's political bequest" (Pollman 2017).

In Japan, despite the rise in women's societal and civic engagement following democratization, both employment and family policies revolved around traditional stereotypes. Men were expected to embody traits of strength and reliability as breadwinners, while women were confined to roles as nurturers, mothers, and dutiful wives within the domestic sphere. In the 1980s, the Liberal Democratic Party (LDP) reinforced these stereotypes, promoting a clear separation of gender roles, positioning men primarily in the workforce and women primarily in the home, rather than advocating for equal opportunities and treatment for all individuals (Usui et al. 2003).

5 Entrepreneurship

According to the United Nations Development Programme (UNDP) Report on Green Business Initiatives by Women, women entrepreneurs are an underexplored market segment with significant potential for high returns and sustainability. Globally, businesses owned by women account for 20–40 percent of small and medium-sized enterprises. However, the potential contribution of women to initiatives such as green enterprises and environmental investments has yet to be noticed, mainly due to biases that hinder women's access to policymaking, financial resources, credit, information, and equitable compensation (Griggs et al. 2014).

6 Organized and Unorganized Sectors: Poverty and Illiteracy Among Women

It isn't easy to operationalize legal and civil rights in organized and unorganized sectors if a person has the economic resources to pay for shelter, food, clothing, healthcare, and education. There is interrelatedness between economic prospects, social opportunities, and legal rights, which come together more visibly in the unorganized sector (Ebrahimi-far 2007).

Gender equality in economic terms is unfavorable even in developed countries. Hour for hour, women earn approximately three-quarters of a man's wage. This is equally true in countries like Belgium (where the female-to-male hourly earnings ratio in manufacturing jobs is 74.5), New Zealand (74.5), Costa Rica (74.0 for monthly earnings), and Kenya (730 also for monthly earnings) (ILO in Anker, 1997). It is estimated that 70 percent of the population living in absolute poverty (earning less than US$1 a day) are women (Griggs et al. 2014).

Many studies as late as the 1990s and even now showed that the impoverishment of women is growing, and their work burdens are expanding. Their contribution to the production system is not less significant, but their role is not valued and properly counted—their talents, skills, knowledge, and insights are not treated as necessary. Consequently, their work has often needed to be paid (Mia et al. 2007).

According to the Department for International Development in London's sustainability and economic development reports, literacy is a crucial metric for assessing global development achievements. In discussions surrounding literacy, post-1980s statistics reveal that the international adult literacy rate is 68 percent for women and 82 percent for men. Despite progress, disparities persist, especially among young women who lag their male counterparts in literacy rates, particularly in regions like South Asia (41 percent) and sub-Saharan Africa (51 percent). The lower literacy rates among women stem from gender inequalities within educational systems, where girls in developing nations are less likely to access schooling compared to boys (with 92 percent enrolment for boys, even at the primary level). The UN Economic and Social Council highlighted in 2001 that families' insufficient investment in female primary education contributes to the stagnation in educational progress (Pollman 2017).

7 Turmoil and Women

In times of war, women are found to have higher chances of being harmed or losing their lives. Statistics from different studies highlight those above. Ninety percent of victims of war as well as refugees are now civilians, and in great likelihood, the bulk of them are women and children. The majority of whom are likely to be women and children. In 1999, De Groot estimated this as high as 80 percent. It was also estimated that refugee females across age groups must embark on foot to collect bare necessities like water and wood in challenging conditions, typically for two hours on average (Sexsmith 2017).

Even in times of environmental crisis like climate change, women across different backgrounds are affected severely.

Individuals residing in developing nations and marginalized communities within developed countries disproportionately feel the impacts of climate change. This phenomenon worsens disparities in accessing healthcare, nutritious food, clean water, and other essential resources. As the Climate Justice Network emphasized, climate change has a precise environmental justice dimension, necessitating an understanding of its interconnections with social issues like racism, gender inequality, and economic inequity. Ironically, the developing world has primarily borne the brunt of climate change consequences and related environmental hazards. Trapped amidst poverty and ecological degradation, burdened with household duties and familial responsibilities, and often overlooked within the informal economy, impoverished rural women in nations like India may find themselves at the brink of physical endurance (Dankleman 2002).

Since the mid-1980s, several studies have revealed that the relationship between communities and their physical environment is far from gender-neutral. It is known and elucidated that women, particularly in rural areas in developing nations, engage with their physical environment extraordinarily and are thus inordinately affected by any and every kind of environmental devastation.

In The State of India's Environment Report of 1985, the Centre for Science and Environment (CSE), headquartered in New Delhi, highlighted that perhaps no other demographic bears the brunt of environmental degradation as acutely as impoverished rural women. Each day, they embark on arduous journeys at dawn, scouring for fuel, fodder, and water. With deteriorating ecological conditions, these journeys only lengthen and become more burdensome (Dankleman 2002).

Through the instances mentioned above of various fields in life, we see that spectacular events and international mentalities have undeniably led women to become vulnerable social, political, cultural, and economic actors (Doyle 1994).

8 The Issue of Women's Empowerment

As the concept of sustainable development and the SDGs explicitly state, the inclusion of the most marginalized groups in society is fundamental in the agenda for development. As vulnerable groups, particularly women, continue to experience inequality in several forms, the achievement of SDGs may become a distant vision. Therefore, we explore a link between the empowerment of women and the attainment of a sustainable world.

The United Nation's Sustainable Development Goals (SDGs) for gender equality (Goal 5) recognizes that connecting gender equality with sustainable development is fundamental, because any image of an impartial and sustainable world must include the rights of women and must note that, relative to their male counterparts, females in diverse and specific scenarios are unevenly affected by economic, social, and environmental crises and conditions (Leach et al. 2016).

Goal 5 in the SDG framework emphasizes the need to accomplish equality between genders and the empowerment of every woman. However, despite being a fundamental right, several noteworthy gaps exist in implementing gender equality. For instance, they have limited access to opportunities for healthcare, education, employment, and even essential participation in the sociopolitical and economic lives of their community (Columbia Center on Sustainable Investment, UNDP, UN Sustainable Development Solutions Network, and World Economic Forum 2016).

Thus, gender equality can only be achieved once women are empowered globally.

Women's empowerment signifies a higher sense of control and power over their lives. It denotes that "rights, responsibilities, and opportunities of individuals will not depend on whether they are born male or female." As per the United Nations Population Fund, when a woman is empowered entirely, she has the authority of her life in the realm of the domestic and outside. Thus, she can be vital in social transformation. Therefore, in order to create equal opportunities for men and women and, in turn, reduce the ever-persistent gender gap and develop and maintain gender equality, women need to be "empowered" (Bhardwaj 2004).

Myriad studies and practical experiences bring to light the connection between sustainable development, gender equality, and the empowerment of women. As per the UNDP 2011 Human Development Report, sustainable development connects policy, economy, society, and the physical environment concerns. It establishes gender equality as a precursor to development, with a role in eliminating poverty, above and beyond being just a human right (Wanjiru 2012).

The Earth Summit 1992 in Rio was the UN's first conference that stated that women needed to be involved at all decision-making stages to achieve environmental sustainability. In this regard, women's empowerment is one of the critical factors in achieving sustainability. Furthermore, to construct women as catalysts for sustainable development, their familial, community, and societal functions must be independent of social, cultural, and religion-based restrictions and traditions that avert their participation in said matters (Pathania 2017).

When women's participation and stake in decision-making avenues increase, it enables an equitable allocation of resources to nutrition, health, education, employment, and social protection. An example would be how an increase in female literacy levels is inversely proportional to child and infant mortality rates and directly proportional to family health. Education also increases their participation in the labor force and household and national income contributions. Their increased earning capacity, in turn, positively affects children's nutrition, health, and educational prospects (Pathania 2017).

Therefore, women's empowerment is critical in reaching sustainable economic growth, social development, and environmental sustainability (Steele 1998).

Just like instances of inequitable treatment of women, there are also descriptions and examples of a more inclusive society for women, where they have emerged not only as leaders but also as significant contributors to sustainable development. Some of these experiences shall be traced in the next section.

9 Women and Sustainable Development

The liberation of women is an essential component of social development and not merely a "women's issue"; and ultimately, gender equality, as Aster Zaoude observed, is considered "a measure of sound governance because when development is not 'engendered,' it can only be endangered" (Mia et al. 2007). The presence of women in public offices has steadily affected development in countries, making them safer spaces and more inclusive in decision-making. It has significantly impacted the quality of life and sustainability (Brand 1996). In this section, the examples cited explain how equity in development can lead to sustainability in the same process and how women's voices can amount to faster growth advancement of countries.

9.1 Political Leadership in Sweden and Latin America

Sweden, as a country, has successful public policies to progress gender parity, involvement, and education for all genders, even though gender equality has not yet been stated as a concrete policy goal. Women's overt presence in the public sphere/political platforms, such as their membership in the parliament, was identified as the guiding force behind these egalitarian policies (Gustafsson and Kolam 2008).

Historically, in Sweden, immigration policies were constrained, and so was the availability of the labor force. Inevitably, females filled the rungs of blue- and white-collar jobs. This phenomenon led to a shift in women's positions in public and domestic life, culminating in a higher insistence on gender equality and establishing a ground for the "Nordic gender regime." This is marked by the notion that both men and women should be entrusted with nurturing children and making a living, not just one of the two responsibilities (Bergqvist and Jungar 2003).

Since the mid-1970s, the ratio of female members in the parliament has exponentially increased (stands at 45 percent at present); so has the number of female ministers (Statistics Sweden 2006). The Social Democratic Party is being led in contemporary times by Mona Sahlin, and Maud Olofsson spearheads the Centre Party. In addition, a host of other female leaders have taken positions of governance in Swedish politics and strive for formidable policies for women's empowerment in several private and public domains. Combatting violence against women and measures to ease childcare are seen as two fundamental mechanisms for those above. Therefore, the same percentage of women is currently beneficially employed as men (Gustafsson and Kolam 2008). This has relegated it to the position of the most habitable country for women to live in (Lopez-Claros 2007).

In similar illustrations of women's empowerment through leadership positions, we look at Michelle Bachelet of Chile. She was the first female President of Latin America who served

two terms in office (2006 and 2014) after being the Minister of Health and Defence. Another such forerunner is Costa Rica's Laura Chinchilla, who served as the Minister of Public Security and Vice President of the country before eventually taking office as the President (2010–2014).

These female leaders symbolize a vigorous structural and cultural change that can fuel greater equality and engagement in all spheres of civic life. Such standardization of women in leadership positions in the public arena over time may be the most convincing marker of Latin America's developments in political gender equality (Gustafsson and Kolam 2008).

9.2 Sustainability and Women in India

In the Indian context, women have taken leadership positions and directly emphasized sustainable development through numerous initiatives. Even in tragedy, they have shown remarkable resilience. For example, in the unfortunate event of the partition, women found a way to solidify their identity in a country that was no longer their stable home. In the said event, the Indian state provided avenues to support women.

The Indian state, following the partition, undertook the ardent task of the rehabilitation of uprooted women. Various transit camps, relief centers, rehabilitation homes, and vocational training centers were set up for this purpose. A "Women's Section" under the guidance of prominent women social workers was set up by the Ministry of Relief and Rehabilitation.

To facilitate women's vocational training, the Women Section aligned with the various bodies mentioned above. These encompassed the Tata Institute of Social Sciences, the vocational centers of the Ministry of Labour in Delhi, the Kasturba Gandhi Memorial Trust, and Lady Hardinge Medical College, New Delhi. An Employment Bureau was also operationalized in conjunction with the employment exchange of the Labour Ministry to hire women when they completed their training (Bhardwaj 2004).

In March 1949, a survey documented that five hundred women had gained employment through the Employment Bureau. In a review of the resettlement work by the Director General of Resettlement and Employment in May 1949, the records showed 98,314 registrations with the Directorate of Employment Exchanges. Of these, the number of people placed in employment was 24,443; of these, the number of displaced women was 4962, i.e., 20 percent. These included school *mais* (maids), women tailors, and calico printers. The directorate of training registered 278 women (displaced) who were undergoing training at a center at Curzon Road run by the Ministry of Labour (Bhardwaj 2004).

Extensive research was conducted in 495 communities in India, which revealed that as political leaders, women become exemplary to female adolescents, raising the latter's aspirations for both public and private levels of achievement. In areas governed by women, better services in the public interest are assured, and issues such as bribery become obsolete. A similar analysis in India disclosed that women in public office significantly bring down the prejudice and negative biases inherent in society about women and positions of authority (Pollman 2017).

A story of local communities utilizing resources at the grassroots from the author's hometown has been another instance of women's empowerment leading to sustainable development (especially in a crisis). It tells a story of resilience and gender equity through the emergence of self-help groups (SHGs). SHGs are India's most potent instruments for supporting women moving from survival to sustainability. The self-help group (SHG) movement in India came as the sturdiest silver lining in the grassroots amidst the COVID-19 pandemic.

It is one of the most robust incubators of female strength and entrepreneurship and an avenue for modifying the socially constructed definition/idea of gender in rural communities. Women in rural areas are now able to generate individual earning sources. While many young semi-literate women have homegrown skills, they lack assets, and restrictive social conventions often prevent them from setting up independent businesses before the pandemic. However, in 2020, when COVID-19 caused difficulties and societal/civic pressure, women-led SHGs effectively brought about social well-being in their families and communities (Qazi 2021).

The primary data collected by the author in the mentioned locations—Deochutal and Tetelia—are given below, which are adopted villages of her institution of affiliation. Data was collected using a qualitative approach, with convenience and snowball sampling.

The researcher/author used focus group discussions and personal interviews as methods and an interview guide as a tool. The sample consisted of all the women in the said communities. They volunteered to speak to the researcher/author and describe their experiences during the pandemic.

The suburbs of Deochutal and Tetelia are two communities near each other, near Charaideo Village, near National Highway 37, Guwahati, Assam, India. These communities are composed of 30 and 25 households, respectively. Approximately 90% of the households have male members working in private factories near the area or engaged in agriculture, fisheries, cattle-/poultry-rearing, and selling fried snacks (*nimki*, *bhajia*, *pakora*) by the roadside. These communities have resources like lower primary schools, wells, common grazing lands, temples, and ration shops.

The women of these communities are exceptional in their coexistence with men despite the lack of formal education or the prevalence of female dropouts in the areas.

Many women from these communities are engaged in the production of traditional garments: Assamese Gamusa, Mekhela Sador, Dhoti, Lungi, Bihuwan, and Boro Dokhona (Figs. 1 and 2).

Fig. 1 Women at Deochutal weave the traditional *gamusa* (a locally woven symbol of Assamese identity and representation of Assamese culture)

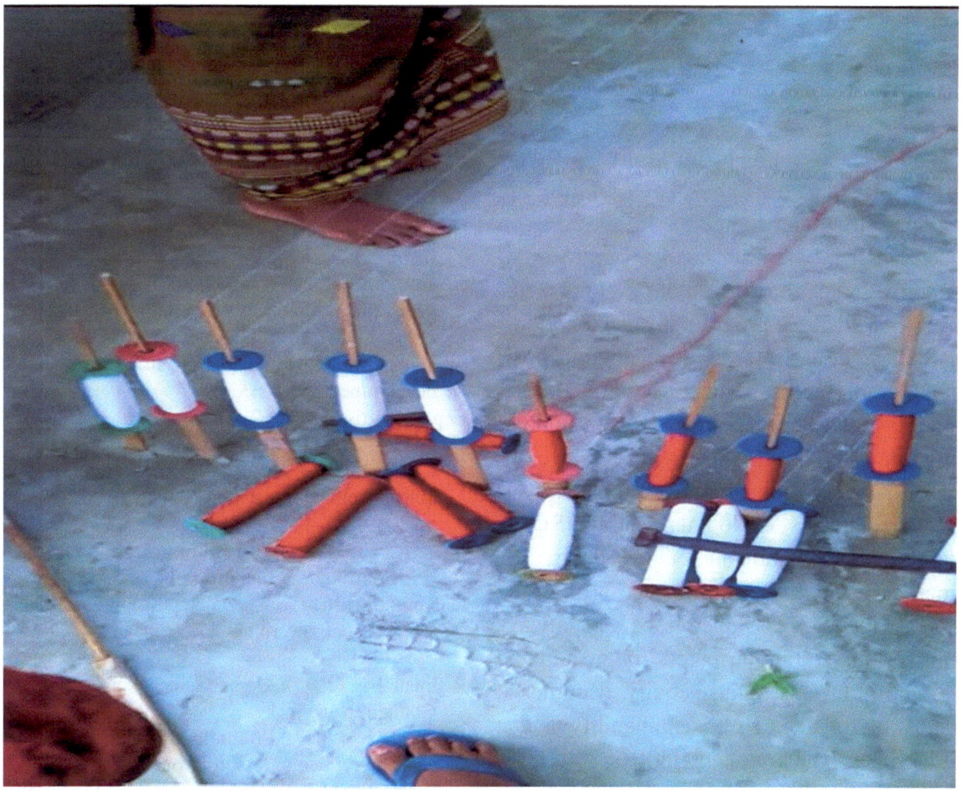

Fig. 2 The *gamusa* is made of organic material, using a self-made equipment *tathal*

A total of 30 women from these communities drive the entire process. These women are masters of the art, as they engage themselves in a fruitful yet tedious craftsmanship from scratch.

In their production of the garments above, the women avoid using fabric but create the produce solely from threads. This process requires exceptional handloom and weaving skills, determination, and focus.

Their work begins with selecting the material, a combination of Madras yarn, Padmini yarn, Kopahi (cotton), plain yarn (also known as Kasa huta), and Thailand yarn, and tracking the availability of the same and buying it from tailoring outlets; to weaving this fabric in different permutation combinations using the "tathal" or locally made weaving machine; to setting up stalls in the weekly market near their communities; to handpicking their best creations; to fixing the prices at a rate affordable to the local population as well as buyers from the city; and finally, to selling these products (range of 250–2500 rupees).

While a handful of women choose to make these for the community, the others are small market entrepreneurs. They contribute to the family's earnings, work with an eco-friendly approach, and form a self-help group in the community. This is an anecdote of resilience, independence, and hard work that kept their families afloat during the pandemic (Bhattacharjee 2023).

10 Conclusion

Even though sustainable development goals are yet to be achieved, including women in various spheres of life can accelerate the process (Low and Sengupta 1996). As seen above, in instances where women are involved more in all sectors of the economy, polity, the domain of law, and sociocultural life, there is considerable progress in such locations. This means that their voices—in terms of leadership positions, employment opportunities, infrastructural adjustments, valued work contribution, and decision-making powers at home—can open creative pathways to development in sustainable ways. Women's contributions always carry an element of sustainable development and the SDGs, as the primary data from Deochutal and Tetelia reflects.

Creating women empowerment in countries through the measures mentioned above across the globe has always bestowed fruitful results and contributed to sustainable development in manifold ways. Thus, equity becomes a framework or pathway for achieving sustainability and a goal itemized in the SDGs (Buckingham-Hatfield 2002).

To conclude, given below is a table deriving from the chapter that lists the fields discussed here that reflect how women are not represented in an equitable and just manner in domains of society. Following that, some suggestions are provided on ensuring the same.

References

Bergqvist C, Jungar AC (2003) Adaptation or diffusion of the Swedish gender model? In: Hantrais L (ed) Gendered policies in Europe: reconciling employment and family life. Palgrave, Hampshire, pp 160–179

Bhardwaj A (2004) Partition of India and Women's experiences: a study of women as sustainers of their families in post-partition Delhi. Soc Sci 32(5/6):69–88

Bhattacharjee J (2023) Post development through reselience: the journey of women SHGs and entrepreuners in the COVID-19 era. In: Sharma DA (ed) Prayash. Women Cell, Pandu College, Guwahati, pp 91–99

Black K, Guest D, Bagnol B, Braaten YB, Laven A (2019) Gender, health, and smallholder farming. In: Walton M (ed) One planet, one earth. Sydney University Press, Sydney, pp 105–128

Brand J (1996) Sustainable development: the international, national and local context for women. Built Environ Women Environ 22(1):58–71

Buckingham-Hatfield S (2002) Gender equality: a prerequisite for sustainable development. Geography 87(3):227–233

Columbia Center on Sustainable Investment, UNDP, UN Sustainable Development Solutions Network, and World Economic Forum (2016) Mapping mining to the sustainable development goals: an atlas. Sustainable Development Solutions Network, New York

Dankleman I (2002) Climate change: learning from gender analysis and women's experiences of organising for sustainable development. Gend Dev 10(2):21–29

Doyle E (1994) War and madness. Fortnight 94(334):37–39

Ebrahimi-far T (2007) Women, sustainable development and dialogue among civilizations. Pak Horiz Women Concern Int Relat 60(4):93–107

Griggs D, Smith M, Rockstrom J, Ohman C, Gaffney O, Glaser G et al (2014) An integrated framework for sustainable development. Ecol Soc 19(4):48–73

Gurung HB (2003) Education for sustainable development in Nepal. Institute for Global Environmental Strategies, Kathmandu

Gustafsson G, Kolam K (2008) Political women's leadership in Sweden: developments and challenge. Signs 34(1):27–32

Leach M, Mehta L, Prabhakaran P (2016) Sustainable development: a gendered pathways approach. In: Gender equality and sustainable development, pp 1–33

Lopez-Claros DA (2007) The global competitiveness report 2006–07. World Economic Forum, Geneva

Low B, Sengupta M (1996) Guest editorial: women and sustainable development, a reflection. In: Population and environment, pp 105–109

Mia KA, Ali K, Yasin M, Yasmin, Yash M (2007) Gender equality- rhetoric or a PAncea? Indian J Polit Sci 68(4):739–750

Pathania SK (2017) Sustainable development goal: gender equality for women's empowerment and human rights. Int J Res Granthaalayah 5(4):1–11

Pollman LA (2017) Women's leadership in Latin America. Atlantic Council, New York

Sexsmith K (2017) Promoting gender equality in foreign agricultural investments: lessons from voluntary sustainability standards. International Institute for Sustainable Development (IISD), Geneva

Statistics Sweden (2006) Country report: statistics Sweden. Miljorakenskaper, Sweden

Steele F, Amin S, Naved RT (1998) The impact of an integrated micro-credit programme on women's empowerment and fertility behavior in rural Bangladesh
Usui C, Rose S, Kageyama R (2003) Women, institutions, and leadership in Japan. Asian Perspect 27(3):85–123
Wanjiru L (2012) Gender, climate change, and sustainable development. Fordham Environ Law Rev 23(2):1–6

Other Sources

Qazi M (2021, March 11) Retrieved February 19th, 2022, from www.theleaflet.in: https://www.theleaflet.in/self-help-groups-women-become-the-change-they-want-to-see-in-the-world

Sustainable Consumption and Production in Emerging and Developed Economies: A Bibliometric Analysis

Hariharan C., Dhanalakshmi T., and Naganandini S.

Abstract

Sustainable consumption and production have been identified as critical prerequisites for sustainable development. Because of differences in economic situations and sociocultural aspects, sustainable consumption and production necessitate a multifaceted approach in emerging and developed countries. To date, few attempts have been made to comprehensively assess sustainable consumption and production and its direction in emerging and established countries. This report will review the sustainable consumption and production literature published in international scientific journals between 1998 and 2023. This article will examine sustainable consumption and production patterns in developed and developing economies by addressing three carefully selected questions. This discovery and others are discussed and explored in more detail in this work, identifying gaps and potential areas for future research in the existing material.

Keywords

Sustainable consumption and production · Developed nations · Developing nations · Sustainable supply chain management · Sustainable development

Hariharan C. (✉)
Department of Management Studies, SRM Institute of Science and Technology, Tiruchirappalli, Tamil Nadu, India

Dhanalakshmi T.
Nehru Institute of Health Sciences, Coimbatore, Tamil Nadu, India

Naganandini S.
Department of Management Studies, Nehru Institute of Technology, Coimbatore, Tamil Nadu, India

1 Introduction

The scientific world realized in the 1970s that unsustainable expansion was dangerous and that there would soon be economic and environmental crises, which the Club of Rome had dubbed "limits to growth" (Meadows et al. 1972). It has become clear that sustainable development (SD) is essential to human existence (Seiffert and Loch 2005). They integrated the production and consumption systems with SD (Akenji and Bengtsson 2014; Lozano et al. 2015). Globally, there has been an increase in concern over sustainable consumption and production (SCP), as evidenced by major international conferences addressing SCP and climate change. According to the Report of United Nations, Plan of Implementation of the World Summit on Sustainable Development (2002), industrialized countries should promote SCP, with advantages accruing to all nations. Ten years later, the United Nations Conference on Sustainable Development (Rio+20) proposed a 10-year framework of programs on SCP, reinforcing SCP as essential to SD.

These activities inspired various stakeholders, including governmental agencies, nongovernmental organizations, and academic institutions, to include SCP in their strategic planning. Prior research on SCP has examined aspects of consumption and production while focusing on social inclusion, environmental preservation, and economic growth. Few reviews have addressed SCP literature from different perspectives, even though several have been done (Brizga et al. 2014; Pallaro et al. 2015; Roy and Singh 2017). These studies concentrated on business domains, automobile industry factors, and SCP policy. Furthermore, a few papers (Caeiro et al. 2012; Grabs et al. 2016; Liu et al. 2016) have focused only on sustainable consumption. Nevertheless, comprehensive studies contrasting SCP's difficulties and advancements in developed and emerging countries are still required (Table 1).

The principal objectives of this work are to (a) carry out a thorough assessment of the literature in the SCP research

J. A. Parray et al. (eds.), *Environmental Landscape and Sustainable Biodiversity for Healthy Green Growth*, Sustainable Landscape Planning and Natural Resources Management, https://doi.org/10.1007/978-3-031-75405-0_12

Table 1 Previously published review papers

Authors (year)	Time	Field	N	Comments
Roy and Singh (2017)	1990–2016.07	Both	71	Area of focus: business research Questions: What are the main study themes found in the SCP literature? Are you summarizing the dominant business orientation?
				What more explanations are possible for these main themes to summarize the theme-specific critical implications that should direct scholarly work?
Pallaro et al. (2015)	2004–2014	Both	42	Focus area: the automotive industry What are the primary obstacles and factors that support sustainable production and consumption in the automobile sector? Which ideas, approaches, and resources have been applied for sustainable consumption and production? What else has to be done to get the car industry closer to more all-encompassing sustainable practices?
Brizga et al. (2014)	1990–2010	Both	N.A.	Area of interest: creation and application of SCP policies. Research questions include SD tactics and goals, policy instruments, the organizational and legal frameworks that support SCP governance, and features of SCP in addition to sectoral policies and programs
Caeiro et al. (2012)	1979–2011	C	49	Focus area: household SCP. The study questions are as follows: metrics of household sustainable consumption and a methodological approach to HSC indicators

(continued)

Table 1 (continued)

Authors (year)	Time	Field	N	Comments
Grabs et al. (2016)	N.A.	C	93	Focus area: community-based efforts towards sustainable consumption Study questions: Which systems and elements encourage and facilitate the adoption of sustainable consumption habits at the community level? Which processes and group variables provide the most support for maintaining grassroots support operations? Which social structures and procedures enable grassroots projects to be implemented successfully?
Liu et al. (2016)	2003–2013	C	N.A.	Focus area: China's sustainable consumption Research questions: What is the method for addressing Chinese sustainable consumption initiatives based on different implementation strategies? Which sustainable consumption behaviors are most prevalent in Chinese society?

Note: The third column indicates that the examined papers covered only sustainable production (P), sustainable consumption (C), or both. N is the number of documents that have been reviewed

domain, with a focus on well-defined taxonomies, and (b) apply the findings to a comparative analysis, emphasizing significant SCP differences and focal points between developed and developing countries. This is how the paper develops: The research approach is described in Sect. 2, descriptive and content analyses are covered in Sect. 3, a thorough comparative analysis is provided in Sect. 4, and potential directions for future research based on gaps in the data are examined in Sect. 5. Limitations and concluding remarks are covered in Sect. 6.

2 Research Design

2.1 Research Questions

Table 1's literature surveys shaped our research questions, melding past insights and scholarly perspectives. Multiple revisions honed three questions, informed by analyses in Sect. 3. Section 4 delves deeper, leveraging insights for robust discussion.

Research question 1: What are the core distinctions in approaching SCP initiatives and methodologies between advanced and emerging economies?

Research question 2: What quantitative and qualitative methodologies and validation strategies have been devised for examining, embracing, and executing SCP frameworks in advanced and emerging economies?

Research question 3: Which sectors participate in SCP deliberations within advanced and emerging economies?

2.2 Research Methodology and Boundaries

The primary objective of the literature analysis phase is to furnish reliable and high-quality insights into the designated topic. Consequently, the scope of this analysis encompasses peer-reviewed journal articles published in English and indexed in the Scopus database. Given that SCP was initially delineated in the Oslo Symposium of 1994, publications sought for this study commence from 1994, with the earliest journal papers discovered dating back to 1998. Thus, the research period spans from January 1998 to December 2023, during which a meticulous selection process yielded 90 articles meeting the specified criteria.

To ensure consistency and robustness in the literature analysis, it is imperative to establish appropriate limitations, as Seuring and Müller (2008) noted. For the present study, these limitations are as follows:

1. The research scope is confined to "sustainable consumption and production," excluding articles focusing solely on either "sustainable consumption" or "sustainable production." Articles addressing specific industries such as agri-food, chemical, or energy production and mobility, housing, clothing, and nutrition consumption were also excluded. For instance, despite a special issue on the "sustainability of seafood production and consumption," articles were not considered due to their exclusive focus on consumption systems within the seafood industry.
2. Articles needing a discussion on sustainability in the context of consumption and production were excluded, aligning with the aim of this paper to review works integrating sustainability with consumption and production operations.
3. Articles discussing consumption and production across industries without considering sustainability and those covering more general subjects like waste management and sustainable supply chain management should have been included.
4. The review database should have included articles about metabolism, public health, technical science, and microcosmic chemical systems.
5. This study did not include non-journal materials such as governmental documents, regulations, SCP projects, corporation activities, and web content.

2.3 Coding Process

This study included three main dimensions—SCP in established and emerging countries, research methodology and validation approaches, and industry sector classifications using Mayring's (2004) coding process model. The work of Tseng et al. (2013) serves as an example of the coding procedure utilized to classify specific papers (as explained in Sect. 2.2).

1. SCP in developed and developing economies: Tseng et al. (2013) are primarily concerned with SCP practices in India; therefore, they fall under the "developing economies" category even though the writers are from several other nations, including China, India, and the UAE.
2. Approaches to validation and research methodology: Tseng et al. (2013) fall within the category of "multi-criteria decision modeling" for research approaches when they use the analytical hierarchy process (AHP). Furthermore, in terms of validation techniques, its application to a real-world case study makes it eligible for the "real-world application/case studies" category.
3. Industry sector classification: Tseng et al. (2013) fall under the "manufacturing" category since their study evaluated the obstacles to SCP adoption within an Indian plastic product manufacturing organization.

2.4 The Coding Process's Rigor

Seuring and Müller's (2008) double-check guidelines were employed to mitigate subjectivity in coding. Both authors independently coded, resolving discrepancies with a third and fourth author. Inter-coder reliability, assessed via Cronbach's (1951) method, yielded a coefficient alpha of 0.89, surpassing the 0.70 threshold for acceptable reliability.

3 Descriptive and Content Analysis

This section employs a dual analysis approach comprising descriptive and content analysis. Descriptive analysis delineates fundamental research features within the SCP domain, including yearly publication rates and primary journals (refer to Sect. 3.1). Additionally, content analysis deciphers published literature content via systematic classification taxonomies, unveiling present and future trends (Hsieh and Shannon 2005). Economic statuses are assessed to discern varied thematic focuses across economies, categorized as developed or developing according to the United Nations (UN 2012) classification. To address the three research queries in Sect. 2.1, three content analysis classification taxonomies are delineated: (a) SCP in developed and developing economies, (b) research methodologies and validation approaches, and (c) associated industries. Based on insights from prior literature reviews, these taxonomies furnish ample data for a comparative analysis of SCP considerations in Sect. 4. Sections 3.2, 3.3, and 3.4 elaborate on each classification.

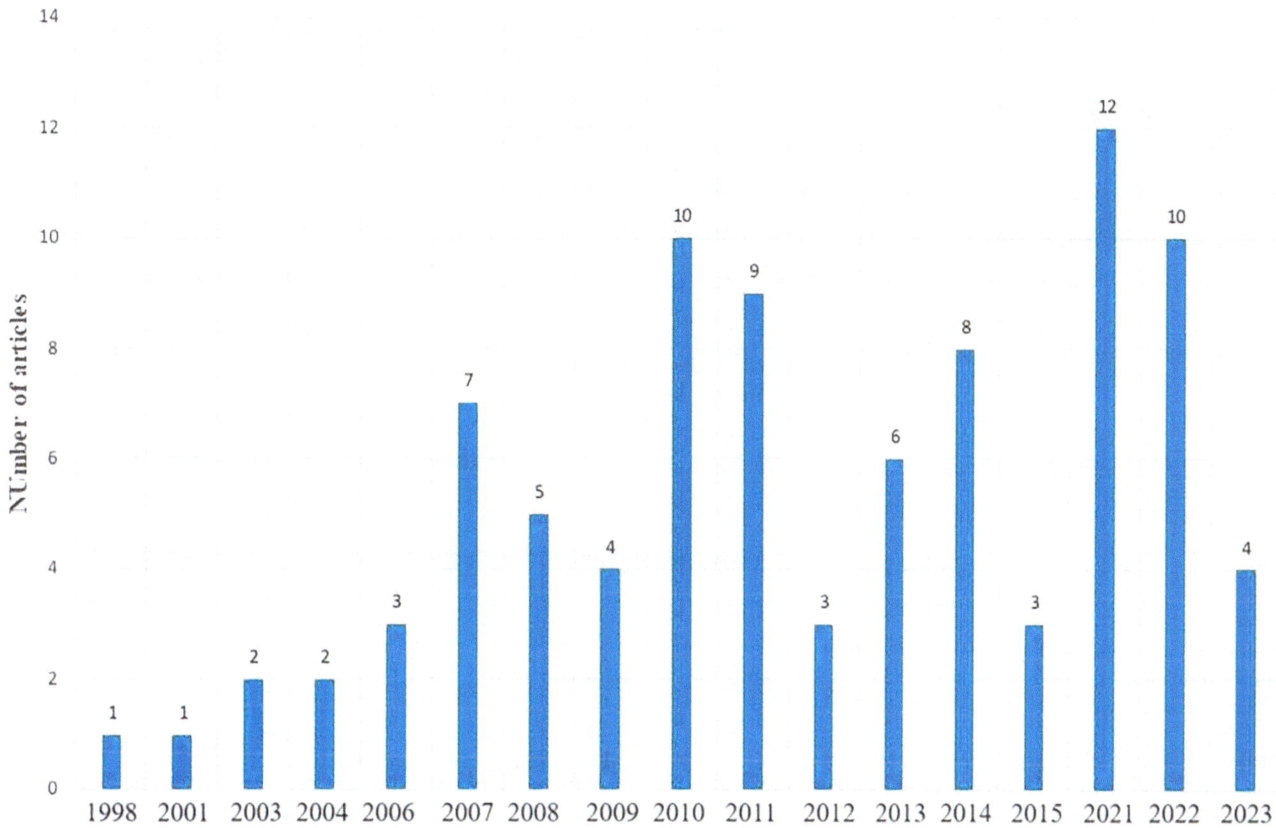

Fig. 1 Annual publications during the period under investigation

3.1 Publications Per Year and Leading Journals

Figure 1 illustrates the yearly publication frequency in the SCP field across the Scopus database, highlighting trends in knowledge production. Publications have notably surged since 2007, comprising 85.6% of the total, with an average of 7.0 per year from 2007 to 2022. This upward trajectory reflects a growing interest in SCP-focused research. Peaks in publication quantity, such as in 2021 (12), often coincide with special issues sponsored by journals like the Journal of Industrial Ecology and the Journal of Cleaner Production, showcasing concentrated efforts in SCP-related exploration (Fig. 1).

It is interesting to see how articles are distributed among journals. Journals of engineering and economics contained the majority of the articles. There were 90 papers in total, published in 45 different publications. Remarkably, the Journal of Cleaner Production published a third of these articles. Others appeared in various journals, including Ecological Economics, the International Journal of Production Economics, and Natural Resources Forum (Fig. 2).

3.2 SCP in Both Developed and Emerging Markets

Due to sociocultural, demographic, and economic variables, different countries have different strategies for sustainable consumption and production (SCP) (Spaargaren 2011). SCP practices in developed nations and those in developing/underdeveloped economies are the two usual categories into which research publications are divided. This classification aims to identify critical areas of attention and contrast them to identify similarities and weaknesses (refer to Sects. 4.1 and 4.2). It is crucial to remember that economies are categorized according to the nations or areas researched, not by the authors' associations. Tables 3 and 4 show the results for the developing/underdeveloped and developed economies, respectively. A few sources are redundant because different authors have examined different economies. Furthermore, specific articles address SCP practices in more general economies, such as the European Union (EU), without concentrating on particular nations; these outcomes are summarized in Table 2.

Out of the 90 papers that were surveyed, 22 of them looked at SCP in emerging or undeveloped economies (see

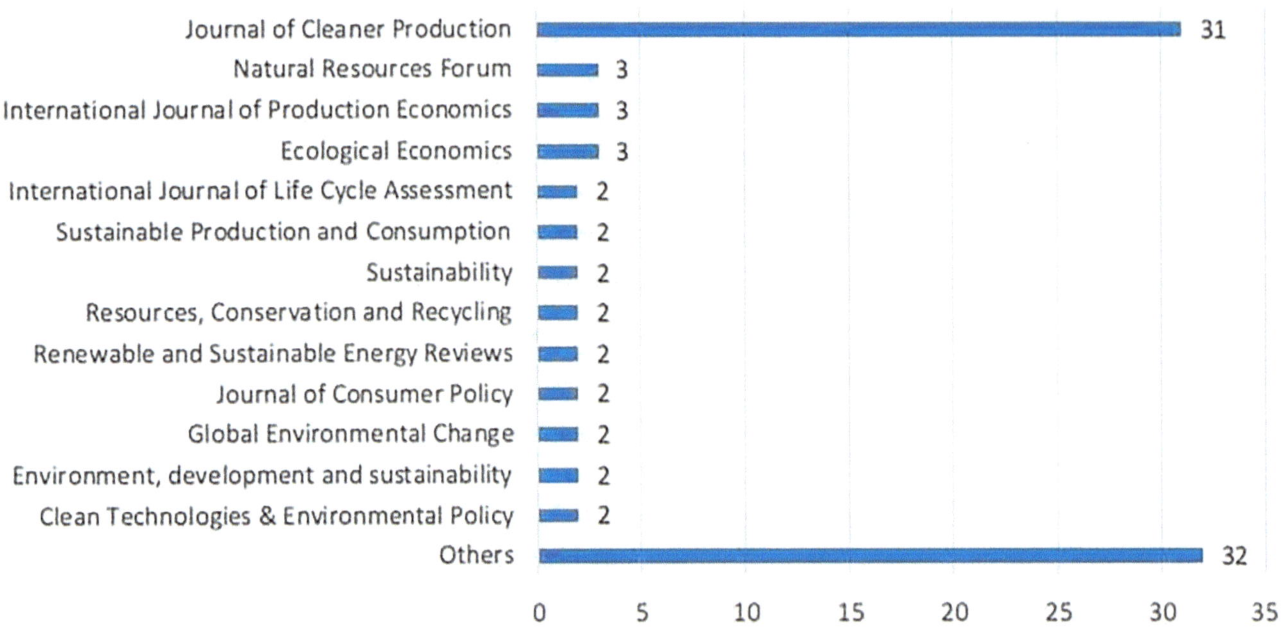

Fig. 2 Number of articles from the bibliography published in prestigious journals

Table 2 SCP in developing/underdeveloped economies

Emerging/ underdeveloped economy	References	Count
India	Mangla (2017), Dubey et al. (2016), Luthra et al. (2017), Soni et al. (2016), Luthra et al. (2016), Hoff et al. (2014)	6
China	Ely et al. (2016), Vergragt et al. (2016), Schroeder (2014), Chiou et al. (2013), Liu et al. (2010)	5
Turkey	Yılmaz and Uslu (2007), Soyhan (2009), Bilen et al. (2008)	3
Malaysia	Adham et al. (2015), Wong et al. (2016)	2
Thailand	Mungkung et al. (2012), Thongplew et al. (2017)	2
Indonesia	Moreno-Peñaranda et al. (2015)	1
Tikopia	Mertz et al. (2010)	1
Mexico	Corral (2003)	1
Nigeria	Hoff et al. (2014)	1
Total		22

Table 3 SCP in developed nations

Developed economy	References	Count
Finland	Berg and Hukkinen (2011b), Lehtoranta et al. (2011), Berg (2011), Berg and Hukkinen (2011a), Honkasalo (2011), Niinimäki and Hassi (2011), Risku-Norja and Mäenpää (2007)	7
UK	Dewick and Foster (2018), Berg (2011), Azapagic et al. (2016), Yakovleva and Flynn (2004), Seyfang (2004)	5
Germany	Lorenz and Veenhoff (2013), Hoff et al. (2014), Grözinger et al. (2010)	3
Spain	Staniškis (2012), Cazcarro et al. (2016), Partidário et al. (2007)	3
Lithuania	Staniškis et al. (2012), Staniškis (2012)	2
Sweden	Petry et al. (2011), Berg (2011)	2
Netherlands	Petry et al. (2011)	1
Australia	Clay et al. (2007)	1
Austria	Schönhart et al. (2009)	1
Bulgaria	Staniškis (2012)	1
Canada	Petry et al. (2011)	1
Czech Republic	Dobes (2016)	1
Estonia	Staniškis (2012)	1
Romania	Deselnicu et al. (2014), Lakatos et al. (2018)	2
Total		25

Note: There are no duplicate articles in the overall number

Table 3). About 86% of the articles were about Asia, with China (5 articles) and India (6 articles) coming in first and third, respectively (3 articles). China and India are well known due to their substantial populations and quick economic development. Table 4 demonstrates that out of 90 papers, 25—24 of which focused on Europe—addressed SCP in developed economies. Finland had the most (7 articles), followed by the UK (5 articles), Germany (3 articles), and Spain (3 articles).

Table 1 shows 46 publications covering SCP behaviors beyond national borders in a more general economic framework. Brizga et al. (2014) examined the problems and advancements of SCP in post-Soviet nations. SCP issues in European countries are covered in eight articles, with particu-

Table 4 Sorting the techniques/instruments used according to various economies

Methodology	Developed economies	Developing economies	General research	Count
PPIS review and analysis	Dewick and Foster (2018), Liobikienė and Dagiliūtė (2016), Kielin-Maziarz (2013), Lorenz and Veenhoff (2013), Staniškis et al. (2012), Berg (2011), Berg and Hukkinen (2011a), Tukker et al. (2008), Barber Honkasalo (2011), Nash (2009), Kovács (2011), Niinimäki and Hassi (2011), Clay et al. (2007), Seyfang (2004)	Adham et al. (2015), Schroeder (2014), Zhao et al. (2008), Zhao and Patrick (2010), Soyhan (2009)	Zisopoulos et al. (2017), Lebel and Lorek (2008), Geels et al. (2015), Fedrigo and Hontelez (2010), Akenji and Bengtsson (2014), Brodhag (2010), de Haen and Réquillart (2014), Stevens (2010), Kuhndt et al. (2008), Clark (2007), Barber (2003)	15/5/11 total 31
Empirical study/ expert theoretical review/survey	Govindan (2018), Petry et al. (2011), Lakatos et al. (2018), Dendler (2014), Staniškis (2012), Partidário et al. (2007), Yakovleva and Flynn (2004), Deselnicu et al. (2014), Berg and Hukkinen (2011b), Lehtoranta et al. (2011), Schönhart et al. (2009)	Thongplew et al. (2017), Ely et al. (2016), Moreno-Peñaranda et al. (2015), Vergragt et al. (2016), Yakovleva and Flynn (2004), Tseng et al. (2013), Bilen et al. (2008), Mertz et al. (2010)	Pialot et al. (2017), Vinkhuyzen and Karlsson-Vinkhuyzen (2014), Notarnicola et al. (2017), Gandenberger et al. (2011), Sala et al. (2017), Spangenberg et al. (2010), Haake and Jolivet (2001), Maxwell and Sheate (2006), Church and Lorek (2007), Maxwell et al. (2006), Mulder (1998)	11/8/11 total 30
Mathematical modeling and MCDM	Risku-Norja and Mäenpää (2007), Azapagic et al. (2016)	Luthra et al. (2017), Luthra et al. (2016), Chiou et al. (2013) Mangla et al. (2017)	Ülkü and Hsuan (2017), Bai et al. (2018), Zhu et al. (2006)	2/4/3 total 9
Life cycle management and carbon/water foot printing-based methods	Cazcarro et al. (2016), Camillis and Goralczyk (2013), Hoff et al. (2014), De	Soni et al. (2016), Mungkung et al. (2012), Hoff et al. (2014), Liu et al. (2010)	Ridoutt and Pfister (2010), Parent et al. (2013)	3/4/2 total 8
Statistical data analysis	Grözinger et al. (2010)	Dubey et al. (2016), Brizga et al. (2014), Wong et al. (2016), Corral (2003)	Gilli et al. (2017), Cohen and Muñoz (2016)	1/4/2 total 7
Other tools	Dobes (2016), Burja (2009)		Jonkutė and Staniškis (2016), Schinkel and Spiegel (2017), Cohen (2010)	2/0/3 total 5
Total				34/25/32 total 90

Note: The number format in the final column is "a/b/c total d." The number of articles about developed economies is indicated by the letters a and b, respectively. The number of papers on the broad study, which does not concentrate on one nation or economy, is indicated by the c. The d means the total quantity of publications in each row

lar attention paid to policy briefs, implementation, action plans, routes, labeling programs, and green buying. Three articles examine Asian trends, policy frameworks, and sustainable green designs. Initiatives for SCP promotion in North America were reviewed by Barber (2007). Peeters and Deketelaere's (2006) findings are echoed in the remaining 33 articles, which provide broad SCP research. European scholars co-authored 31 of these papers, highlighting Europe's critical role in encouraging SCP participation and initiatives.

3.3 Methods of Research and Strategies for Validation

Appropriate methods of synchronizing production and consumption objectives are essential to achieving sustainable development (Stevens 2010). The research articles in this section are arranged according to the approaches utilized in developed and emerging economies for sustainable consumption and production (SCP). Section 4.3 presents a comprehensive comparison based on findings, presenting six different methodological categories (Table 5).

Approximately 34.4% of articles in the examined set employed a review and analysis technique to explore various public-private innovation systems (PPISs) developed and evaluated by diverse stakeholders. These systems aim to foster a more cohesive approach to improving sustainability performance. Primarily, these articles targeted policymakers, governments, and regulatory bodies. Empirical studies, expert theoretical reviews, and surveys constituted the second most utilized methodological approach in the sustainable consumption and production (SCP) domain, accounting for 33.3%. Nine out of 90 papers attempted to address SCP research gaps through mathematical modeling and multi-criteria decision-making. Additionally, methodologies such as life cycle management and carbon/water footprinting comprised around 8.89% of approaches within SCP research. Statistical data analysis was the least employed methodol-

Table 5 SCP areas of concentration revealed in the top-ranked developed and developing nations

Focus area	China, India, and Turkey	UK, Finland, and Germany
Materials and products	Luthra et al. (2017), Mangla et al. (2017), Luthra et al. (2016)	Honkasalo (2011), Niinimäki and Hassi (2011)
Energy and industrial ecology	Soyhan (2009), Bilen et al. (2008), Soni et al. (2016), Yılmaz and Uslu (2007)	Azapagic et al. (2016), Lehtoranta et al. (2011)
Leadership	Dubey et al. (2016), Schroeder (2014)	Berg (2011), Seyfang (2004), Berg and Hukkinen (2011a), Berg and Hukkinen (2011b)
Agri-food	Ely et al. (2016), Hoff et al. (2014)	Dewick and Foster (2018), Risku-Norja and Mäenpää (2007), Hoff et al. (2014), Lorenz and Veenhoff (2013)
Housing and lifestyle/workstyle	Vergragt et al. (2016), Liu et al. (2010)	Grözinger et al. (2010)

ogy, representing approximately 7.8% of research activities, likely due to the focus on PPISs, empirical studies, expert reviews, and surveys.

Furthermore, the validation approaches of these methodologies across different economies were scrutinized. The analysis aimed to shed light on how these methodological tools were validated. Results revealed three main types of validation approaches. Over 54% of papers focused on theoretical grounds, defining theoretical gaps and empirical claims. Around 37.8% utilized case studies to demonstrate the applicability of proposed models or approaches. While numerical test beds were employed in some instances, they were deemed insufficient for validation without practical application to real-world scenarios.

3.4 Arrangement According to Industry Sector

SCP literature primarily examines various industrial sectors, with articles classified by validated approaches in specific industries, illuminating sectorial influences on SCP across economies. Utilizing the North American Industry Classification System (NAICS), the focus mainly spans the agricultural and manufacturing sectors. Notably, 48.9% of articles are in general industries, while 31.1% concentrate on manufacturing, attributed to production operations' intrinsic sustainability relevance and historical regulatory scrutiny. The production of food is the focus of manufacturing. With 16.7% of articles, agriculture also attracts interest. Studies by Adham et al. (2015), Brizga et al. (2014), Cohen and Muñoz (2016), Dobes, Liu et al. (2016), Tukker et al. (2010), and Zhao et al. (2008) demonstrate how many academics investigate SCP across industries, encouraging external verification and expansion of the results.

4 Discussion of Research Findings

This section compares findings from Sect. 3, addressing RQs 1, 2, and 3. Sections 4.1 and 4.2 analyze content from 3.2. Sections 4.3 and 4.4 build on 3.3 and 3.4.

4.1 Key Distinctions Between Developed and Developing Economies' SCPs

SCP systems include both "bottom-up" initiatives started by businesses and "top-down" measures for sustainability spearheaded by legislators (Akenji and Bengtsson 2014; Tseng et al. 2013). The "bottom-up" strategy incorporates SCP pledges into corporate operations, whereas the "top-down" approach entails government economic actions. The emphasis on SCP practices changes with the state of the economy. Due to previous disregard for sustainability, economic expansion takes precedence in developing economies. Here, the focus is primarily on "bottom-up" initiatives, with little government attention towards SCP.

Conversely, developed economies engage in "top-down" and "bottom-up" approaches, implementing both supply and demand interventions. Despite economic constraints, developing economies prioritize economic development over environmental concerns, often adopting a "grow first, clean up later" mentality. However, increasing awareness of ecological pressures prompts efforts to balance economic growth with environmental protection. "Bottom-up" initiatives such as SCP strategies, adoption of cleaner technologies, and environmental product design are prevalent.

In contrast, developed economies focus on altering overconsumption patterns and reducing material and energy intensity through interventions on both supply and demand sides. National SCP programs encourage nationwide participation in achieving SCP goals. Despite these efforts, challenges persist, particularly in SCP governance functionality. Developed economies grapple with the environmental impacts of established production technologies while addressing overconsumption and inefficiencies. Achieving SCP necessitates a multifaceted approach, blending "top-down" policies with "bottom-up" initiatives for holistic sustainability.

4.2 The Most Dynamic Developed and Emerging Nations' SCP Emphasis Areas

The results of Sect. 3.2 are further explored in this section, which also looks at the particular focal areas of SCP practices in the most active developed or emerging economies. Finland, the UK, and Germany are the top three developed economies, whereas China, India, and Turkey are the most active emerging nations in SCP. Previous studies by Tukker et al. (2008) showed that 70–80% of attention and effort was directed towards the most critical aspects of SCP: food, housing, energy use, and mobility. The five focal areas that were determined through study in these six nations are listed in Table 6. Although the results support Tukker's analysis, other recent studies have looked into leadership, products, and materials in SCP.

Prioritizing food, housing, and transportation, SCP in Finland expanded to include goods and materials (Honkasalo 2011). Although Finnish consumers are free to select how they want to live, they are also aware of how their decisions may affect the environment (Niinimäki and Hassi 2011). On the other hand, India is concentrating on determining the factors that encourage and hinder the adoption of SCP, especially in the plastic and automotive industries (Luthra et al. 2016, 2017; Mangla et al. 2017). Increased household consumption in China causes a rebound effect that raises CO2 emissions, while technology developments reduce energy and material intensity (Liu et al. 2010; Staniškis 2012). A notable deficiency in emerging economies' literature from China is the absence of global SCP discourse (Schroeder 2014).

To overcome organizational barriers to SCP in India, senior management's function as internal agents must synchronize with external pressures, such as governmental and community impacts (Dubey et al. 2016). However, more enforcement of governmental SCP rules is required, which impedes development (Dubey et al. 2016; Schroeder 2014). While developed economies like the UK and Finland have vital SCP programs on paper, government commitment to these initiatives must be followed through (Berg 2011). Therefore, even though SCP development in wealthy and developing economies differs, both require more outstanding government commitment than current practices.

4.3 Methods and Instruments Used in the SCP Study

Policies, programs, and institutional supports (PPISs) are a study approach used in around 35% of published literature. This highlights the critical role that government regulations, strategic plans, and programs play in promoting sustainable consumption and production systems. Studies reveal that PPISs for sustainable consumption and production (SCP) activities fluctuate throughout nations and economic zones because of variations in financial circumstances, cultural aspects, and demographics (Adham et al. 2015; Liobikienė and Dagiliūtė 2016; Schroeder 2014). Unfortunately, established economies—especially the UK and Finland—focus most of the research on SCP practices, with emerging economies receiving less attention than developed economies. This tilt, which reflects basic variations in SCP between economic contexts, can be linked to more research and industrial investment in developed nations (Sect. 4.1).

Even with a wealth of research on the political and theoretical aspects of encouraging SCP adoption, initiatives and laws frequently place only loose requirements on corporate and industrial groups. Quantitative modeling of consumption and production activities is necessary to transition to new sustainability regulations and governmental policies. This should motivate the adoption of more quantitative methodologies after legislation is passed (de Haen and Réquillart 2014). Nevertheless, it is clear from Sects. 3.3 and 4.2 that not enough thought has been given to quantitative approaches to evaluate how policy and strategy reforms may affect SCP practices. This is unfortunate because quantitative methods are essential for making well-informed decisions in business settings (Tseng et al. 2018).

Additionally, while scholars propose guidelines and frameworks to enhance SCP implementation policies, adequate governmental and management support impedes the adoption of these advancements, leading to sustainable consumption patterns and efficient production practices (Luthra et al. 2016). Notably, PPISs seldom integrate sustainability with industrial practices, educational institutions, national governments, local authorities, and the public/consumers in developing economies, where SCP remains a secondary concern despite early introduction. Pilot studies conducted in some developing economies over recent years have facilitated the collection of valuable experiences and lessons, informing the establishment of national regulations and standards to promote SCP across new industrial sectors and regions (Schroeder 2014).

Efficiently managing scarce resources with sociocultural considerations in production is the primary aim of sustainable output, while sustainable consumption aims to heighten consumer awareness for eco-friendly purchasing. The inter-

Table 6 SCP PPIS focuses on developed and developing economies

Type	References	Focus	Developed economy	Developing economy
Policy	Liobikienė and Dagiliūtė (2016)	EU's adoption of SCP strategy	√	
	de Haen and Réquillart (2014)	SCP policies for food systems	General research	
	Schroeder (2014)	China's SCP system of government		√
	Honkasalo (2011)	Whole SCP policies	√	
	Brodhag (2010)	Social responsibility & sustainable tourism policies under SCP	General research	
	Zhao and Patrick (2010)	Policies for the sustainable evolution of SCP patterns		√
	Nash (2009)	Action plan for sustainable industrial policy & SCP	√	
	Soyhan (2009)	Turkey's energy strategy		√
	Zhao et al. (2008)	Asia-Pacific SCP regional guidelines for sustainable urban development		√
Program	Kielin-Maziarz (2013)	EU SCP plan of action	√	
	Staniškis et al. (2012)	SCP operations in Lithuania		√
	Berg (2011)	The SCP programs in Finland, Sweden, & the UK	√	
	Berg and Hukkinen (2011a)	The SCP promotion initiative in Finland	√	
	Tukker et al. (2008)	The SCP programs' ten-year structure for nations with developed economies	√	
	Clay et al. (2007)	Sustainability Victoria's initiative to lessen its influence on the environment	√	
	Clark (2007)	Activities of UNEP SCP		√
	Barber (2003)	Global conference on the implementation of the sustainable development plan	General research	
Initiatives	Zakari et al. (2022)	Ensuring Africa has sustainable production and consumption patterns	√	
	Zakari, A. Adham et al. (2015)	Malaysian efforts concerning SCP		√
	Akenji and Bengtsson (2014)	SCP in the objectives for sustainable development	General research	
	Barber (2007)	Initiatives for SCP in the USA & Canada	√	
Strategy	Singh et al. (2023)	The contribution of energy efficiency, green innovation, and financial inclusion to environmental performance	√	
	Singh, A. Dewick, and Foster (2018)	Strategies for procurement & interactive relationships between buyers and suppliers	√	
	Zisopoulos et al. (2017)	An energy-based conceptual framework for an agri-food industry that is resource-efficient	√	
	Geels et al. (2015)	Strategic analytical stances in SCP debates	General research	
	Lorenz and Veenhoff (2013)	Techniques for Germany's sustainable food production and consumption	√	
	Niinimäki and Hassi (2011)	Strategies for designing products in specialized markets	√	
	Kovács (2011)	Sustainable food production and consumption	√	
	Fedrigo and Hontelez (2010)	European SCP blueprint	√	
	Stevens (2010)	Using sustainability methods to address systemic and market shortcomings	General research	
	Lebel and Lorek (2008)	SCP frameworks in projects for fair trade	General research	
	Kuhndt et al. (2008)	Global endeavors about SCP	General research	
	Seyfang (2004)	The British approach to SCP	√	

play between these facets necessitates exploration, often through empirical and theoretical analysis, including moral leadership theories. Developing dynamic models like system dynamics simulation is crucial for comprehending long-term sustainable consumption and production behaviors.

4.4 Trends in SCP Use Across All Industrial Domains

Table A3 reveals that SCP practices are more frequently addressed in the manufacturing and agriculture industries.

Scholars from developing economies focus narrowly on SCP within their countries, seeking local solutions. In contrast, those from developed economies take a broader approach, examining domestic and international SCP issues. Notably, manufacturing industries in developing economies garner significant attention due to their role in driving economic growth, particularly in countries like China and India, where industrialization is increasing, albeit accompanied by environmental and social challenges stemming from manufacturing waste.

As highlighted by Govindan (2018), population growth drives increased food production needs alongside higher consumption levels, leading to more incredible food waste in both developed and developing economies. In developed economies, attention has shifted towards sustainable food consumption and production in the agri-food and service industries. However, there needs to be more focus on sustainable food practices despite food abundance. Developing indicators for sustainable agri-food practices is crucial in these economies.

Addressing the challenge of reducing food miles and minimizing food loss and waste in the agri-food industry necessitates logistical adjustments in the global food transportation network, a task developed economies are better equipped to undertake due to technological advancements. However, these advantages serve as barriers in developing economies.

Furthermore, the transition from manufacturing to service-oriented economies has begun in recent decades. SCP practices in service industries are still developing, with limited literature discussing SCP developments in this sector. Exploring and developing business models to promote SCP in service industries is necessary.

5 Opportunities for Research in the SCP Sector

Sections 3 and 4 compare and discuss potential future directions. SCP opportunities vary with economies and sociocultural factors, guiding efficient interventions.

5.1 Possibilities for SCP in Emerging Markets

Ninety articles highlight Europe's leadership in sustainable consumption and production (SCP). However, SCP's complexity and gradual nature raise doubts about replicating European practices in developing economies like China and India. Proposed frameworks demand intricate accounting, robust monitoring, governmental enforcement, and corporate responsibility. Corruption, income inequality, and poverty are challenging for many developing nations. Developed economies receive more governmental support for SCP research, but understanding barriers and drivers in developing countries is crucial. Further exploration is needed to gauge their adoption and adaptation of SCP frameworks.

5.2 Possibilities in Developed Economies for SCP

SCP practices in developed economies have reached maturity compared to their counterparts in developing nations. Strategic research on product-process-system integration (PPIS) has garnered attention, driven by the proliferation of national and international PPIS initiatives guiding SCP implementation. Evaluating the environmental and economic impacts of these initiatives is crucial. In Ireland, the Sustainability Evaluation Metric for Policy Recommendation (SEMPRe) project aims to quantitatively assess sustainability policies using sustainable development (SD) indices. Enhancing sustainability scoring with multi-criteria decision-making (MCDM) methods is proposed. Leveraging data technologies in developed economies can further research opportunities. Integrated SCP sustainability indicators should span industries like agriculture, transportation, plastics, and electronics, benefiting from MCDM, fuzzy inference modeling, and life cycle analysis. Stochastic and multi-objective approaches are vital for addressing demand, product availability, and consumer behavior uncertainties.

5.3 Possibilities in Developing as well as Developed Economies for SCP

Trade-off analysis among various stakeholders and target audiences in sustainable consumption and production (SCP) practices reveals intricate dynamics shaping the sustainability landscape. Stakeholders encompass a diverse range, including academics, government bodies, and business entities, each playing a pivotal role in fostering SCP initiatives. Governmental intervention often involves implementing stringent regulations and policies to enforce SCP standards, while businesses strive to integrate green technologies and sustainable strategies into their operations. Concurrently, academics focus on educating consumers about responsible purchasing practices (Schinkel and Spiegel 2017).

Beyond stakeholder identification lies the imperative of understanding their behavioral inclinations, a facet crucial for innovation in sustainability across developed and developing economies. This necessitates acknowledging the potential trade offs wherein benefits for one stakeholder may come at the expense of another. Transitioning towards SCP demands a nuanced analysis of the complex interplay among stakeholders, necessitating dynamic methodologies like system dynamics and agent-based models to delineate feedback mechanisms and long-term sustainable behaviors (Tseng et al. 2018).

Similarly, sustainable behavior analysis of consumers and producers underscores the need for holistic product life cycle management approaches. Governments must enforce the integration of production and consumption practices to dissuade consumers from solely seeking low-cost products, thus aligning consumption patterns with sustainability goals (Scherer and Verburg 2017). Initiatives such as eco-labeling and educational campaigns can enhance consumer awareness and prompt shifts towards SCP behaviors. Consequently, there is a pressing need for scholarly and governmental scrutiny to gauge the efficacy of such initiatives in influencing production behaviors, thereby necessitating the development of multi-agent systems embedded with analytical models for comprehensive analysis (Ghadimi and Heavey 2013).

Moreover, while large corporations dominate SCP efforts, small and medium enterprises (SMEs) constitute a significant yet overlooked segment. Despite SMEs' substantial ecological and social footprint, they receive scant attention in sustainability discourse (Jenkins 2006). Given their prevalence in developed and developing economies, it is imperative to empower SMEs with sustainable practices. Robust methodologies and business models must be devised to integrate SCP considerations into SME operations, thereby leveraging their collective impact to drive sustainable patterns on a broader scale (Garetti and Taisch 2012).

6 Conclusion and Limitations

Sustainable consumption and production (SCP) are pivotal for advancing sustainable development, a cornerstone underscored by the UN's Sustainable Development Goal (SDG) #12. This ethos has garnered significant attention across both developed and developing nations. Discussions on SCP dynamics vary across economies, influenced by cultural diversity, economic growth stages, and political frameworks. Developing countries often need help with resource inadequacies to fulfill basic needs, while developed ones contend with entrenched production technologies hindering significant environmental shifts. Such disparities necessitate deeper exploration, yet comprehensive research in this domain needs to be more extensive.

This study embarks on a thorough comparative analysis, drawing insights from a systematic literature review, to delineate SCP considerations between developed and developing economies. It delineates fundamental disparities and examines prominent actors in SCP practices across these economic categories. The analysis furnishes nuanced perspectives, highlighting areas of convergence and divergence essential for informed policy formulation and strategic planning.

Nevertheless, this study acknowledges limitations. Its reliance on peer-reviewed literature may cause SCP initiatives to be noticed in countries more inclined towards academic dissemination. For example, lauded for its sustainability prowess, Switzerland lacks peer-reviewed publications elucidating its SCP strategies. Such gaps underscore the need for diverse data sources to understand global SCP endeavors comprehensively.

In conclusion, while this study offers valuable insights, it underscores the imperative for broader data acquisition methodologies to capture the full spectrum of SCP initiatives worldwide. By amalgamating rigorous research with real-world experiences, it aspires to steer future SCP endeavors towards more equitable and sustainable trajectories in developed and developing economies.

References

Adham KN, Siwar C, Bhuiyan MAH (2015) An overview of Malaysian government initiatives on sustainable consumption and production practices. Int J Sustain Dev 9(6):23–32

Akenji L, Bengtsson M (2014) They are making sustainable consumption and production the core of sustainable development goals. Sustain For 6(2):513–529

Azapagic A, Stamford L, Youds L, Barteczko-Hibbert C (2016) Towards sustainable production and consumption: A novel decision-support framework integrating economic, environmental and social sustainability (DESIRES). Comput Chem Eng 91:93–103

Bai C, Shah P, Zhu Q, Sarkis J (2018) Green product deletion decisions: an integrated sustainable production and consumption approach. Ind Manag Data Syst 118(2):349–389

Barber J (2003) Production, consumption and the world summit on sustainable development. Environ Dev Sustain 5(1):63–93

Barber J (2007) Mapping the movement to achieve sustainable production and consumption in North America. J Clean Prod 15(6):499–512

Berg A (2011) Not roadmaps but toolboxes: analysing pioneering national sustainable consumption and production programs. J Consum Policy 34(1):9–23

Berg A, Hukkinen JI (2011a) Beyond effectiveness: the uses of Finland's national program to promote sustainable consumption and production. J Clean Prod 19(16):1788–1797

Berg A, Hukkinen JI (2011b) The paradox of growth critique: narrative analysis of the Finnish sustainable consumption and production debate. Ecol Econ 72:151–160

Bilen K, Ozyurt O, Bakırcı K, Karslı S, Erdogan S, Yılmaz M, Comaklı O (2008) Energy production, consumption, and environmental pollution for sustainable development: A case study in Turkey. Renew Sust Energ Rev 12(6):1529–1561

Brizga J, Mishchuk Z, Golubovska-Onisimova A (2014) Sustainable consumption and production governance in countries in transition. J Clean Prod 63:45–53

Brodhag C (2010) A differentiated approach for sustainable consumption and production policies. Nat Res Forum 34(1):63–70

Burja A (2009) Using Green Public Procurement (GPP) for sustainable consumption and production. J Eur Environ Plan Law 6(3):319–338

Caeiro S, Ramos TB, Huisingh D (2012) Procedures and criteria to develop and evaluate household sustainable consumption indicators. J Clean Prod 27:72–91

Cazcarro I, Duarte R, Sánchez-Chóliz J (2016) Downscaling the grey water footprints of production and consumption. J Clean Prod 132:171–183
Chiou Y-C, Lan LW, Chang K-L (2013) Sustainable consumption, production, and infrastructure construction for operating and planning intercity passenger transport systems. J Clean Prod 40:13–21
Church C, Lorek S (2007) Linking policy and practice in sustainable production and consumption: an assessment of the role of NGOs. Int J Innov Sustain Dev 2(2):230–240
Clark G (2007) Evolution of the global sustainable consumption and production policy and the United Nations Environment Programme's (UNEP) supporting activities. J Clean Prod 15(6):492–498
Clay S, Gibson D, Ward J (2007) Sustainability Victoria: influencing resource use towards zero waste and sustainable production and consumption. J Clean Prod 15(8):782–786
Cohen B (2010) A guidance framework for mainstreaming resource efficiency and sustainable consumption and production in a developing country. Environ Dev Sustain 12(6):1051–1068
Cohen B, Muñoz P (2016) Sharing cities and sustainable consumption and production: Towards an integrated framework. J Clean Prod 134:87–97
Corral CM (2003) Sustainable production and consumption systems—cooperation for change: assessing and simulating the willingness of the firm to adopt/develop cleaner technologies. The case of the In-Bond industry in northern Mexico. J Clean Prod 11(4):411–426
Cronbach LJ (1951) Coefficient alpha and the internal structure of tests. Psychometrika 16(3):297–334
De Camillis C, Goralczyk M (2013) Towards more robust measures for sustainable consumption and production policies: proposal of a new fiscal framework based on a life cycle approach. Int J Life Cycle Assess 18(1):263–272
de Haen H, Réquillart V (2014) Linkages between sustainable consumption and sustainable production: some suggestions for foresight work. Food Secur 6(1):87–100
Dendler L (2014) Sustainability Meta Labelling: an effective measure to facilitate more sustainable consumption and production? J Clean Prod 63:74–83
Deselnicu DC, Vasilescu AM, Purcarea AA, Militaru G (2014) Sustainable consumption and production in the footwear sector. Leath Footwear J 14(3):159–180
Dewick P, Foster C (2018) Focal organisations and eco-innovation in consumption and production systems. Ecol Econ 143:161–169
Dobes V (2016) An integrated diagnostic tool for identifying and quantifying the opportunities for sustainable consumption and production in industry. J Clean Prod 135:567–576
Dubey R, Gunasekaran A, Childe SJ, Papadopoulos T, Wamba SF, Song M (2016) Towards a theory of sustainable consumption and production: Constructs and measurement. Resour Conserv Recycl 106:78–89
Ely A, Geall S, Song Y (2016) Sustainable maize production and consumption in China: practices and politics in transition. J Clean Prod 134:259–268
Fedrigo D, Hontelez J (2010) Sustainable consumption and production. J Ind Ecol 14(1):10–12
Gandenberger C, Garrelts H, Wehlau D (2011) Assessing the effects of certification networks on sustainable production and consumption: The cases of FLO and FSC. J Consum Policy 34(1):107–126
Garetti M, Taisch M (2012) Sustainable manufacturing: trends and research challenges. Prod Plan Control 23(2-3):83–104
Geels FW, McMeekin A, Mylan J, Southerton D (2015) A critical appraisal of Sustainable Consumption and Production research: the reformist, revolutionary and reconfiguration positions. Glob Environ Chang 34:1–12
Ghadimi P, Heavey C (2013) A Review of agent-based modeling and simulation applications in supplier selection problem. In: Modelling, and Simulation (EUROSIM), 2013 8th EUROSIM Congress on. IEEE, pp 101–107
Gilli M, Marin G, Mazzanti M, Nicolli F (2017) Sustainable development and industrial development: manufacturing environmental performance, technology and consumption/production perspectives. J Environ Econ Policy 6(2):183–203
Govindan K (2018) Sustainable consumption and production in the food supply chain: A conceptual framework. Int J Prod Econ 195:419–431
Grabs J, Langen N, Maschkowski G, Schäpke N (2016) Understanding role models for change: a multilevel analysis of success factors of grassroots initiatives for sustainable consumption. J Clean Prod 134:98–111
Grözinger G, Matiaske W, Tobsch V (2010) Employee-friendly labor time: a key element to a sustainable pattern of production and consumption. Int J Public Policy 5(4):357–372
Haake J, Jolivet P (2001) Some reflections on the link between production and consumption for sustainable development. Int J Sustain Dev 4(1):22–32
Hoff H, Döll P, Fader M, Gerten D, Hauser S, Siebert S (2014) Water footprints of cities indicators for sustainable consumption and production. Hydrol Earth Syst Sci 18:213–226
Honkasalo A (2011) Perspectives on Finland's sustainable consumption and production policy. J Clean Prod 19(16):1901–1905
Hsieh H-F, Shannon SE (2005) Three approaches to qualitative content analysis. Qual Health Res 15(9):1277–1288
Jenkins H (2006) Small business champions for corporate social responsibility. J Bus Ethics 67(3):241–256
Jonkutė G, Staniškis JK (2016) Realising sustainable consumption and production in companies: the SUstainable and Responsible Company (SURESCOM) model. J Clean Prod 138:170–180
Kielin-Maziarz J (2013) Problems and challenges in the EU's sustainable consumption and production action plan. Int J Public Law Policy 3(3):276–286
Kovács B (2011) Transition towards sustainable food consumption and production in a resource constrained world. EuroChoices 10(2):44–47
Kuhndt M, Tessema F, Herrndorf M (2008) Global value chain governance for resource efficiency building sustainable consumption and production bridges across the global sustainability divides. Environ Res Eng Manag 45(3)
Lakatos ES, Cioca L-I, Dan V, Ciomos AO, Crisan OA, Barsan G (2018) Studies and investigation about the attitude towards sustainable production, consumption and waste generation in line with circular economy in Romania. Sustain For 10(3):865
Lebel L, Lorek S (2008) Enabling sustainable production-consumption systems. Annu Rev Environ Resour 33(1):241–275
Lehtoranta S, Nissinen A, Mattila T, Melanen M (2011) Industrial symbiosis and the policy instruments of sustainable consumption and production. J Clean Prod 19(16):1865–1875
Liobikienė G, Dagiliūtė R (2016) The relationship between economic and carbon footprint changes in EU: the EU's sustainable consumption and production policy implementation achievements. Environ Sci Pol 61:204–211
Liu J, Wang R, Yang J, Shi Y (2010) The relationship between consumption and production systems and their implications for the sustainable development of China. Ecol Complex 7(2):212–216
Liu W, Oosterveer P, Spaargaren G (2016) Promoting sustainable consumption in China: a conceptual framework and research review. J Clean Prod 134:13–21
Lorenz U, Veenhoff S (2013) Integrated scenarios of sustainable food production and consumption in Germany. Sustain Sci Pract Policy 9(2):92
Lozano R, Ciliz N, Ramos T, Blok V, Caeiro S (2015) Bridges for a more sustainable future: joining Environmental Management for Sustainable Universities (EMSU) and the European Roundtable for

Sustainable Production and Consumption (ERSCP) conferences. J Clean Prod 106:1–2
Luthra S, Mangla SK, Xu L, Diabat A (2016) Using AHP to evaluate barriers in adopting sustainable consumption and production initiatives in a supply chain. Int J Prod Econ 181:342–349
Luthra S, Govindan K, Mangla SK (2017) Structural model for sustainable consumption and production adoption—A grey-DEMATEL based approach. Resour Conserv Recycl 125:198–207
Mangla SK, Govindan K, Luthra S (2017) Prioritizing the barriers to sustainable consumption and supply chain production trends using a fuzzy Analytical Hierarchy Process. J Clean Prod 151:509–525
Maxwell D, Sheate W (2006) Enabling sustainable development through sustainable consumption and production. Int J Environ Sustain Dev 5(3):221–239
Maxwell D, Sheate W, Van RDV (2006) Functional and systems aspects of the sustainable product and service development approach for industry. J Clean Prod 14(17):1466–1479
Mayring P (2004) Qualitative content analysis. In: Flick U, von Kardoff E, Steinke I (eds) A companion to qualitative research. SAGE, Glasgow, pp 266–269
Meadows DH, Meadows DL, Randers J, Behrens WW (1972) The limits to growth. N Y 102:27
Mertz O, Bruun TB, Fog B, Rasmussen K, Agergaard J (2010) Sustainable land use in Tikopia: food production and consumption in an isolated agricultural system. Singap J Trop Geogr 31(1):10–26
Moreno-Peñaranda R, Gasparatos A, Stromberg P, Suwa A, Pandyaswargo AH, Puppim de Oliveira JA (2015) Sustainable palm oil production and consumption in Indonesia: what can stakeholder perceptions offer to the debate? Sustain Prod Consump 4:16–35
Mulder KF (1998) Sustainable consumption and production of plastics? Technol Forecast Soc Change 58(1–2):105–124
Mungkung R, Gheewala SH, Kanyarushoki C, Hospido A, Werf HVD, Poovarodom N, Bonnet S, Aubin J, Moreira MT, Feijoo G (2012) Product carbon footprinting in Thailand: a step towards sustainable consumption and production? Environ Dev 3(1):100–108
Nash HA (2009) The European Commission's sustainable consumption, production, and industrial policy action plan. J Clean Prod 17(4):496–498
Niinimäki K, Hassi L (2011) Emerging design strategies in sustainable production and consumption of textiles and clothing. J Clean Prod 19(16):1876–1883
Notarnicola B, Sala S, Anton A, McLaren SJ, Saouter E, Sonesson U (2017) The role of life cycle assessment in supporting sustainable agri-food systems: a review of the challenges. J Clean Prod 140:399–409
Pallaro E, Subramanian N, Abdulrahman MD, Liu C (2015) Sustainable production and consumption in the automotive sector: integrated review framework and research directions. Sustain Prod Consump 4(2):47–61
Parent J, Cucuzzella C, Revéret J-P (2013) Revisiting the role of LCA and SLCA in the transition towards sustainable production and consumption. Int J Life Cycle Assess 18(9):1642–1652
Partidário PJ, Lambert J, Evans S (2007) Building more sustainable solutions in production-consumption systems: the case of food for people with reduced access. J Clean Prod 15(6):513–524
Peeters M, Deketelaere K (2006) EU climate change policy: the challenge of new regulatory initiatives. Edward Elgar Publishing
Petry RA, Fadeeva Z, Fadeeva O, Hasslöf H, Hellström Å, Hermans J, Mochizuki Y, Sonesson K (2011) Educating for sustainable production and consumption and sustainable livelihoods: learning from multi-stakeholder networks. Sustain Sci 6(1):83–96
Pialot O, Millet D, Bisiaux J (2017) "Upgradable PSS": Clarifying a new sustainable consumption/production concept based on upgradablility. J Clean Prod 141:538–550
Ridoutt BG, Pfister S (2010) A revised approach to water footprinting to make transparent the impacts of consumption and production on global freshwater scarcity. Glob Environ Chang 20(1):113–120
Risku-Norja H, Mäenpää I (2007) MFA model to assess economic and environmental consequences of food production and consumption. Ecol Econ 60(4):700–711
Roy V, Singh S (2017) Mapping the business focus in sustainable production and consumption literature: review and research framework. J Clean Prod 150:224–236
Sala S, McLaren SJ, Notarnicola B, Saouter E, Sonesson U (2017) In quest of reducing the environmental impacts of food production and consumption. J Clean Prod 140:387–398
Scherer L, Verburg PH (2017) Mapping and linking supply- and demand-side measures in climate-smart agriculture. A review. Agron Sustain Dev 37(6):66
Schinkel MP, Spiegel Y (2017) Can collusion promote sustainable consumption and production? Int J Ind Organ 53:371–398
Schönhart M, Penker M, Schmid E (2009) Sustainable local food production and consumption: challenges for implementation and research. Outlook Agric 38(2):175–182
Schroeder P (2014) Assessing effectiveness of governance approaches for sustainable consumption and production in China. J Clean Prod 63(2):64–73
Seiffert MEB, Loch C (2005) Systemic thinking in environmental management: support for sustainable development. J Clean Prod 13(12):1197–1202
Seuring S, Müller M (2008) From a literature review to a conceptual framework for sustainable supply chain management. J Clean Prod 16(15):1699–1710
Seyfang G (2004) Consuming values and contested cultures: a critical analysis of the UK strategy for sustainable consumption and production. Rev Soc Econ 62(3):323–338
Singh AK, Raza SA, Nakonieczny J, Shahzad U (2023) Role of financial inclusion, green innovation, and energy efficiency for environmental performance? Evidence from developed and emerging economies in the lens of sustainable development. Struct Chang Econ Dyn 64:213–224
Soni V, Singh SP, Banwet DK (2016) Sustainable coal consumption and energy production in India using life cycle costing and fundamental options analysis. Sustain Prod Consump 6:26–37
Soyhan HS (2009) Sustainable energy production and consumption in Turkey: a review. Renew Sust Energ Rev 13(6):1350–1360
Spaargaren G (2011) Theories of practices: Agency, technology, and culture: Exploring the relevance of practice theories for the governance of sustainable consumption practices in the new world-order. Glob Environ Chang 21(3):813–822
Spangenberg JH, Fuad-Luke A, Blincoe K (2010) Design for Sustainability (DfS): sustainable production and consumption interface. J Clean Prod 18(15):1485–1493
Staniškis JK (2012) Sustainable consumption and production: how to make it possible. Clean Technol Environ Policy 14(6):1015–1022
Staniškis JK, Arbačiauskas V, Varžinskas V (2012) Sustainable consumption and production as a system: experience in Lithuania. Clean Techn Environ Policy 14(6):1095–1105
Stevens C (2010) Linking sustainable consumption and production: the government role. Nat Res Forum 34(1):16–23
Thongplew N, Spaargaren G, van Koppen CK (2017) Companies in search of the green consumer: sustainable consumption and production strategies of companies and intermediary organizations in Thailand. NJAS-Wageningen J Life Sci 83:12–21
Tseng ML, Chiu ASF, Tan RR, Siriban-Manalang AB (2013) Sustainable consumption and production for Asia: sustainability through green design and practice. J Clean Prod 40(2):1–5
Tseng M-L, Chiu AS, Liang D (2018) Sustainable consumption and production in business decision-making models. Resour Conserv Recycl 128:118–121
Tukker A, Emmert S, Charter M, Vezzoli C, Sto E, Andersen MM, Geerken T, Tischner U, Lahlou S (2008) Fostering change to sustainable consumption and production: an evidence based view. J Clean Prod 16(11):1218–1225

Tukker A, Cohen MJ, Hubacek K, Mont O (2010) Sustainable consumption and production. Journal of Industrial Ecology 14(1):1–3.

Ülkü MA, Hsuan J (2017) Towards sustainable consumption and production: competitive pricing of modular products for green consumers. J Clean Prod 142:4230–4242

UN (2012) United Nations Country Classification. Retrieved May 26, 2018, from http://www.un.org/en/development/desa/policy/wesp/wesp_current/2012country_class.pdf

United Nations (2002) Plan of implementation of the World Summit on Sustainable Development. Johannesburg, South Africa: United Nations

Vergragt PJ, Dendler L, Jong MD, Matus K (2016) Transitions to sustainable consumption and production in cities. J Clean Prod 134:1–12

Vinkhuyzen OM, Karlsson-Vinkhuyzen SI (2014) The role of moral leadership for sustainable production and consumption. J Clean Prod 63(2):102–113

Wong WP, Soh KL, Chong CL (2016) Differentiated service consumption and low cost production: striking a balance for a sustainable competitive advantage in Malaysia. Int J Prod Econ 181:450–459

Yakovleva N, Flynn A (2004) Innovation and sustainability in the food system: a case of chicken production and consumption in the UK. J Environ Policy Plann 6(3-4):227–250

Yılmaz AO, Uslu T (2007) The role of coal in energy production—Consumption and sustainable development of Turkey. Energy Policy 35(2):1117–1128

Zakari A, Tawiah V, Khan I, Alvarado R, Li G (2022) Ensuring sustainable consumption and production pattern in Africa: Evidence from green energy perspectives. Energy Policy 169:113183

Zhao W, Patrick S (2010) Sustainable consumption and production: trends, challenges, and options for the Asia-Pacific region. Nat Res Forum 34(1):4–15

Zhao J, Liu H, Dong R (2008) Sustainable urban development: a policy framework for sustainable consumption and production. Int J Sustain Dev World Ecol 15(4):318–325

Zhu X, Van Wesenbeeck L, Van Ierland EC (2006) Impacts of novel protein foods on sustainable food production and consumption: lifestyle change and environmental policy. Environ Resour Econ 35(1):59–87

Zisopoulos FK, Overmars L, van der Goot AJ (2017) A conceptual exergy-based framework for assessing, monitoring, and designing a resource efficient agri-food sector. J Clean Prod 158:38–50

Advancements in Technological, Environmental, Economic, and Social Feasibility: A Comprehensive Review

Farah Naaz, Mohd. Farhan, and Mohd. Zeeshan

Abstract

Technological, social, economic, and environmental opportunities are powerful factors of a prosperous society where resources can be preserved and every person can flourish. Technological opportunities can be built on with collective cooperation. *Sustainable development* is also a wide and complex term that represents any kind of development that meets societal requirements while not being detrimental to future generations' ability to meet their needs. Sustainability is the capacity for something to continue to exist in the future.

Keywords

Technology · Innovation · Social · Sustainable · Futuristic approach

1 Introduction

Civilization today is subject to numerous interdependent factors that define human development because it brings together scientific, natural, economic, and social factors. Many scholars and analysts have focused on how managerial decisions on the operations, integration, performance, and resources of a business organization concern the sustainability of resource utilization as well as the advancement of sustainable development (Kautish et al. 2020). To solve global problems and ensure sustainable development, these civilizational and business-level interconnections need to be better understood. The current study explores the relationship between corporate governance mechanisms, social and environmental responsibility, and the efficiency of technologies implemented to achieve sustainable development goals, as advanced by Williams et al. (2019). The following is a summary of the connections between four dimensions (Table 13.1):

1. *Technological dimension:* Technology plays a pivotal role in inventions and development for improving human lives. Digitalization, technovation, biotech, artificial intelligence, and renewable energy are some of the latest technologies that have transformed industries and improved productivity all over the world, in addition to advancements in communication media. On the same note, the social, economic, and environmental aspects are also beneficially affected by the uptake of technology and its advancement.
2. *Environmental dimension:* The environment supports life on Earth and provides life's needed resources and services. People have continued to impose themselves on the environment through factors like industrialization, urbanization, and agricultural growth, putting pressure on the environment as it has continued to be subjected to pollution, environmental degradation, and climate change. Safeguarding the environment requires practices such as minimizing the use of scarce natural resources and developing more-efficient infrastructure, such as renewable energy technology.
3. *Economic dimension:* Main aspects of an economy are determined by advancements in technology and environmental concerns. Reducing poverty and raising people's standard of living requires employing growth through productivity improvements and invention. However, there are often shortcomings in mainline economic approaches and theories, whereby they do not adequately incorporate social justice concerns or consider external environmental impacts. This chapter argues that policies and decisions involving economies need to be influenced by social

F. Naaz (✉)
ICMR-National JALMA Institute for Leprosy and Other Mycobacterial Diseases, Tajganj, Agra, India

M. Farhan
Jamia Millia Islamia University, New Delhi, India

M. Zeeshan
Dr. Bhim Rao Ambedkar University, Agra, India

J. A. Parray et al. (eds.), *Environmental Landscape and Sustainable Biodiversity for Healthy Green Growth*, Sustainable Landscape Planning and Natural Resources Management, https://doi.org/10.1007/978-3-031-75405-0_13

Table 13.1 The sustainable development paradigm

Element	Criteria
Economic Sustainability	Growth
	Development
	Productivity
	Trickle Down
Social Sustainability	Equity
	Empowerment
	Accessibility
	Participation
	Sharing
	Cultural Identity
	Institutional Stability
Environmental Sustainability	Ecosystem Integrity
	Carrying Capacity
	Biodiversity

and environmental aspects to reorient the economy toward sustainable economics.

4. *Social dimension:* Social aspects such as justice, equity, and individual welfare are essential for the growth of a society and improving quality of life. Opportunities for stakeholder involvement in decision-making processes also have key roles in societal betterment through opportunities for work, housing, healthcare, and education, among other necessities. Advancements in technology can help influence positive social change because impoverished communities gain better access to resources that satisfy their basic needs. Although technology can improve various aspects of society, it can also exacerbate inequality and disrupt cultures. Technological advancements, for instance, can contribute to a country's gross domestic product by enhancing efficiency and unlocking opportunities for trade. But if technological progression is left uncontrolled, it can disrupt society and harm the environment. In the same way, we know that social stability and sustainable long-term economic development depend on the wise use of natural resources. Recourses are depleted, pollution occurs, and greenhouse gases are emitted through the course of most economic activities. These factors then negatively affect the health of individuals, their options for earning a living, and overall human welfare. However, once the research observed that humans are exerting forces on various environments and ecological systems in a negative manner, environmental studies shifted focus, and any harmful human actions came to be deemed as unsustainable. The world has shifted toward what has been referred to as sustainable development (Mebratu 1998). In addition, human behavior and technical preparedness can pose problems as they steer this progress. Technological advancements and environmental protection processes are socially constructed in terms of their direction, speed, and mechanisms, which are influenced by societal, cultural, and institutional values. Mainstreaming social justice in inclusive economic policies can protect and improve the environment and human quality of life if we choose an economic program conducive to both.

To ensure that these four components work in concert, a strategy focusing on social justice and protecting the environment during technological developmental side is needed. The proposed concepts for this chapter are informed by those mentioned in the 2030 Agenda for Sustainable Development of the United Nations, which help to unite all of the relevant aspects and maximize their overall feasibility. Such conditions can be realized only by acknowledging and utilizing the synergy between technological, environmental, economic, and social interactions, where communities can effectively manage complicated situations in society to ensure an integrated and sustainable environment for everyone.

2 Technological Advancements

2.1 Evolution of Key Technological Advancements

Over the past few decades, technical progress has impacted and altered various segments of human life. Universities are central to increasing innovation in science and technology (García-Vega and Vicente-Chirivella 2020). Although regional economies are fully capable of supporting universities' innovative scientific research, incorporating social justice in universities' scientific and technological outcomes is meaningful to the economic aspects that affect most of society. Doing so would present an opportunity to enhance the high-quality development of regional economies. The following are among the most significant technological developments.

(a) *Artificial intelligence (AI):* This is a branch of computer science that involves the design of AI applications, which are adapted to solve problems that are typically solved by humans, including learning, reasoning, and decision-making, among others. More specifically, enhancements have been made in robotics, machine learning, and natural language processing. Some of the fields where artificial intelligence is used include healthcare, where it is used to diagnose diseases; analytics; self-driving cars; and smart home devices, such as Amazon's Alexa.

(b) *Renewable energy:* There is increasing awareness of the sustainability of renewable energy technologies, and this has been greatly boosted by its development. Major renewable energy production processes include the use of hydroelectric power, biofuels, wind power, and solar cells. By using cleaner fuel sources and fossil fuel alternatives, these technologies assist in combating global warming by reducing greenhouse gas emissions.

(c) *Biotechnology:* This is a general class of technology where living organisms, biodegradable products, and their structures are employed to develop products or services. As a field of study and industry, it has directly improved precision farming, genetically modified foods, human health via gene-editing techniques, personalized medicines, and ecosystems (via techniques such as bioremediation).

2.2 Impact on Productivity, Efficiency, and Innovation

Almost every industry has been transformed by certain technological advancements, especially in terms of productivity, efficiency, and innovation.

(a) *Efficiency:* Through the improvement of operations, reduction of inefficiencies, and better utilization of resources, such innovations as automation, the Internet of Things, and innovative manufacturing processes have enhanced efficiency. Predictive maintenance algorithms are used by artificial intelligence to make energy use more efficient and reduce downtime.

(b) *Productivity:* Due to the use of various technological devices, particularly in social media, computers, and data processing, work has undergone a great transformation. Cloud computing, the mobilization of gadgets, and opportunities for working from home have made technological and knowledge-based work available from anywhere at any time.

(c) *Innovation:* Many argue that technologies give rise to other technologies and the latter to even more technologies; thus, innovations are continuously produced. Technological advancements in each sector, be it medical, energy, transport, or communication systems, are developed through open repositories, collaborative communities, and transdisciplinary science aiming to make unique innovations and share information.

2.3 Case Studies Illustrating Technological Feasibility

The following are three specific cases where innovation in technology has been used successfully:

(a) *Artificial intelligence in healthcare:* Watson of IBM helped oncologists make therapy decisions to improve patient outcomes and reduce medical mistakes.

(b) *Renewable energy deployment:* Pervasive technological advancement together with favorable policies has led to a global shift toward renewable energy sources, as seen through large-scale installed solar photovoltaics in such countries as China and Germany.

(c) *Biotechnology in agriculture:* Genetically modified (GM) crops have been accepted globally because they enhance production, reduce the use of pesticide, and enhance crop resistance to drought, salinity, and carcinogens. Some of such crops include pest-resistant cotton and broadleaf-tolerant soybeans.

2.4 Ethical Issues and Difficulties in Technology Adoption

However, several ethically challenging issues and controversies remain with regard to all the benefits that technology has brought about:

(a) *Data security and privacy:* This is especially the case given that a number of AI-driven technologies are coming into our lives and that AI-enabled models are prone to such challenges as algorithmic bias, data leakage, and data privacy violations. Anticipating the potential risks for business and society at large, they remain committed to fostering innovation while protecting people's rights and trust in data management.

(b) *Economic inequality and job displacement:* These technologies have the potential to disrupt labor markets, meaning that income disparity will widen and more people will lose their jobs. Employee training, the promotion of learning throughout employees' lifetimes, and creating fair opportunities for embracing technologies mitigate these impacts significantly.

(c) *Environmental impacts:* The production and recycling of batteries, electric devices, and other products also impact resource consumption and pollute the environment, though the usage of renewable sources of energy offers a cleaner form of power than do conventional energies. Much like the seven revised sustainability principles, circular-economy approaches are also needed as green innovations evolve to become technologies in their own right.

3 Evaluation of Strategies for Reducing Environmental Impacts

This section evaluates the measures that could be taken to minimize impacts on the environment.

3.1 Environmental Sustainability

Reducing impacts on the environment means measuring the feasibility of solutions that provide a way to reduce the hazardous effects of human activities on ecosystems/natural resources. Flamini in his estimates, according to the 1993 Noble prize winner Robert Solow, emphasized that the stock level of the economy has to be maintained to actualize prosperity. The case of sustainability highlights that natural resources and artificial intelligence are employed in production processes and commonly preserve a particular stock of natural resources. Andersen points out that renewable resources should be seen as "money in the bank"—here is where capital remains, and only revenue is drawn to spend it, so many people believe that they should give it to their descendants. However, future generations won't be able to enjoy a sustainable environment if limited resources have been depleted. According to Martinet in 2012, this scenario amounts to so-called "cake-eating." Environmental sustainability requires using four key strategies.

(a) *Pollution control:* Environmental protection means following procedures or steps to check and minimize detrimental effects on ecosystems, rules and other constraints, and advancements in eradicating pollutants in the air, water, and soil.
(b) *Waste management:* Environmental protection and the proper disposal of wastes must be encouraged and promoted in waste minimization, the proper management of waste, and recycling programs.
(c) *Resource efficiency:* We must advocate for resource efficiency, more specifically the efficient use of materials and energy, reductions in emissions and consumption, and waste minimization.
(d) *Sustainable transportation:* This strategy features some proposals for reducing greenhouse gas emissions and air pollution by using new and clean means of transportation technology, including electric cars and efficient transport systems, among others.

3.2 The Function of Sustainable Practices and Renewable Energy Technologies

Being effective in minimizing the effects of burning fossil fuels and reducing global warming, renewable energy, along with the implementation of sustainable practices, cannot be excluded when protecting the environment. The important factors of sustainable practices include the following:

(a) *Renewable energy sources:* Heating and electricity can contribute positively to the environment, especially as the efficiency of some common sources of energy, such as solar energy, wind energy, hydroelectricity, and geothermal energy, increases.
(b) *Energy efficiency:* Investments are needed in energy-efficient designs for new buildings and renovations to old buildings, energy-efficient manufacturing processes, energy-efficient technology and appliances, efficient transportation, and awareness campaigns that target energy consumers and aim to modify their behavior.
(c) *Sustainable agriculture:* Sustainable agriculture requires using such strategies as geo-agriculture, precision production to improve stewardship of the soil ecosystem, water conservation, the minimal use of chemicals, and minimizing disruption to habitats.
(d) *Green building designs:* Effective technology and efficiency help to build infrastructure that is resource efficient and has small impacts its surrounding environment.

3.3 Programs to Conserve Biodiversity and Restore Ecosystems

Some of the major projects that have been employed particularly to safeguard species and ecosystems and promote their flourishing are guided by biodiversity preservation and ecosystem restoration.

(a) *Protected areas:* Overseeing conservation parks, including national, nature, protected, and marine environment conservation areas for ecosystem services, supports biodiverse resources and the conservation of critical ecosystem spaces.
(b) *Habitat restoration:* Restoring damaged ecosystems by using ecosystem engineering, afforestation, and the rehabilitation of the habitats improve the function, diversity, and resilience of biotic and abiotic factors.
(c) *Species conservation:* Endangered species management, breeding, and artificial rearing activities and methods for restoring habitats aim to prevent the extinction of various species and to improve measures that restore balance to species populations and maintain biodensity.
(d) *Sustainable land-use planning:* To prevent habitat degradation, the loss of biotic diversity must factored into land use, population growth, and planning how best to sustainably use bioresources.

3.4 Policies Encouraging Environmental Sustainability

Legal provisions, incentives, and other measures can help to encourage actions that do not damage the environment or that will slow down the rate of degradation and can help to discourage environmentally detrimental activities. Several important policy can be enacted and followed to encourage environmental sustainability, which can be categorized into four policy domains.

(a) *Environmental regulations:* These regulations ensure compliance with and the enforcement of environmental laws, regulations, policies, and standards for the disposal of waste, pollution, land use, and the conservation of natural resources.

(b) *Economic instruments:* To make individuals and organizations pay for the degradation of the environment and for resource depletion, the development of policies that promote sustainability and that provide incentives for developing environmentally sound technology, economic instruments such carbon prices, pollution taxes, subsidies, and tradable permits are employed.

(c) *Conservation programs:* To increase the capacity of addressing environmental challenges and the ability of ecosystems to sustain themselves, state funding, state subsidies, and grants from various nongovernmental organizations can be used to fund initiatives in the form of conservation, restoration, and sustainable development.

(d) *International accords:* Supporting and ratifying international treaties and protocols, including the Ramsar Convention, the Convention on Biological Diversity, and the Paris Agreement, requires international cooperation, goal setting, and managing international environmental issues.

4 Economic Viability

4.1 Examining Models and Indicators of Economic Growth

Economic growth models and indicators are fundamental to measuring the status and efficiency of various economies. According to several studies in the literature (Al Khalifa 2018; Chien et al. 2021a; Danielle and Masilela 2020), a correlation exists between new entrants to technological development and creativity. As competitors incorporate novelty into their existing business models, power struggles result in new structures mainly through the creative destruction concept. Previous studies have also depicted economic growth as "accumulation by creation," which has established that business entities resort to the application of innovation more often than even to countering their competition (Dlalisa and Govender 2020; Hussain et al. 2019; Nelson and Mattes 2017). As highlighted by Hussain et al. (2020) and NgoNdjama et al. (2020), these two dimensions determine the economic aspects and, therefore, the nature and issues of the business. A recent analysis has demonstrated that R&D investment has some advantages and risks in young firms but is closely related to the United Nations' Sustainable Development Goals (SDGs) (Chien et al. 2021b; Koloba 2020). The following are the three essential elements of such analysis.

(a) *Gross domestic product (GDP):* Gross domestic product consists of the overall aggregate of the value of output, which includes both manufactured goods and services, from a country within a given period. Entailing the measurement of relative per capita income, it is widely used for analyzing the levels of economic development of different counties and is an essential indicator that depicts the pace of economic growth.

(b) *Economic growth models:* These models can be understood as paradigms for understanding the antecedent conditions and dynamics of economic growth, such as the accumulation of capital, technology development, or institutional factors. Some of the common models are the Solow growth model, the neoclassical growth model, and endogenous growth theory.

(c) *Sustainable growth:* The likely evaluative indicators to be used in assessing sustainability involve not only macroeconomic indicators but also environmental, social, and governmental indicators. According to its definition, *sustainable development* aims to improve the quality of life of the human population within a framework of minimizing negative environmental impacts.

4.2 Strategies for Economic Development Led by Innovation

Innovative GDP-growth strategies lay emphasis on creativity, new venture development, and technological advancement. When considering sustainable development from this angle, some people find the term contradictory because "development" is linked to ongoing expansion in business, economies, agriculture, and consumption. However, as this study postulates, organisms' growth can be continuous, though this is not practicable (Jackson 2009; Robinson 2004; Rubin 2012; Daly 1990). The following are among these strategies' essential components.

(a) *Research and development (R&D):* Companies spend funds on R&D to develop new technologies that unleash creativity, inventions, and innovations and that increase the number of outputs in various sectors. Research money comes from grants, tax credit schemes, subsidies, and key technology partnerships between governments and the private sector.

(b) *Knowledge-based economies:* Some countries shifted their economic paradigm from industrial manufacturing to a knowledge-based economy for its growth, high level of value addition, absorption capacity, and digitization.

(c) *Cluster development:* Organizations, entrepreneurs, universities, and industries require the creation of complex commercial ecosystems and key information variety, hence positioning themselves in innovation clusters or technology parks for knowledge spillover or knowledge creation and collaborative innovation.

4.3 Entrepreneurship, Job Creation, and Inclusive Economic Policies

This includes putting an economy back onto a growth track, creating employment opportunities, and putting into practice economic policies that are balanced and agreeable by people in the same region and in different regions. The essential factors here are as follows.

(a) *Supporting entrepreneurship:*
   I. Unlocking access to finance helps independencies, startups, and businesses.
   II. Creating opportunity for incubation and counseling for independencies, startups, and Small and medium size Enterprises (SMEs).
   III. Changing regulations to support independencies startups and SMEs.
(b) *Techniques for creating jobs:* Policies must promote the generation of employment, especially in sectors that we foresee will experience vibrant growth in the future, such as technology, health, farming, and renewable energy. Ensuring that all the various categories of people in various fields are included can be enhanced by targeting population groups such as minorities, women, and youths.
(c) *Inclusive economic policies:* Countermeasures are needed to overcome crises, more specifically countermeasures that address the equality concerns of affected citizens and industries that would benefit from economic policies. Policy actions can also help to deal with the financial sector, health and education, social security, and measures to rectify inequality.

4.4 Wealth Distribution and Social Justice Issues

Financial resources can be used for the betterment of different aspects of society, such as wealth distribution and social justice, which cannot be achieved without the following factors.

(a) *Wealth distribution:* The distribution of wealth and income aims to offset inequality, and policies must be formulated to increase opportunity and socioeconomic mobility, to effect economic change.
(b) *Poverty alleviation:* Various social protection programs and policies have aimed to elevate lower-income people's quality of life.
(c) *Access to opportunities:* Equal employment opportunities, consumer opportunities, and other categories of opportunities should be realized for people of color, women, and other minorities to allow every person to achieve their potential, thus stimulating the economy with a highly skilled population.

5 Social Feasibility

5.1 Analysis of Inclusion, Justice, and Equity

In providing equal opportunities and prospects for everyone in a country and to ensure that they have all their rights, social equity, justice, and inclusion are critical concepts. Important components of this analysis are as follows.

(a) *Equity:* Assessing where people seem to fit in the social hierarchy for opportunities, resources, and advantages helps to identify disparities in society based on income, gender, color, ethnicity, age, disability, or place of residence. This transition requires the development of specific management strategies and equitable policy aiming at nondiscrimination against any group to eradicate disparities.
(b) *Justice:* Social justice needs to be couched in the legal frameworks of institutions that govern them, like rules, laws, and regulatory policies as well as procedures concerning conflict resolution and the enforcement of laws. Due process, human rights, and justice are the major ways through which fairness and accountability in a society can be upheld.
(c) *Inclusion:* Inclusion first requires investigations into the extent to which oppressed and minority populations are included in governmental administration and policy, social institutions, and overall decision-making. Thus, spaces where those at the periphery of society are heard, promoting diversity, and actually factoring in the needs of minority groups are essential requirements which must not be ignored by global leaders when establishing and maintaining an equitable and inclusive society.

5.2 Access to Basic Services, Education, and Healthcare

Basic services can be summarized as social needs that are fundamental to people's welfare, social development, and economic success.

(a) *Healthcare access:* Evaluations must be made on the accessibility of citizens to receive medical treatment, therapy, or preventive healthcare as per their necessity at comparatively low costs and irrespective of their geographical status or economic stratum.
(b) *Education access:* Evaluations must be made on the barriers to and gaps in enrolment, persistence, and attainment by exploring education opportunities from elementary education to university education. Four key areas are identified as barriers to education, namely money, society, culture, and fairness, and students

deserve the opportunity to overcome all barriers and have equal learning conditions.

(c) *Basic services:* Evaluations must be made on the breakthroughs in products and services by analyzing the increased availability of affordable and improved health, housing, food security, water, hygiene, and social welfare for human standards of living. Consequently, what we have seen is that modifications to infrastructure, guarantees for service delivery, and recommendations on the introduction of appropriate welfarism and restricted needs-based services still need to be made.

5.3 Community Engagement, Empowerment, and Participation

Community engagement, empowerment, and participation make appropriate social relations, democracy, and active societies possible.

(a) *Community engagement:* Community engagement aims to increase the participation of stakeholders, local intuitions, civil society organizations, and local communities in the decision-making process pertaining to policies and programs. People are more supportive when they are involved in and take responsibility in defending their project, embracing an undertaking more warmly.

(b) *Community empowerment:* People and communities need support for them to be equipped with the knowledge, skills, resources, and opportunities needed for them to choose their own directions in life. Hegemonic institutional action—the idea of a leading agency, self-governance, and self-determination—is indispensable to breaking the cycle of marginalization and empowering underprivileged and marginalized populations.

(c) *Community participation:* Such participation involves combining and persistently endorsing a lack of full ownership over social and material resources and community participation in governmental bodies as well as in all other matters of interest for society. Communities must be constructed and organized without the exclusion of any element and need equal public access to information, and decision makers need to live near the communities that they have authority over.

5.4 Preservation of Cultural Diversity and History

People generally aim to perverse their social identity, live in democracy, and participate in their culture and history.

(a) *Diversity of cultures:* The social differences in color, language, beliefs, and preferences through which various communities identify themselves and live in today's complex global civilization must be acknowledged and embraced. Cultural diversity therefore calls for the protection of cultural rights, policy enshrining diversity, antidiscrimination programs, and cultural decolonialization.

(b) *Heritage preservation:* Culture, legacy, and traditions are kept or passed down with the aim of ensuring that upcoming generations have access to the knowledge, beliefs, skills, or material objects of preceding generations. Heritage management involves various activities across communities, such as the preservation of cultural and historical resources, the promotion of heritage that belongs to various peoples, and the practice of heritage education.

(c) *Intercultural dialogue:* Cultural intercommunication, intercourse, and relations between people of different cultures promotes an understanding of and tolerance to diverse cultural differences. Especially in multicultural societies, intercultural discussion proves to be essential for contributions from society and building a peaceful society.

6 Comprehensive Analysis and Its Significance to Sustainable Development

Sustainable development is a wide and complex term that represents any kind of development that meets societal requirements while not being detrimental to future generations' ability to meet their needs. Sustainability is the capacity for something to continue to exist in the future (Basiago 1999). In the literature on development, most scholars, researchers, and practitioners (Milne and Gray 2013; Tjarve and Zemīte 2016; Mensah and Enu-Kwesi 2018; Thomas 2015) understand it in the following ways in order to secure and sustain a healthy economic, ecological, and social structure for human development. In the effort toward sustainable development, including the economic, social, cultural, and physical needs of both present generations and future generations is critical; these needs also help us to better understand the numerous linkages and interactions between so many economic, ecological, and social factors. A critical analysis must guide policymaking, decision-making, and actions to promote sustainable outcomes.

(a) *Holistic awareness:* Relevant stakeholders can gain a lot of knowledge on the opportunities and challenges of sustainable development if they only study the issue. Thus, decisionmakers are able to determine the areas of synergy, win win situations, and potential tradeoffs in development objectives by taking into consideration their economic, environment, and social impacts. This provides a broad perspective to help eliminate tactful and limited measures that may exclude significant facets that are relevant to sustainability.

(b) *Integrated solutions:* For the effective and successful utilization of integrated solutions, solutions must target various sectors at once for sustainable development. Disentangling appropriate strategies and measures that positively harness synergy for social, environmental, and economic facets proves more manageable when properly examined. For instance, purchasing renewable energy technologies reduces carbon emissions and increases economic returns, income, and energy security. With such benefits in mind, those formulating policies can direct resources at ideas that bring the biggest total benefit.

(c) *Long-term view:* Sustainable development is a concept in which the supplies of resources for present and future generations are met. A comprehensive analysis provides full evaluation of the long-term ramifications of development endeavors. This includes the idea of sustainability, which might refer to the sustainability of communities, economic systems, and/or ecosystems. Assessing history and scenario planning also enable decision makers to anticipate and avoid future risks and otherwise-unknown situations.

(d) *Stakeholder engagement:* Societies are made up of multisector players who must play active roles in sustainable development, including governments, corporations, Civil society organizations (CSOs), and local communities. A qualitative assessment reveals relevant prospects and conflicts of interest, and it can be used to find common ground to communicate with the stakeholders. Such communication can enable decision makers to take every aspect into consideration whenever they devise sustainable development strategies.

(e) *Adaptive management:* Change, especially in learning and continuous improvement, is vital in development given that management processes are cyclical and dynamic. By providing feedback on the progress of different plans and the results of different actions taken, a thorough analysis enables the evaluation of the efficacy of and the management of the proposed actions after their implementation. In the long run, this cycle results in creating more-robust goals and reaching those goals because it encourages innovative thinking, risk taking, and learning from past experiences.

7 Multidisciplinary Synergy

Intersection of Technology, the Environment, Economies, and Society Technological, environmental, economic, and social factors intersect in the integration of various sectors.

(a) *Technological innovations:* Technological advancement has brought with it significant consequences on a range of factors, including the environment, economies, and social coherence. Renewable energy technologies, for instance, decrease carbon emissions, create business prospects, and extend consumer access to clean energy services.

(b) *Environmental sustainability:* Environmental issues affect human activities, the development of society, economic status, and technology. Environmental policies give impetus to the deployment of clean technology, innovatively revolutionize industries and improve the well-being of humans by purifying the atmosphere and water sources.

(c) *Economic viability:* Economic viability is a crucial socioeconomic factor that affects social support measures, environmental legislation, and the development of new technologies. Technology creates environmental approaches that reduce adverse effects on the surrounding environment, stimulates economic development, and creates jobs in growing fields like bio-based farming and green energy systems.

(d) *Social equity:* The effectiveness of environmental justice depends on conspicuous social aspects, such as using advanced technology, which has impacts on economic results. Minimizing environmental hazards, raising people's economic status, fairly distributing wealth, and providing equal opportunities to exploit technology are essential in determining social justice for vulnerable groups.

8 Integrated Methods and Synergistic Advancement

This kind of combining not only ensures that one method always feeds into another but also keeps the focus on the synergistic advancement of various integrated methods. Complex problems should be solved and development goals reached by integrating approaches. Such approaches would integrate technologies with the environment while factoring in social and economic aspects.

(a) *Systems thinking:* The systems-thinking approach to developing inventions and policy analyses interconnects social, economic, and environment factors into one framework; provides feedback, ratios, and tradeoffs; and balances the value of these factors.

(b) *Multistakeholder collaboration:* Such collaboration must include governments, businesses, civil society, and local people all taking part in consultative procedures leading up to decision-making. All these voices together can help consolidate knowledge or resources from every stakeholder or devise solutions that incorporate the views and needs of all these stakeholder, filling in gaps that each separately does not have.

(c) *Policy integration:* The systematic interconnection between policies and planning at the sectoral and government levels can integrate technological, environmental, economic, and social goals. This involves dismantling sectoral barriers to yield an overarching sectoral organizational framework through which sectors can be integrated to work in a coherent way.

(d) *Innovative ecosystems:* Innovative ecosystems develop settings for the promotion of knowledge sharing and applied and interactive professional, occupational, and academic collaborations between different professions, disciplines, and vocations, including practice, research, industry, and policy areas. There are programs supporting today's type of modern initiatives in the context of business, including those of innovation centers, incubators, and accelerators that can be the nuclei of interdisciplinarity shifts in perspective.

Collaborative Solutions for Handling Complex Global Difficulties Strategies, processes, and roles are involved in coping with multifaceted international challenges. Partnership approaches in this context relate to a process or a number of activities that involve everybody and need concerted efforts to tackle complex cross-boundary issues that affect several nations and people in different departments and, consequently, require cooperation for welfare.

(a) *Global partnerships:* Global governmental partnerships can be used to solve worldwide problems like pandemics, climate change, or species extinction. International relations and agreements are enable global undertakings (often backed by legal frameworks).

(b) *Implementing cross-sectoral initiatives:* These initiatives address various questions at the local, regional, and global levels and adopt an all-encompassing view of the technological, environmental, economic, and social aspects of interconnected problems. Other forms of context resilience plans include circular economies, smart cities, and ecofriendly agriculture.

(c) *Community-based approaches:* Local populations should be involved when determining what problems are as well as in decision-making processes. This is further supported by solutions that utilize local assets, comparative advantages, talents, and resources to meet a community's needs.

(d) *Capacity building:* Capacity building develops institutional capacity augmentation in terms of both technical and human resources and endowments to yield interdisciplinary innovation and integration. The primary goals of capacity-building interventions are to increase sharing capacity and to increase the capacity of partnerships, with an emphasis on enhancing competencies that are complex and multiple.

9 Future Aspects

In the future, biotechnology, IoT, and artificial intelligence (AI) are anticipated to redefine multiple industries and human lives. These are some of the critical technologies that will help to address the pressing challenges facing the world today in key sectors like healthcare. Thus, as research and development continues, we should prepare ourselves for innovations that will enhance productivity, efficiency, and creativity among several sectors to further enhance technological development and interconnectivity.

The desire for environmental friendliness also suggests that efforts to revive natural habitats can boost the circular economy and harness green energy, which will be among the most important of future endeavors. In the future, more efforts will focus on enhancing sustainable practices, reducing emissions, and promoting the conservation of species as societies come to understand how crucial solving climate change and protecting our current resources are. A green economy promises innovation, employment, and durable infrastructure that are capable of withstanding the drastic shifts in climate.

On the economic level, the element of viability can only be realized through advocating for socially integrated development standards and equitable wealth distribution. Approaches for the development of even more economic potential come from innovative environments and entrepreneurial culture and imply social equity and human-oriented principles. The level of new industries affecting people's lives by solving ecological and social problems while contributing to economic development is increasing; examples include green technology, changes in the digital environme-nt, and sustainable finance.

Moreover, fundamental to social feasibility is guaranteeing equal opportunities, services, outcomes, and rights in the future. The future steps toward social justice will focus on reforms in strengthening social protections, increasing access to healthcare and education, and supporting the less privileged groups of society. By promoting diversity and multicultural societies, future generations can ensure that no one is left behind during technological development.

10 Conclusion

The technological advancements demonstrate that addressing complex global issues necessitates a multidisciplinary approach and continuous innovation across all sectors. Continuous integration of technological, environmental, economic, and social feasibility into solutions will contribute to a more sustainable and equitable future.

References

Al Khalifa F (2018) Hope, utopia, and creativity in higher education: pedagogical tactics for alternative futures. Taylor & Francis

Basiago AD (1999) Economic, social, and environmental sustainability in development theory and urban planning practice: the environmentalist. Kluwer Academic Publishers, Boston

Chien F, Ananzeh M, Mirza F, Bakar A, Vu HM, Ngo TQ (2021a) The effects of green growth, environmental-related tax, and eco-innovation towards carbon neutrality target in the US economy. J Environ Manag 299:113633

Chien F, Kamran HW, Nawaz MA, Thach NN, Long PD, Baloch ZA (2021b) Assessing the prioritization of barriers toward green innovation: small and medium enterprises Nexus. Environ Dev Sustain 24:1897

Daly HE (1990) Toward some operational principles of sustainable development. Ecol Econ 2:1–6

Danielle NEL, Masilela L (2020) Open governance for improved service delivery innovation in South Africa. Int J eBusin eGov Stud 12(1):33–47

Dlalisa SF, Govender DW (2020) Challenges of acceptance and usage of a learning management system amongst academics. Int J Ebusin Egov Stud 12(1):63–78

Hussain HI, Kamarudin F, Thaker HMT, Salem MA (2019) Artificial Neural Network to model managerial timing decision: non-linear evidence of deviation from target leverage. Int J Comput Intell Syst 12(2):1282–1294

Hussain HI, Kamarudin F, Mohamad Anwar NA, Nassir AM, Sufian F, Tan KM (2020) Impact of country's governance dimensions on bank revenue efficiency: overview on Middle East, Southeast Asia, and South Asia countries. Transf Busin Econ 19(1):191–228

Jackson T (2009) Prosperity without growth. Earthscan, London

Kautish P, Khare A, Sharma R (2020) Values, sustainability consciousness and intentions for SDG endorsement. Mark Intell Plan 38(7):921–939

Koloba HA (2020) Purchase intention towards environmentally friendly products among consumers in South Africa. Applying the theory of planned behaviour. Int J Busin Manag Stud 12(1):34–49

García-Vega M, Vicente-Chirivella Ó (2020) Do university technology transfers increase firms' innovation? Eur Econ Rev 123(4):103388

Martinet V (2012) Economic theory and sustainable development: what can we preserve for future generations? Routledge, New York

Mebratu D (1998) Sustainability and sustainable development: historical and conceptual review. Environ Impact Assess Rev 18:493–520

Mensah J, Enu-Kwesi F (2018) Implication of environmental sanitation management in the catchment area of Benya Lagoon, Ghana. J Integr Environ Sci 16:23

Milne MJ, Gray R (2013) W(h)ither ecology? The triple bottom line is the global reporting initiative and corporate sustainability reporting. J Bus Ethics 118(1):13–29

Nelson C, Mattes A (2017) Lego, creative accumulation and the future of play multidisciplinary perspectives on play from birth and beyond. Springer, pp 265–279

NgoNdjama JD, Dhurup M, Joubert PA (2020) Exploring the relationship between organisational identification and organisational citizenship behaviour among employees in a university of technology in South Africa. Int J Busin Manag Stud 12(1):84–100

Robinson J (2004) Squaring the circle? Some thoughts on the idea of sustainable development. Ecol Econ 48:369–384

Rubin J (2012) The end of growth. Random House, Toronto

Solow R (1993) It is an almost practical step toward sustainability. Resour Policy 19:162–172

Thomas CF (2015) Naturalizing sustainability discourse: paradigm, practices and Pedagogy of Thoreau, Leopold, Carson and Wilson: Ph.D Thesis: Arizona State University

Tjarve B, Zemīte I (2016) The role of cultural activities in community development. Acta Universitatis Agriculturae et Silviculturae Mendelianae Brunensis 64(6):2151–2160

Williams A, Whiteman G, Parker JN (2019) Backstage interorganizational collaboration: corporate endorsement of the sustainable development goals. Acad Manag Discov 5(4):367–395

Innovative Approaches to Managing Natural Resources and Halting Biodiversity Loss

Shahnaz Anjum, Bisma Farooq, and Madiha Farooq

Abstract

The worldwide ecological disaster has led to a need for innovation to accelerate crisis response. Effective management of natural resources is crucial to prevent biodiversity loss and address environmental problems. This chapter summarizes the key concepts and achievements of innovative resource management and provides a quick overview of the area. This study uses a multidisciplinary approach to analyse how innovation plays a crucial role in addressing the complex link between resource management and biodiversity conservation. This article explores how innovation may enhance resource efficiency, minimize environmental impact, and boost resilience in socio-ecological systems through sustainable resource management and biodiversity conservation. This chapter offers a variety of creative strategies that appear to have the potential to achieve the two main objectives of protecting biodiversity and managing resources sustainably. These strategies include precision farming, renewable energy technology, community-based conservation programs, and biotechnological treatments. By combining cutting-edge research and best practices, comprehensive insights are presented into how innovation may spur revolutionary change and open the door to a more sustainable and biodiverse future.

Keywords

Natural resources · Biodiversity conservation · Socio-ecological systems · Precision farming

1 Introduction

Natural resources provide food, water, energy, and a variety of other essentials for human survival, making them essential to life as we know it. However, overexploitation of these resources has caused significant environmental degradation and biodiversity loss, putting human civilizations and ecosystems at risk. This chapter looks at how innovation may help reduce biodiversity loss and manage natural resources sustainably. In the face of mounting environmental challenges, sustainable natural resource management and biodiversity loss prevention have become critical prerequisites for preserving the health of human societies and ecosystems (Reid et al. 2005; SCBD 2020). The convergence of rising resource extraction and dwindling biodiversity emphasizes the critical need for novel methods to dealing with these interconnected issues (MEA 2005). This introduction provides a comprehensive overview of the importance of adopting innovation to manage natural resources and prevent biodiversity loss, highlighting the vital role that interdisciplinary collaboration and cutting-edge technologies play in achieving these objectives (Foley et al. 2011). Sustainable resource management is necessary to protect our planet's ecological integrity and ensure the delivery of critical ecosystem services on which human civilizations rely (UNWWDR 2020). However, in the face of ever-increasing human requirements and rapidly changing environmental circumstances, traditional resource management approaches have repeatedly proven insufficient (Tallis et al. 2008). As a result, innovative solutions that may reconcile the competing aims of social justice, environmental preservation, and economic growth are urgently required.The crux of this dilemma

S. Anjum (✉)
Department of Botany, Lovely Professional University, Phagwara, Punjab, India

B. Farooq
Department of Biomedical Engineering, Satyabhama Institute of Science and Technology, Chennai, India

M. Farooq
Departement of Microbiology, Asian International University, Imphal, Manipur, India

J. A. Parray et al. (eds.), *Environmental Landscape and Sustainable Biodiversity for Healthy Green Growth*, Sustainable Landscape Planning and Natural Resources Management, https://doi.org/10.1007/978-3-031-75405-0_14

is the transformative power of innovation, which encompasses new governance frameworks, policy reforms, and technological advancements, to totally alter how we utilize and manage natural resources (Ostrom 2009). Technological advancements like artificial intelligence, renewable energy, and remote sensing provide never-before-seen possibilities to monitor environmental changes, optimize resource efficiency, and create sustainable alternatives to extractive industries (Pretty and Bharucha 2014). Market-based mechanisms, regulatory frameworks, and participatory decision-making procedures are examples of policy innovations that can promote collaborative governance arrangements at various scales, internalize environmental externalities, and encourage sustainable behaviours (Pagiola et al. 2005). Furthermore, a paradigm alteration towards comprehensive, ecosystem-based methods that emphasize biodiversity conservation and restoration alongside sustainable resource management is required to solve the interwoven crises of resource depletion and biodiversity loss (MEA 2005). In addition to being a source of intrinsic value, biodiversity—the complex web of life that includes species, ecosystems, and genetic diversity—also supports the resilience and functionality of ecosystems, which are ultimately necessary for human civilizations to operate (Kareiva and Marvier 2012). However, anthropogenic activities have led to an alarming reduction in biodiversity worldwide, with significant consequences for ecosystem services, food security, and human well-being. These activities include habitat destruction, overexploitation, pollution, and climate change (SCBD 2020).

In this context, creative measures are desperately needed to prevent biodiversity loss and environmental damage (IRENA 2021). These could include cutting-edge technologies such as genomics, synthetic biology, and conservation biotechnologies, which have the potential to reduce the drivers of biodiversity decline and increase ecosystem resilience, as well as innovative conservation interventions such as habitat restoration, species reintroduction, and landscape-scale conservation initiatives (Redford and Adams 2009). Additionally, innovative financial strategies, market incentives, and community-driven conservation efforts can raise funds, enhance local stakeholder interests, and promote conservation activities across various socio-ecological contexts (Muradian et al. 2010). Natural resource management innovation and biodiversity conservation are interconnected demands that necessitate bold, interdisciplinary solutions based on social justice, ethics, and science (Berkes 2009a). By using innovation's revolutionary potential, we may create a sustainable future that balances human needs with the need to maintain and replenish the natural capital that serves as the foundation of all life (Ried et al. 2005).

2 Innovation in Sustainable Resource Management

Innovation is the key to addressing the challenges associated with responsible natural resource management. From technological advances to innovative management strategies, innovation offers a wide range of solutions to increase resource efficiency, decrease waste, and mitigate environmental consequences. Farmers, for example, may now increase crop yields, reduce environmental harm, and use less pesticides as a result of the advancement of precision agricultural equipment. Similarly, advances in renewable energy technologies, such as hydroelectric, solar, and wind power, provide fossil fuel alternatives while cutting greenhouse gas emissions and mitigating climate change, which is a major contributor to species extinction. Furthermore, innovative techniques of managing water resources, such as collecting rainwater and reusing wastewater, reduce water shortages while protecting freshwater environments.

2.1 Technological Innovations

Technological advancements, which are at the heart of the resource management revolution, enable more effective resource utilization, waste reduction, and environmental impact mitigation. Precision agricultural technologies, which maximize inputs such as water, fertilizer, and pesticides while decreasing environmental pollution and resource depletion, are a well-known example. These technologies utilize data analytics, remote sensing, and GPS-guided equipment (Pretty and Bharucha 2014). Similarly, advances in renewable energy technologies, such as wind, solar, and bioenergy systems, provide viable alternatives to fossil fuels, lowering greenhouse gas emissions and contributing to the mitigation of climate change, a major cause of resource scarcity and environmental degradation (IRENA 2021). Furthermore, advances in water management, such as water recycling, desalination, and drip irrigation, assist conserve freshwater resources, enhance water efficiency, and safeguard aquatic ecosystems (UNWWDR 2020).

2.1.1 Policy Innovations

Policy innovations are equally vital as technological advancements in promoting an environment conducive to sustainable resource management. This necessitates the development and implementation of institutional structures, market incentives, and regulatory frameworks that promote sustainable behaviours, internalize environmental costs, and align financial incentives with environmental objectives. Carbon pricing systems, such as carbon taxes and cap-and-trade schemes, are notable examples of how to support the transition to low-carbon technologies and energy sources while also incentivizing emission reductions (Hepburn et al. 2019). Similarly, payments for ecosystem services (PES) programs incentivize communities and landowners to preserve or restore ecosystem services such as carbon sequestration, biodiversity preservation, and watershed protection (Pagiola et al. 2005).

2.1.2 Governance Innovations

In the subject of sustainable resource management, innovative ideas must be converted into tangible actions and outcomes via effective governance. This requires establishing frameworks for multi-stakeholder partnerships, flexible administration, and participatory decision-making that enable efficient coordination, interaction, and learning across many actors and institutions. Co-management agreements, for example, have been effective in promoting sustainable fisheries, forest management, and protected area governance. These agreements call for collaboration in natural resource management among government entities, communities at large, and other stakeholders (Ostrom 2009). Similarly, community-based natural resource management (CBNRM) systems promote social equity, livelihood endurance, and environmental stewardship by allowing local residents to actively participate in decision-making, resource monitoring, and enforcement activities (Berkes 2009a). The broad category of sustainable resource management innovations includes technological, regulatory, and governance innovations that aim to enhance resource efficiency, reduce environmental harm, and improve resilience in socio-ecological systems. By using innovation's revolutionary power, we can build a future in which human needs are met while also conserving and replenishing natural resources. Sustainable resource management necessitates innovative governance structures and policies, in addition to cutting-edge technology. Protected area establishment, sustainable fisheries management practices, and ecosystem services payment programs are just a few examples of ways to promote conservation and sustainable use of natural resources. Innovation is a driving factor behind transformative change in sustainable resource management, bringing new methods to addressing the multiple issues of resource depletion, degradation of the environment, and climate change. This section investigates several aspects of innovation in the management of sustainable resources, such as technological breakthroughs, changes in legislation, and governance structures, with an emphasis on enhancing resource efficiency, minimizing environmental effects, and cultivating resilience in socio-ecological systems.

3 Halting Biodiversity Loss Through Innovation

The abundance of life on Earth, often known as biodiversity, is critical for ecosystem resilience and health. However, as a result of human activities such as pollution, overexploitation, habitat loss, and climate change, biodiversity is vanishing at previously unprecedented rates. The alarming worldwide decline in biodiversity needs novel solutions to address the wide range of complex reasons threatening the sustainability of the environment and the survival of various species. This section looks at how innovation may assist to prevent biodiversity loss. It addresses a wide range of technological, ecological, and socioeconomic solutions for managing biological diversity in a sustainable manner, as well as conservation and restoration efforts.

3.1 Biotechnological Innovations

Biotechnology has the potential to significantly contribute to biodiversity conservation since it provides novel ways and tools for addressing important conservation concerns. Genetic engineering advancements such as gene editing and biotechnology provide new techniques to managing invasive species, avoiding animal diseases, and increasing ecosystem and endangered species resilience (Redford and Adams 2009). Gene drive technology, for example, can eliminate an invading species by altering its capacity to reproduce (Esvelt et al. 2014). Furthermore, biotechnology breakthroughs such as nuclear transfer of somatic cells, cryopreservation of DNA material, and technology for assisted reproduction are critical for the ex situ preservation of endangered species (Lacy 2016). By retaining genetic diversity and returning captive-bred animals into the wild, these strategies serve to stabilize dwindling populations and enhance their chances of survival in changing environments.

3.2 Ecological Restoration

To reverse ecosystem deterioration and accelerate biodiversity recovery, novel ecological restoration technologies are required. For example, rewilding is a novel concept that seeks to promote biodiversity and ecosystem resilience by returning keystone species, reinstating biological processes, and establishing connected habitats (Soule and Noss 1998). Rewilding projects have the ability to preserve biodiversity and restore ecological functions at the landscape scale as they mimic natural processes and promote self-sustaining ecosystems. Similarly, cutting-edge restoration technologies such as ecological engineering and assisted migration provide novel ways to assist ecosystems in adapting to changing climatic conditions and mitigating the consequences of environmental degradation on biodiversity (Hobbs et al. 2009). Environmental engineering is the deliberate modification of the environment to restore degraded habitats and aid biodiversity recovery, whereas supported migration is the intentional relocation of populations or species to more appropriate environments to increase their probability of survival in the face of global warming.

4 Community-Based Initiatives for Conservation

Community-based conservation initiatives combine conventions, local knowledge, and community participation to manage resources from nature sustainably and maintain biodiversity. Local communities are given the ability to participate in decision-making, asset monitoring, and conservation efforts through community-managed reservations, indigenous protected areas, and shared management agreements. This encourages stewardship and increases biodiversity conservation outcomes (Berkes 2009a). Furthermore, innovative funding strategies such as biodiversity offsets, grassroots efforts, ecotourism, and compensation for ecosystem services (PES) programs provide monetary incentives for conservation and help sustain the way of life of the local community while protecting biodiversity-rich areas (Muradian et al. 2010). Community-based strategies promote coexistence between humans and animals and contribute to the long-term viability of conservation operations by incorporating social, cultural, and economic considerations into conservation policy. It will need audacious and creative strategies to address the underlying causes of environmental deterioration and encourage the preservation and sustainable use of biological variety in order to stop the loss of biodiversity. Through the use of biotechnology, ecological restoration, and community-based conservation, we can effectively steer towards a future that is more robust and biodiverse, characterized by the coexistence of humans and environment.

Community-based programs aim to improve biodiversity conservation and foster resilience in socio-ecological systems by utilizing the knowledge, skills, and cultural values of indigenous and local populations.

1. *Empowerment of Local Stakeholders*

By involving local stakeholders in the procedures of decision-making, resource surveillance, and preservation efforts, community-based projects offer them more authority. Community participants have a greater feeling of possession and accountability as a result of this collaborative approach, which produces more enduring and successful results (Berkes 2009a).

2. *Integration of Traditional Knowledge*

The unique customs and knowledge contribute significantly to biodiversity conservation and sustainable resource management. The integration of conventional systems of knowledge with modern conservation approaches is thought critical for community-based programs to boost conservation efforts' resilience and adaptability (Berkes 2009b).

3. *Promotion of Stewardship Values*

Community-based initiatives promote cultural norms and stewardship principles, emphasizing biodiversity preservation and equitable utilization of resources. These initiatives contribute to the long-term survival of conservation efforts by promoting a greater connection to the land and a sense of responsibility to future generations (Chapin et al. 2010).

4. *Enhancement of Livelihoods*

Community-based conservation initiatives usually combine attempts to protect biodiversity with strategies to improve livelihoods, such as non-timber forest product collection, ecotourism, and sustainable agriculture. These initiatives address the root causes of biodiversity loss while also supporting local livelihoods through financial incentives and opportunities for alternative sources of income (Kremen et al. 2012).

5. *Building Resilience in Socio-ecological Systems*

Community-based activities play an important role in increasing the resilience of socio-ecological systems by strengthening social networks, promoting flexibility, and supporting ecosystem-based natural resource management strategies. These initiatives strengthen communities' resilience to environmental changes and external shocks by using local knowledge and resources (Olsson et al. 2004).

Community-based initiatives are important in natural resource management because they build local communities, incorporate traditional knowledge, promote stewardship values, improve livelihoods, and develop resilience in socio-ecological systems. By recognizing the importance of local involvement and participation, innovative approaches to natural resource management can result in more equitable and sustainable results for people and the environment.

5 Conclusion

The management of resources from nature via innovation emphasizes the importance of innovation in accomplishing the dual imperatives of environmental sustainability and biodiversity protection. Through leveraging the transformational power of technology breakthroughs, policy changes, and innovations in governance, stakeholders may devise innovative ways to optimize resource use, reduce environmental effects, and promote resilience in socio-ecological systems. A wide range of creative strategies, from biotechnological treatments and community-based conservation programs to precision agriculture and renewable energy technology, show promise for accomplishing sustainable development objectives and preserving biodiversity for future generations. But in order to fully utilize innovation, deliberate efforts must be made to get over obstacles and difficulties such as multidisciplinary cooperation, equality concerns, scalability constraints, and legislative support. Through cultivating collaborations, advocating for variety, and establishing conducive conditions for creativity, interested parties may open up fresh avenues for sustainable growth and wildlife preservation. In the end, "Managing Natural Resources Through Innovation and Halting Biodiversity Loss" promotes a paradigm change in favour of more forward-thinking, integrative, and adaptable strategies that put the health of the earth and people first. We can all look forward to a more robust, just, and biodiverse future if we work together and find creative solutions.

References

Berkes F (2009a) Evolution of co-management: role of knowledge generation, bridging organizations and social learning. J Environ Manag 90(5):1692–1702

Berkes F (2009b) Indigenous ways of knowing and the study of environmental change. J R Soc N Z 39(4):151–156

Chapin FS, Carpenter SR, Kofinas GP, Folke C, Abel N, Clark WC, Olsson P, Smith DM, Walker B, Young OR, Berkes F (2010) Ecosystem stewardship: sustainability strategies for a rapidly changing planet. Trends Ecol Evol 25(4):241–249

Esvelt KM, Smidler AL, Catteruccia F, Church GM (2014) Concerning RNA-guided gene drives for the alteration of wild populations. elife 3:e03401

Foley JA, Ramankutty N, Brauman KA, Cassidy ES, Gerber JS, Johnston M, Mueller ND, O'Connell C, Ray DK, West PC, Balzer C (2011) Solutions for a cultivated planet. Nature 478(7369):337–342

Hepburn C, Adlen E, Beddington J, Carter EA, Fuss S, Mac Dowell N (2019) The technological and economic prospects for CO2 utilization and removal. Nature 575(7781):87–97

Hobbs RJ, Higgs E, Harris JA (2009) Novel ecosystems: implications for conservation and restoration. Trends Ecol Evol 24(11):599–605

International Renewable Energy Agency (IRENA) (2021) Renewable power generation costs in 2020. IRENA, Abu Dhabi

Kareiva P, Marvier M (2012) What is conservation science? Bioscience 62(11):962–969

Kremen C, Iles A, Bacon C (2012) Diversified farming systems: an agroecological, systems-based alternative to modern industrial agriculture. Ecol Soc 17(4):44

Lacy RC (2016) The role of genetics in ex situ conservation. In: Soorae RE (ed) Global re-introduction perspectives: 2016. Further case studies from around the globe. IUCN/SSC Re-introduction Specialist Group & Environment Agency-Abu Dhabi, pp 115–121

Millennium Ecosystem Assessment (2005) Ecosystems and human well-being: synthesis. Island Press

Muradian R, Corbera E, Pascual U, Kosoy N, May PH (2010) Reconciling theory and practice: an alternative conceptual framework for understanding payments for environmental services. Ecol Econ 69(6):1202–1208

Olsson P, Folke C, Berkes F (2004) Adaptive comanagement for building resilience in social-ecological systems. Environ Manag 34(1):75–90

Ostrom E (2009) A general framework for analyzing sustainability of social-ecological systems. Science 325(5939):419–422

Pagiola S, Arcenas A, Platais G (2005) Can payments for environmental services help reduce poverty? An exploration of the issues and the evidence to date from Latin America. World Dev 33(2):237–253

Pretty J, Bharucha ZP (2014) Sustainable intensification in agricultural systems. Ann Bot 114(8):1571–1596

Redford KH, Adams W (2009) Synthetic biology and conservation of nature: wicked problems and wicked solutions. PLoS Biol 7(8):e1000248

Reid WV, Mooney HA, Cropper A, Capistrano D, Carpenter SR, Chopra K, Dasgupta P, Dietz T, Duraiappah AK, Hassan R, Kasperson R (2005) Ecosystems and human well-being-synthesis: a report of the millennium ecosystem assessment. Island Press

Secretariat of the Convention on Biological Diversity (SCBD) (2020) Global biodiversity outlook 5. Secretariat of the Convention on Biological Diversity, Montreal, Canada

Soulé ME, Noss RF (1998) Rewilding and biodiversity: complementary goals for continental conservation. Wild Earth 8(3):18–28

Tallis H, Kareiva P, Marvier M, Chang A (2008) An ecosystem services framework to support both practical conservation and economic development. Proc Natl Acad Sci 105(28):9457–9464

United Nations World Water Development Report (UNWWDR) (2020) Water and climate change. UNESCO, Paris

Green Growth in Sustainable Environment and Future Outlook

Bisma Farooq, Asma Nazir, Madiha Farooq, and Shahnaz Anjum

Abstract

This conceptualization of green growth responds to the exigency of transitioning towards more sustainable economic frameworks, accentuating the imperative of conserving natural resources and safeguarding the welfare of present and future populations. It underscores the imperative of disassociating economic advancement from ecological degradation, cognizant that perpetuating unsustainable practices is untenable in the protracted trajectory. Instead, green growth advocates for a holistic development approach that amalgamates environmental considerations into economic policies and strategies, nurturing resilience and sustainability. By scrutinizing global case studies and examples, insights emerge regarding how nations and communities operationalize green growth strategies to realize economic prosperity while safeguarding environmental integrity and advancing social equity

Keywords

Green growth · Sustainability · Models · Green standards · Sustainable power

B. Farooq (✉)
Department of Biomedical Engineering, Sathyabama Institute of Science and Technology, Chennai, India

A. Nazir
Department of Biochemistry, Srinagar University of Kashmir, Srinagar, India

M. Farooq
Department of Microbiology, SKUAST, Srinagar, Kashmir, India

S. Anjum
Department of Botany, Lovely Professional University, Phagwara, Punjab, India

1 Introduction

Green growth within the context of sustainable development epitomizes a paradigmatic shift in economic paradigms, encapsulating the intrinsic nexus among economic affluence, ecological integrity, and societal justice. Conventional models of economic progression have often favored short-term gains at the expense of long-term planetary and intergenerational repercussions. Conversely, green growth endeavors to reconcile economic expansion with environmental preservation and societal inclusiveness, envisioning a future wherein prosperity is attained while respecting the ecological constraints of our biosphere (Grunwald 2018).

This conceptualization of green growth responds to the exigency of transitioning towards more sustainable economic frameworks, accentuating the imperative of conserving natural resources and safeguarding the welfare of present and future populations. It underscores the imperative of disassociating economic advancement from ecological degradation, cognizant that perpetuating unsustainable practices is untenable in the protracted trajectory. Instead, green growth advocates for a holistic development approach that amalgamates environmental considerations into economic policies and strategies, nurturing resilience and sustainability (United Nations Environment Programme (UNEP) 2011).

Delving into the multifaceted dimensions of green growth necessitates an exploration of its ramifications across various sectors, including energy, agriculture, transportation, and urban planning. Each of these domains presents distinctive challenges and opportunities for advancing sustainability and fostering inclusive economic progress. By scrutinizing global case studies and examples, insights emerge regarding how nations and communities operationalize green growth strategies to realize economic prosperity while safeguarding environmental integrity and advancing social equity (Brown 2012).

Furthermore, an exhaustive examination of the policy frameworks and institutional mechanisms requisite to

J. A. Parray et al. (eds.), *Environmental Landscape and Sustainable Biodiversity for Healthy Green Growth*, Sustainable Landscape Planning and Natural Resources Management, https://doi.org/10.1007/978-3-031-75405-0_15

buttress green growth endeavors is indispensable. Effective governance structures, regulatory frameworks, and fiscal instruments are pivotal for cultivating an enabling milieu for sustainable development. Moreover, international collaboration and alliances play a pivotal role in addressing transboundary environmental exigencies and propelling the global green growth agenda forward (OECD 2011).

In probing the socioeconomic dimensions of green growth, consideration must be given to its ramifications on employment dynamics, income distribution, and poverty mitigation. Sustainable economic endeavors have the potential to engender novel employment prospects, particularly in burgeoning green sectors such as renewable energy, ecotourism, and sustainable agriculture. However, ensuring the inclusivity of these opportunities necessitates proactive measures to foster skills enhancement, entrepreneurialism, and societal integration.

Additionally, green growth holds promise in ameliorating income disparateness by affording avenues for small-scale entrepreneurs and marginalized communities to participate in the green economy. By empowering women, indigenous populations, and other vulnerable cohorts to partake in sustainable livelihoods, green growth can contribute to equitable outcomes and social cohesion. At the crux of green growth lies the ethos of sustainability, epitomizing the imperative of meeting present needs without jeopardizing the capacity of future generations to fulfill their own exigencies. This mandates a profound transformation in consumption patterns, production modalities, and lifestyle preferences towards more sustainable commodities. Education and awareness initiatives play a pivotal role in fostering an ethos of sustainability and instilling responsible consumption and production behaviors among individuals and collectives (Wilkinson and Pickett 2009).

In summation, green growth represents a transformative approach to economic development, harmonizing the imperatives of economic prosperity, environmental sustainability, and societal equity. By embracing this paradigmatic transition and embracing comprehensive strategies for sustainable development, nations can cultivate resilient and inclusive economies that flourish within the ecological confines of our planet. Through concerted endeavors at local, national, and global echelons, a trajectory towards a more sustainable and prosperous future for all can be charted.

2 Principles of Green Growth

Central to the ethos of green growth are several foundational principles that underpin its implementation:

Efficient resource utilization: Green growth places paramount importance on optimizing resource utilization to minimize waste and environmental impact while maximizing economic output. This entails streamlining production processes, reducing energy consumption, and promoting recycling and reusability practices.

Investment in sustainable technologies: A cornerstone of green growth involves directing investments towards the development and adoption of environmentally sustainable technologies. This strategic allocation of resources is pivotal for driving innovation and facilitating the transition towards a low-carbon economy, encompassing areas such as renewable energy, clean transportation, and eco-friendly manufacturing (United Nations Environment Programme (UNEP) 2011).

Promotion of innovation: Innovation stands as a key driver in realizing the objectives of green growth. This encompasses the continual development of novel technologies, business models, and policy frameworks that prioritize sustainability and resource efficiency, fostering economic growth and enhancing competitiveness.

Social equity: A fundamental tenet of green growth is the commitment to ensuring equitable distribution of economic benefits across all segments of society. This entails addressing social disparities, poverty alleviation, and ensuring universal access to essential services such as clean water, sanitation, and energy (Brown 2012).

An in-depth exploration of these guiding principles necessitates a comprehensive examination of their intricacies, implications, and interdependencies across diverse sectors and geographical contexts. By delving into the multifaceted dimensions of resource efficiency, sustainable technology investment, innovation promotion, and social equity, a deeper understanding of green growth's transformative potential emerges.

Resource efficiency: A detailed analysis of resource efficiency strategies reveals a range of approaches aimed at optimizing resource usage across industries. From embracing circular economy principles to adopting sustainable procurement practices, initiatives focused on resource efficiency are integral to green growth strategies.

Investment in sustainable technologies: The domain of sustainable technologies encompasses a broad spectrum of innovations, including renewable energy, energy storage, sustainable transportation, and eco-efficient manufacturing. By exploring technological advancements and investment opportunities within these areas, a roadmap for fostering a robust green technology ecosystem can be delineated (Grunwald 2018).

Promotion of innovation: Innovation serves as a catalyst for driving the transformative agenda of green growth, enabling the development and dissemination of solutions to environmental challenges. Through case studies, best practices, and policy interventions, the pivotal role of innovation in advancing sustainability goals becomes evident.

Social equity: Achieving social equity within the framework of green growth involves dismantling barriers to access and opportunity. This requires examining policy interventions, community-led initiatives, and grassroots movements aimed at promoting social inclusivity and reducing disparities (Wilkinson and Pickett 2009).

3 Definition and Standards of Green Development

Green development can be characterized as an improvement method that aims to focus on economic development while guaranteeing the protection of natural resources and ecological services that are significant for human prosperity (United Nations Development Programme (UNDP) 2015). It recognizes that financial development should not be sought after at the cost of irreversible damage to the environment and, on second thought, advocates for a change in economic practices and innovations (World Bank Group 2016; Sachs 2015).

Standards of Green Development

1. Resource efficiency: The guideline of asset proficiency highlights the significance of advancing asset use to limit squandering and the ecological impacts while expanding financial results. This includes taking on practices that decrease energy utilization, advance reusing, and limit material wastage throughout the creation and utilization process (United Nations Environment Programme (UNEP) 2011).
2. Investment in green technologies: Green development accentuates the need to put resources into advances that are harmless to the ecosystem and feasible. This incorporates environmentally friendly power sources, such as solar, wind, and hydroelectric power, as well as perfect transportation, eco-accommodating assembling cycles, and energy-effective advancements by supporting the development of green innovations and green development plans to speed up the change to a low-carbon economy (Brown 2012).
3. Fostering innovation: Development is a focal principle of green development, driving innovation progress, plans of action, and strategies that advance supportability. This includes growing new answers for ecological difficulties, such as contamination decrease, environmental change moderation, and typical assets on the board. By cultivating a culture of development, green development tries to open new doors for financial development while tending to squeeze natural worries (Grunwald 2018).
4. Social inclusiveness: Green development focuses on friendly value and comprehensiveness, guaranteeing that the advantages of monetary improvement are shared fairly among all fragments of society for fundamental requirements like clean water, disinfection, and energy for minimized networks, tending to social imbalances, advancing fair circulation of assets, and opening doors of opportunities (Wilkinson and Pickett 2009).

4 Critical Parts of Green Development

Green development includes a few essential parts that together structure a far-reaching system for practical turn of events:

1. Sustainable power

   Progressing to environmentally friendly power sources is a foundation of green development, offering a maintainable option in contrast to petroleum derivatives and relieving environmental change. Environmentally friendly power advances, such as sunlight-based photovoltaics, wind turbines, hydroelectric power, and geothermal energy, bridle average assets without exhausting limited holds or radiating ozone-harming substances. Interests in environmentally friendly power framework and exploration are fundamental for increasing sustainable power organization and accomplishing a carbon-impartial energy framework (International Energy Agency (IEA) 2020a)
2. Economical agribusiness

   Manageable horticulture practices are fundamental for safeguarding standard biological systems, rationing soil and water assets, and guaranteeing food security for future generations. Supportable horticulture procedures, like natural cultivating, agroforestry, regenerative agribusiness, and accuracy cultivating, limit ecological effects while amplifying efficiency and flexibility. By advancing soil well-being, biodiversity, and biological system administrations, reasonable farming adds to environmental change alleviation, variation, and practical land management (Food and Agriculture Organization of the United Nations (FAO) 2019).
3. Roundabout economy

   The idea of a roundabout economy intends to dispose of waste and contamination by designing items, materials, and frameworks to be fixed, remanufactured, or reused endlessly. A round economy limits asset extraction, diminishes ecological contamination, and cultivates monetary flexibility and development. Critical standards of a circular economy incorporates planning out waste problems. By changing from a direct "take-make-arrange" model to a roundabout economy, green development advances reasonable utilization and creation of designs, focusing on asset effectiveness and natural stewardship (Ellen MacArthur Foundation 2019).

4. Green framework

Interests in green frameworks are fundamental for improving versatility in environmental change, advancing natural sustainability, and supporting practical metropolitan turn of events. Green framework envelops many drives, including green structures, reasonable transportation frameworks, sustainable power productions, hydropower projects, and everyday living space reclamation projects. By incorporating green foundations into metropolitan advancements, and green development aims to create reasonable, solid, and strong networks that limit imapcts on ecosystems and improve personal satisfaction (United Nations Human Settlements Programme (UN-Habitat) 2019).

5 Advantages of Green Development

The reception of green development methodologies offers a number of advantages across monetary, natural, and social aspects:

1. Financial development

Green development can animate financial development by driving advancement, making new business sectors, and producing work that opens doors in arising green ventures. Interests in environmentally friendly power, energy effectiveness, and reasonable foundation undertakings can spike monetary development while lessening ecological effects. Green infrastructure, like environmentally friendly power, energy-effective advances, and manageable farming, can result in the creation of employment opportunities, development, and intensity. By advancing green development and business ventures, monetary strength is strengthened and flourishes in a rapidly changing global economy (Organisation for Economic Co-operation and Development (OECD) 2019a).

2. Natural maintainability

By the advancement of economic practices and innovations, the green development can mitigate environmental change, lessens contamination, and protects water, land, and biodiversity. Sustainable power sources, for example, solar and wind power, offer clean alternatives to petroleum products, reducing ozone-depleting substance emissions and air contamination. Feasible horticulture practices, like natural cultivation and agroforestry, advance soil well-being, biodiversity, and carbon sequestration, adding to environmental change alleviation and practical land the board. The circular economy drives waste decrease and reusing, limit asset extraction, and advance asset effectiveness. By focusing on ecological supportability, green development safeguards the environment, protects biodiversity, and natural resources for people in the future (Intergovernmental Panel on Climate Change (IPCC) 2018).

3. Social consideration

Green development projects can enhance access to essential utilities like clean energy, water, and sanitation, especially in underserved areas. Green development can help alleviate poverty, reduce inequality, and improve general well-being by emphasizing social ideals and inclusivity. Investments in energy access initiatives and renewable energy infrastructure can give impoverished communities access to clean, reasonably priced energy services, enhancing their economic, educational, and health prospects. In addition to boosting resilience to climate change, sustainable agricultural initiatives can empower smallholder farmers, improve food security, encourage rural development, and provide employment. Recycling and upcycling are two examples of circular economy projects that can generate revenue and promote community involvement, fostering social cohesion and empowering people. Green development supports human rights, human development, and poverty alleviation by tackling social inequalities and promoting inclusive growth (United Nations Development Programme (UNDP) 2020).

6 Attempts and Hindrances to Green Development

Despite its potential, green development faces different obstacles that impact its reception and execution. These elements can fundamentally influence the advancement towards a more economical and harmless ecosystem. Below, we investigate some critical moves and hindrances to green development (European Environment Agency (EEA) 2019; World Economic Forum (WEF) 2020; International Institute for Sustainable Development (IISD) 2018).

Attempts Towards Green Development

1. Policy support: One of the essential moves towards green advancement is the detailing and execution of solid arrangements at the nearby, public, and worldwide levels. These approaches might incorporate impetuses for environmentally friendly power reception, guidelines on dis-

charges and contamination, and sponsorships for manageable practices (United Nations Environment Programme (UNEP) 2018).
2. Investment in sustainable energy: Expanding interest in sustainable power sources, such as solar, wind, and hydroelectric power, is a massive move towards a green development. Legislatures, organizations, and people are progressively perceiving the significance of transitioning away from nonrenewable energy sources to diminish fossil fuel byproducts and environmental impacts (International Energy Agency (IEA) 2021).
3. Adoption of feasible practices: The reception of manageable practices across different areas, including horticulture, assembling, transportation, and development, is fundamental for green development. This incorporates feasible cultivating techniques, energy-proficient assembling cycles, and green structure plans (World Bank Group 2019a).
4. Technological innovation: Advancements in environmentally friendly power, energy capacity, waste management, and clean transportation contribute to the transition towards a more sustainable economy (International Renewable Energy Agency (IRENA) 2020).
5. Public mindfulness and education: Expanding public awareness and education about the significance of ecological maintainability and the advantages of green advancement can drive positive change. Teaching people and institutes about sustainable ways of life, sustainable utilization of resources, and environmental protection can prompt more acceptance of green practices (United Nations Educational 2017).

7 Blocks to Green Development

1. Lack of political will: One of the main blocks to green improvement is the absence of political will and obligation to execute practical approaches. Generally speaking, momentary political interests might offset long-haul natural contemplations, prompting insufficient activity on environmental change and maintainability (World Bank Group 2019b).
2. Economic challenges: Monetary difficulties, for example, the high initial expenses of changing to environmentally friendly power and sustainable advancements, can hinder the green developments (International Monetary Fund (IMF) 2018).
3. Technological hurdles: Despite advancements in green technologies, significant barriers such as high costs, limited flexibility, and infrastructure constraints can impede their widespread adoption. Further research and development efforts are required to overcome these barriers and render green technologies more accessible and affordable (International Energy Agency (IEA) 2020b).
4. Lack of awareness and education: Insufficient awareness and education regarding environmental issues and the benefits of green development can hinder progress. Without widespread comprehension and support for sustainability initiatives, endeavors to promote green development may encounter resistance or apathy from the public and policymakers (United Nations Environment Programme (UNEP) 2019a).
5. Institutional challenges: Institutional boundaries, like regulatory formality, clashing guidelines, and restricted coordination between government organizations and partners, can dial back the execution of green advancement drives. Smoothing out processes and further developing cooperation among important partners are fundamental for beating these difficulties (Organisation for Economic Co-operation and Development (OECD) 2020).

While there are huge moves towards green development, different blocks exist that can obstruct progress. Tending to these difficulties requires coordinated endeavors from states, organizations, networks, and people to defeat political, financial, mechanical, instructive, and institutional hindrances and speed up the change towards a more sustainable and harmless ecosystem in the future.

8 Advantages of Green Development Advancement

The future benefits of green development are vast, touching on economic, environmental, and social aspects. As societies increasingly realize the importance of transitioning to sustainable practices, green development offers numerous advantages that can contribute to a more prosperous, adaptable, and equitable future. In this comprehensive exploration, we delve into the potential benefits of green development advancement and their implications for various sectors of society (World Bank Group 2018; United Nations Environment Programme (UNEP) 2019b; International Labour Organization (ILO) 2018).

Economic Advantages

Stimulating economic growth: Green development holds the potential to catalyze economic growth through the stimulation of innovation, the creation of new market landscapes, and the proliferation of job opportunities. Strategic investments in renewable energy, energy efficiency, sustainable infrastructure, and green technologies can invigorate economic expansion and bolster competitiveness within global markets (International Renewable Energy Agency (IRENA) 2017).

Fostering employment opportunities: The transition towards green development is poised to engender a diverse array of employment prospects spanning various sectors. Opportunities may materialize in fields such as renewable energy deployment and maintenance, energy-efficient construction practices, sustainable agricultural methodologies, waste management systems, and ecotourism ventures. Green jobs often entail higher levels of innovation, skill development, and job satisfaction (International Labour Organization (ILO) 2020).

Attracting capital inflows: Initiatives geared towards green development can attract investments from both public and private sectors, including venture capital entities, impact investors, and sustainable finance institutions. There is a growing recognition among investors of the economic viability inherent in sustainable ventures and technologies, leading to a surge in investments directed towards renewable energy projects, green infrastructure initiatives, and sustainable enterprises (United Nations Environment Programme (UNEP) 2018).

Mitigating energy expenditures: The adoption of renewable energy sources and the enhancement of energy efficiency measures have the potential to mitigate energy costs for businesses, households, and governmental entities alike. Sustainable energy technologies such as solar and wind power have attained heightened cost competitiveness vis-à-vis fossil fuels, resulting in reduced electricity expenses and heightened energy self-sufficiency (International Energy Agency (IEA) 2020c).

Heightening resilience to economic turbulence: Embracing green development strategies, such as investments in renewable energy and sustainable infrastructure, can bolster resilience to economic upheavals and volatility within energy markets. Nations can mitigate the adverse impacts of price fluctuations and supply disruptions by diversifying energy portfolios and diminishing reliance on imported fossil fuels.

1. *Environmental advantages:*
    1. *Mitigating climate change*: One of the most consequential environmental advantages of green development lies in its capacity to ameliorate climate change. Through the curtailment of greenhouse gas emissions and the adoption of renewable energy alternatives, green development endeavors to mitigate the trajectory of global warming and its attendant repercussions, encompassing phenomena such as extreme weather occurrences, rising sea levels, and disturbances to ecosystems.
    2. *Conserving natural resources*: Green development advocates for the prudent utilization and preservation of natural assets, encompassing water bodies, terrestrial expanses, forested regions, and biodiversity. Through the adoption of sustainable agricultural methodologies, meticulous forest stewardship, and endeavors aimed at ecosystem rejuvenation, green development endeavors to safeguard these invaluable natural resources, thereby guaranteeing their sustenance for posterity (Organisation for Economic Co-operation and Development (OECD) 2019b).
    3. *Reducing pollution*: The overarching goal of green development initiatives is to diminish pollution and environmental degradation through enhancements in air and water purity, reductions in waste production, and transitions towards cleaner manufacturing methodologies. Through the mitigation of pollution, green development endeavors to safeguard both human well-being and the integrity of ecosystems.
    4. *Preserving ecosystem services*: Ecosystem services, encompassing vital provisions such as unpolluted air, pristine water sources, fertile soils, and crucial pollination processes, serve as indispensable cornerstones for sustaining life on our planet. Green development underscores the imperative of safeguarding these ecosystem services through the preservation of natural habitats, the rehabilitation of deteriorated ecosystems, and the advocacy for sustainable land management practices.
    5. *Protecting biodiversity*: Biodiversity stands as a pivotal factor in bolstering ecosystem resilience, ensuring food security, and nurturing human well-being. Strategies within the realm of green development, including the preservation of habitats, sustainable forestry practices, and the safeguarding of wildlife populations, serve to uphold biodiversity and forestall the loss of species. Moreover, the conservation of biodiversity not only fosters ecotourism but also lends support to endeavors aimed at preserving cultural heritage (Intergovernmental Panel on Climate Change (IPCC) 2018).

Social Advantages

- *Enhancing public health*: Green development initiatives play a pivotal role in enhancing public health by curbing exposure to air and water pollution, fostering active modes of transportation and recreational pursuits, and bolstering access to clean energy sources and sanitation facilities. By mitigating environmental hazards, green development endeavors serve as a bulwark against diseases and contribute to the overall enhancement of human well-being.

- *Augmenting quality of life*: The adoption of sustainable urban planning principles, the proliferation of green spaces, and the establishment of eco-friendly infrastructure converge to elevate the quality of life for urban denizens. Access to verdant parks, recreational amenities, and efficient public transportation networks fosters not only physical and mental health but also a sense of community cohesion and enhances overall livability (Food and Agriculture Organization of the United Nations (FAO) 2020).
- *Eradicating poverty*: Green development initiatives harbor the potential to alleviate poverty and foster social equity through the creation of employment opportunities, the enhancement of access to essential services, and the empowerment of marginalized segments of society. Strategic investments in renewable energy, sustainable agricultural practices, and rural development initiatives can catalyze socioeconomic upliftment, thereby reducing disparities and fostering inclusive growth (World Health Organization (WHO) 2018).
- *Fortifying resilience to climate change*: Vulnerable communities are disproportionately susceptible to the adverse impacts of climate change, ranging from extreme weather events to prolonged droughts and food insecurity. Green development strategies, such as the implementation of climate-resilient infrastructure, the formulation of robust disaster preparedness measures, and the establishment of social safety nets, serve to bolster resilience and adaptive capacity in the face of climate-induced risks.
- *Advancing sustainable development*: At its core, green development endeavors to advance the cause of sustainable development by harmonizing economic, environmental, and social imperatives. Through the integration of these multifaceted considerations, green development offers a pathway towards a future characterized by equity, resilience, and prosperity—a future wherein the needs of the present are met without compromising the ability of future generations to fulfill their own (Millennium Ecosystem Assessment 2005).

9 Conclusion

The anticipated dividends of advancing green development are manifold and far-reaching, spanning realms of economic vitality, ecological resilience, and societal welfare. Positioned at the nexus of these spheres, green development holds the potential to catalyze inclusive and sustainable progress by igniting economic expansion, engendering novel avenues for employment, and galvanizing capital inflows. Through its concerted efforts in climate change mitigation, natural resource conservation, and the promotion of public health, green development emerges as a bastion of planetary safeguarding, enriching the present and safeguarding the legacy of generations yet to come (United Nations Environment Programme (UNEP) 2019c).

In embracing green development as an intrinsic pillar of developmental ethos, we lay the groundwork for a future brimming with promise—a future marked by heightened resilience, equitable opportunities, and enhanced quality of life for all inhabitants of our shared Earth. Through meticulous examination and strategic deployment of green development strategies, we pave the path towards a more sustainable and prosperous world, wherein the principles of ecological stewardship and social equity reign supreme.

Expanding upon the multifaceted facets of green development necessitates a comprehensive exploration of its manifold dimensions, implications, and potentialities. By delving into the intricacies of economic dynamism, environmental fortification, and societal well-being fostered by green development paradigms, we unearth a rich tapestry of opportunities and imperatives. By harnessing the transformative potential of sustainable practices and innovations, green development initiatives stand poised to invigorate economies, spur innovation, and foster resilience in the face of global economic challenges. Moreover, by catalyzing the emergence of green industries and markets, green development cultivates a fertile ground for entrepreneurship, job creation, and inclusive prosperity.

In essence, the journey towards a greener, more sustainable future is one fraught with challenges, yet ripe with opportunities for transformative change. By embracing green development as a fundamental value for human progress, we embark upon a collective endeavor to forge a world that is not only prosperous and resilient but also just and equitable—a world wherein the principles of sustainability, stewardship, and solidarity serve as guiding beacons illuminating the path towards a brighter tomorrow.

References

Brown LR (2012) World on the edge: how to prevent environmental and economic collapse. W. W. Norton & Company

Ellen MacArthur Foundation (2019) Completing the picture: how the circular economy tackles climate change

European Environment Agency (EEA) (2019) Environmental sustainability: key issues and trends in the European Union

Food and Agriculture Organization of the United Nations (FAO) (2019) The state of food and agriculture 2019: Moving forward on food loss and waste reduction

Food and Agriculture Organization of the United Nations (FAO) (2020) The state of the world's forests 2020

Grunwald M (2018) The new new deal: the hidden story of change in the Obama era. Simon & Schuster

Intergovernmental Panel on Climate Change (IPCC) (2018) Special report on global warming of 1.5°C
International Energy Agency (IEA) (2020a) World energy outlook 2020
International Energy Agency (IEA) (2020b) Energy technology perspectives 2020: special report on clean energy innovation
International Energy Agency (IEA) (2020c) Renewables 2020: analysis and forecast to 2025
International Energy Agency (IEA) (2021) Renewables 2021: analysis and forecast to 2026
International Institute for Sustainable Development (IISD) (2018) Barriers to a green economy: a systematic review of the evidence
International Labour Organization (ILO) (2018) World employment and social outlook: greening with jobs
International Labour Organization (ILO) (2020) World employment and social outlook: trends 2020
International Monetary Fund (IMF) (2018) World economic outlook, April 2018: cyclical upswing, structural change
International Renewable Energy Agency (IRENA) (2017) Renewable energy and jobs—annual review 2017
International Renewable Energy Agency (IRENA) (2020) Innovation outlook: renewable energy in Latin America and the Caribbean 2020
Millennium Ecosystem Assessment (2005) Ecosystems and Human well-being: synthesis
OECD (2011) Towards Green Growth: Monitoring Progress
Organisation for Economic Co-operation and Development (OECD) (2019a) Green growth indicators 2019
Organisation for Economic Co-operation and Development (OECD) (2019b) Climate change and economic resilience: insights from experimental economics
Organisation for Economic Co-operation and Development (OECD) (2020) OECD environmental performance reviews: Sweden 2020
Sachs JD (2015) The age of sustainable development. Columbia University Press
United Nations Development Programme (UNDP) (2015) Greening development: enhancing capacity for environmental management and governance
United Nations Development Programme (UNDP) (2020) Human development report 2020: the next frontier - human development and the Anthropocene
United Nations Educational, Scientific and Cultural Organization (UNESCO) (2017) Education for sustainable development goals: learning objectives
United Nations Environment Programme (UNEP) (2011) Towards a green economy: pathways to sustainable development and poverty eradication
United Nations Environment Programme (UNEP) (2018) Global trends in renewable energy investment 2018
United Nations Environment Programme (UNEP) (2019a) Global environment outlook: summary for policy makers
United Nations Environment Programme (UNEP) (2019b) Global environment outlook 6: healthy planet, healthy people
United Nations Environment Programme (UNEP) (2019c) Global biodiversity outlook 5
United Nations Human Settlements Programme (UN-Habitat) (2019) Towards a green urbanism: learning from sustainable urban development experiences in Europe
Wilkinson R, Pickett K (2009) The spirit level: why more equal societies almost always do better. Allen Lane
World Bank Group (2016) Inclusive green growth: the pathway to sustainable development
World Bank Group (2018) Green growth and development: a framework for action
World Bank Group (2019a) Sustainable agriculture: key to ending hunger and boosting shared prosperity
World Bank Group (2019b) Political economy analysis for resilience in MENA countries: environment and climate change
World Economic Forum (WEF) (2020) Global risks report 2020
World Health Organization (WHO) (2018) Ambient air pollution: a global assessment of exposure and burden of disease

Honeybees: An Adaptive and Resilient Solution for Sustainable Global Food Security

Dhanashree A. Kirdat, Bhalchandra B. Waykar, and Ashish R. Gawali

Abstract

Sustainable development aims to satisfy current demands without risking the capacity of future generations to meet their own. Various policies and measures influence food security and sustainable development, including productivity, sustainability outcomes, price volatility, public-private partnerships, and agriculture. The price volatility of commodities contributes to national food insecurity, as nearly one-third of people rely on agricultural production. Beekeeping is crucial in increasing yield, as bees pollinate flowers, resulting in a one-fourth increase in crop yields. Furthermore, bees provide ecosystem services for sustainable horticulture, agriculture, and food and nutrition security. The quality and quantity of agricultural outputs are enhanced by honeybee pollination services, which raise market value and growers' earnings. In the United States, pollination generates USD 16 billion annually, with 12 billion due to honeybee accessibility. Culturing stingless honeybees for pollination can be an effective solution as they can withstand severe droughts and low temperatures and require minimal maintenance. In addition to their economic value and contribution to sustainable global food security, these indigenous pollinators support healthy ecosystems needed for clean air, stable soil, clean water, and diverse wildlife.

Keywords

Global economy · Sustainable development · Pollination · Stingless honeybees · Agricultural food security

D. A. Kirdat (✉) · B. B. Waykar · A. R. Gawali
Department of Zoology, Dr. Babasaheb Ambedkar Marathwada University, Chhatrapati Sambhajinagar, India

1 Introduction

In the global economy, resources and wealth are transferred beyond national boundaries. It extends business activities beyond national and international borders, enabled by advancements in communications, technology, and the social, political, and environmental spheres. Likewise, global agriculture is the basic foundation of many countries and interacts with the country's economic growth. The share of the economy that goes to agriculture in developing countries is about half or one-fourth that of employment; although agriculture makes up 4.3% of the global GDP, it employs 26.4% of the workforce worldwide; this percentage needs to be increased for global food security. Sustainable global food security should track the new trends in food behavior consumption, new sensory aspects or evaluation, food supply chain changes, new product development strategies (plant-based foods, cultured meats, etc.), or technologies to fulfill the rising global population (Abdulkadyrova et al. 2016). Experts warn that if food deficiency is not addressed, it could lead to the social collapse in many countries. The world's population is expected to grow by 1.9 billion people by the year 2050, which indicates that the limited resources needed to grow our food are being increasingly strained by human activity. The current paradigm of a growing population dependent on finite resources is unsustainable, making it even more crucial to ensure that we aim for a "mobile" resource economy. Asia, home to 56% of the world's population, has the highest percentage of food-insecure people worldwide (315.2 million as of 2017). With 10.8 million tonnes of grain in 2017, the region has the second-largest food shortfall (Mc Carthy et al. 2018).

In the discussion of the food system, two main concepts are food security and food sustainability. Food security and nutrition must be linked to sustainability transitions in sustainable food systems. Sustainability and food security are related on many levels. An agriculture-centered research

J. A. Parray et al. (eds.), *Environmental Landscape and Sustainable Biodiversity for Healthy Green Growth*, Sustainable Landscape Planning and Natural Resources Management, https://doi.org/10.1007/978-3-031-75405-0_16

policy is needed for comprehensive food security transformation (El Bilali et al. 2019).

The agricultural sector may be negatively impacted by the threats of diseases and natural disasters, irreversible damages due to human acts, etc., which is why extreme caution must be exercised. Food security is a multi-layered concept that is centred on four main dimensions, according to the FAO and UNICEF: (1) Food availability, (2) Food access, which includes both economic and physical access to food, (3) food utilization based on dietary and cultural requirements; and (4) Food stability, or the stability of its provision. Food insecurity is currently a significant problem in the food chain supply, and the government needs to act quickly to address it (Anis and Norfarizan-Hanoon 2022). Food production, distribution, and consumption are fundamental in climate change, social inequality, land degradation, poverty, hunger and malnutrition, poor diets, and water scarcity (Anis and Norfarizan-Hanoon 2022).

Sustainable agriculture aims to support profitable, environmentally sound farming techniques and technologies that benefit farmers, resources, and communities. It encompasses social, economic, and environmental aspects, focusing on satisfying current demands without sacrificing the capacity of future generations to satisfy their wants. The "Green Revolution" was only a tiny part of the twentieth century's massive transformation of global agriculture, fuelled by science's quick advancements, significant public investments, and policy support for farming. Modern plant breeding, agronomy advancements, inorganic fertilizers, improved seed varieties and irrigation, and modern insecticides have all contributed to these advancements, significantly increasing crop productivity (Rani 2023).

2 The World Food Crisis Is the Result of Human Acts

Food insecurity has been on the rise since 2018 due to a variety of circumstances, such as violence, climatic shocks, and the COVID-19 pandemic's effects. These factors work together to raise food prices and have a detrimental impact on food distribution and production. In 2022, the situation drastically deteriorated as pressure mounted on global food staples and fertilizer prices due to Russia's war in Ukraine (Reilly et al. 2020). The global demand for food and energy is expected to rise by 50% from 2000 to 2030, resulting in an annual increase of 2.8 billion tonnes (Inouye and Ogilvie 2017; Lobell et al. 2009). Food production must increase by 70% to keep up with the projected 9.7 billion people by 2050. The rainfed agriculture contributes to one-third of global food production. Hence, the "Green Revolution" was rooted in the significant increase in grain yields and total factor productivity, which would reduce food costs. At first, industrial agriculture, which uses chemical pesticides, synthetic fertilizers, and hybrid high-yield cereals, was seen as a way to reduce hunger, feed expanding populations, and boost economic prosperity. This concept of industrialized farming has led to the emission of greenhouse gases, air and water pollution, and the destruction of wildlife and their natural habitats (Ahvo et al. 2023). Global warming intensifies biotic and abiotic restrictions on agricultural yields. Abiotic stressors cause plant changes, while biotic forces include pests, anthropogenic evolution, and beneficial species. Abiotic restrictions affect soil pH and physicochemical and climatic pressures, while biotic forces include pests (Cheng et al. 2016). This irreparable and irreversible loss created a significant imbalance in the ecosystem and the loss of natural pollinators, which proved to be one of the reasons for the lower crop yield (Shanahan 2022).

3 The Need to Conserve Bee Pollinators: The Ecosystem Engineers

Honeybees are considered the ecosystem's allogenic engineers in terrestrial environments, as they, directly and indirectly, provide a range of multifaceted services like collecting the nectar and pollen from the flowers, which bring a revolutionary impact on the production yield, benefitting the ecosystems. Bees pollinate crops to help increase the yield quality and quantity and contribute to soil fertility and nitrogen cycling by adding organic nitrogen-rich excrement to the soil, providing inorganic nitrogen for plant absorption. Despite their ecological significance, bees are declining, primarily due to anthropogenic activities (Khan 2023). Disruption in bee biota affects its hive population and the surrounding faunal diversity, especially in fragmented habitats that may eventually vanish in a short or medium period (Thakur 2012). Their population decline will influence gene flow, colonization potential, and dispersal ability. According to 75% of local experts, the primary factors contributing to the deterioration of stingless bees are the increased number of *Apis mellifera* extractive practices and land-use changes (65%)—the slow shift from traditional cultivation to high-value horticulture farming, which includes fruits and vegetables; the use of pesticides, agrochemicals, and insecticides (15%); and fires linked to extended severe droughts (5%). Approximately 70% of the local specialists surveyed believe stingless bees will probably go extinct shortly (Reyes-González et al. 2014). The economic cost of reduced yields and increased crop production expenses from the decline of honeybees alone amounts to $5.7 billion annually (Allen-Wardell et al. 1998). In many regions of China, for instance, bees have virtually disappeared, and artificial hand pollination by humans of various crops, fruit trees, and trees is commonly practiced. Although artificial pollination may partially pollinate using its resources, what nature has done for free now needs enormous financial resources (Atsharyan et al. 2023).

The Industrial Revolution's mechanization of agriculture led to monocultures, destroying natural habitats and decreasing flower diversity, reducing food availability for bees despite new opportunities for crop production. Pesticide use, climate change, and new diseases posed challenges to beekeepers, affecting bee health, colony survival, ecological interactions, flowering time, and synchronizing life cycles with flower resources. When crops aren't flowering, pollinators drawn to these ecosystems need substitute food sources. Urbanization also leads to habitat loss and subdivision, causing the extinction of natural pollinators (Rani 2023). The pollinators faced challenges from highways and towns, contaminated with dust and artificial light. However, preserving habitat variability and flowering can help urban environments serve as significant pollinator reservoirs despite the negative impact of road traffic and artificial light. As our country's citizens, we should preserve woodlands for pollinating insects, plant melliferous trees for agroforestry, and encourage native plants in urban areas. Cities serve as pollinator reservoirs, so it's beneficial to have a variety of native plants that bloom at different seasons and attract local pollinators as bees significantly impact agricultural yields, with global food supply largely dependent on pollination. Pollinators, primarily bees, are crucial for producing over 73% of the world's food supply, according to the United Nations Department of Agriculture and Food (Dawoud 2020).

This is an essential aspect that has to be kept in mind. Accordingly, planning needs to be done for the survival of the entire colony and to avoid their migration, thus reducing the mortality rate of the bees both in rural and urban areas (Kevan et al. 1990). The SAMS (Smart Apiculture Management Services) should be brought into practice to strengthen international collaboration between the EU and developing nations in ICT and sustainable agriculture, focusing on creating an open-source remote sensing system for tracking bee colonies' well-being and productivity, lowering the risk of CCD (colony collapse disorder), and promoting sustainable growth in other sectors like forestry, agriculture, cosmetic, and beauty industries. This will also help the worldwide effort to end hunger by generating jobs, increasing the value of goods and revenue, and sustainable food security (Wakjira et al. 2021). Bee conservation initiatives worldwide are gaining popularity due to their crucial role in ecosystems and contributing to a steady food supply. Governments, groups, and individuals promote pollinator-friendly gardening, organic farming, and bee protection to ensure long-term survival and biodiversity conservation (Komlatskiy and Makarova 2023). Bee pollination is crucial for almost 33 fruit crops, while cross-pollination is required globally for the self-sterile flower varieties for desirable and acceptable yields. In India, approximately 737 lakh honeybee colonies are needed for 12 crops, including 8 fruit crops, to increase crop production by Rs. 3000 crores annually. However, India currently possesses only 17 lakh beehive colonies, requiring an additional 720 lakh to ensure an optimal pollination range (Thakur 2012). HEARD (1999) evaluated the significance of stingless bees as pollinators, with Australian beekeepers utilizing them for pollination purposes (Cortopassi-Laurino et al. 2001).

4 The Underappreciated and Crucial Contribution of Stingless Honeybees to Preserving Biodiversity

Since 1961, the number of managed *A. mellifera* colonies has increased by 85% worldwide, primarily in Asia. Overwinter mortality rates of the reared colonies are high, especially in North America and Europe. Growing agricultural reliance on pollinators and the challenges of managing non-native pollinators have likely led to an increase in alternative pollinator species, such as bumblebees, stingless bees, solitary bees, and flies, which are more effective at pollinating specific crops (Osterman et al. 2021). The eusocial bee species with the widest variety, with about 500 species, are the stingless bees (*Meliponini*), with a wide range of characteristics such as the small size of body and colony, the biology of nesting, the layout of brood cells, production of queens, strategies for foraging, and recruiting procedures (Narbona et al. 2020). This gives the added advantage of more chances for tiny flowers of plants like mango, sapota, green chilies, medicinal herbs, and shrubs to undergo successful pollination. These bees require low maintenance, have perennial colonies that usually consist of hundreds or thousands of workers, are capable of buzz pollination, and can be domesticated as they lack functional sting. The colonies are relocated in artificial wooden boxes, inspected, propagated, fed, and divided into new colonies. They require minimal maintenance and are controlled by enemies, transported at long distances, and otherwise managed (On n.d.). These bees are resilient to harsh weather, protracted droughts, unexpected weather shocks, etc. It was also found that the fungal-based insecticide does not interfere with the ability of the African stingless bee, *Meliponula ferruginea*, to pollinate cucumbers (*Cucumis sativus*) (Omuse et al. 2022). These are the only honeybees that can efficiently pollinate in polyhouses and greenhouse cultivations (Fig. 1).

With the broadest range of niches among all eusocial species, *A. mellifera* and *T. spinipes* are considered super generalists in pollination services (Benites et al. 2023). Stingless honeybees, especially *Tetragona* spp. and *Trigona iridipennis* Smith, are eusocial insects exhibiting various nesting pat-

Fig. 1 Diversified stingless honeybees: the efficient foragers

terns in natural and artificial environments. These corbiculate bees, commonly known as stingless bees (Hymenoptera: Apoidea: Anthophila: *Meliponini*), are indigenous to tropical and subtropical regions worldwide. Due to their colonies comprising thousands to tens of thousands of workers, they act as frequent visitors to the flowers in the surrounding ecosystems they inhabit, similar to other social bees like honeybees (*Apis* spp.) (Bueno et al. 2023).These bees can be the best-suited solution to overcome the severe pollution problems that the world is facing today.

The availability of artificial habitats, coupled with the degradation of natural habitats, may cause the shift away of these bees from natural environments toward anthropogenic ones. Stingless bees prefer habitats such as cracks and crevices of old buildings and mansions, unused earthen pots, plastic drums, hollow pipes, hollow bamboo, tree logs, wooden boxes, clay pots, etc. (Vazhacharickal and Sajan 2016).

It is advised that traditional beekeepers should rear the indigenous stingless bee species, such as *Trigona* and *Tetragonula iridipennis*, to enhance pollination and the quality and quantity of agricultural productivity, address food security concerns, and counteract the decline in native bee populations. It will add to the economy by promoting the use of stingless bees for cross-pollination services. This approach will surely contribute to ensuring steady food sustainability (Fig. 2).

Many stingless bee traits are similar to those of *Apis* spp. honeybees. Among the characteristics that influence their capacity as pollinators are (a) collection and adaptations, which allow them to both pollinate various plant species and adapt to new ones, and (b) floral constancy: the bee often trips only one plant species at a time (Heard 1999). Stingless bees, like honeybees, are easily maintained and carried in hives, allowing them to be introduced to orchards during flowering and subsequently relocated, and are the only efficient pollinators in polyhouses and greenhouses (Rajendran 2019).

It was also found that when flowers of Galia muskmelon were pollinated by the endemic African *Meliponinae* stingless bees as opposed to honeybees in greenhouse cultivation, fruits matured much more quickly. The blooms pollinated by *M. bocandei* and *H. gribodai* had the lowest fruit maturation times. Compared to honeybee pollination, flowers pollinated by *M. bocandei and P. hildebrandti* yielded the highest fruit weight and volume (Noiset et al. 2022).

The pollination services provided by the stingless bee pollinator species in the case of mango proved to be excellent regarding fruit weight and size. Moreover, these services were superior to open pollination and exclusion setups (Chuttong et al. 2022). The research findings tell that the minimum foraging range of stingless honeybees is about 13 meters with a maximum attending around 2.1 km and hence can be applied to small farms, polyhouses and greenhouse gardens, terrace gardens, and open plots (Kuhn-Neto et al. 2009). Still, more research on bee behavior can be conducted, considering more variables related to height, direction, and distance during the various ecological seasons (Fig. 3a, b) (Ciar et al. 2013).

Fig. 2 Stingless honeybees work efficiently in polyhouses

(A) (B)

Fig. 3 Stingless honeybees are considered to be a boon for mango

5 We Need an Hour to Save the Indigenous Bees, But Where Do We Begin?

Wild bee colonies, as well as reared colonies, are declining, raising concerns about pollination services globally. We first need to better understand pollinator ecology in the tropics and subtropics and the interaction between floral resources and pollinators at landscape sizes to lessen the effects of pollinator decreases in these places (Rajendran 2019). Stingless bees, the crucial ecological players, may be potential commercial alternatives shortly (Chuttong et al. 2022). Honeybees are far more economically valuable as agricultural pollinators than honey, wax, and other hive products. Estimates of this value range from $1.6 to $5.7 billion annually for the United States and £137.8 million annually for a few selected crops in the United Kingdom. The yearly global value of agricultural pollination has been approximated to exceed $200 billion [38]. The Swiss Red List for Bees states that these wild indigenous flower-visiting bees were classified as "rare" due to the high-density use of *Apis mellifera* or western honeybee (*L.*) for pollination purposes due to the loss of natural habitat, unavailability of bee-friendly flora and water source, excessive use of pesticides and insecticides, soaring temperatures, deforestation, etc. (Slaa et al. 2006). In the present situation, farmers face challenges in increasing agricultural production with limited land and water supply and the ever-increasing population, resulting in a reduction of per person/hectare land availability for cultivation. As a result, eco-friendly polyhouse farming has gained international attention irrespective of certain constraints and may hold the key to resolving the issues mentioned above (Kuhn-Neto et al. 2009; Ciar et al. 2013). Less productive results are observed when input exceeds output, even with agricultural inputs like high-quality seeds, timely irrigation, and other good farming practices to boost yield.

6 Marketing Strategies

- B2C, B2B.
- Targeting various FPOs, farmers' groups, seed-producing firms, polyhouses, organic and natural farming groups, urban enthusiasts, terrace gardeners, horticulturists, etc.
- Agri-universities, agri-institutes, word of mouth, research papers, articles, social media, and print media.
- International partnership network in IT and SAMS will ensure knowledge exchange and mutual learning.

7 Conclusions

- Develop and maintain diverse pollinator habitats in rural and urban areas to enhance the beauty of the landscape.
- Plant strategically selected mixtures of bee-friendly floral varieties.
- Promote apitourism to impart knowledge about indigenous bees and their contribution to sustainable food security.
- Implement artificial insemination techniques to create new bee queens and other caste bees that are more resistant to sudden environmental changes (in the case of *Apis mellifera*).
- Divide colonies, as stingless bees show the presence of more than one queen cell in the colony hive.
- Rear colonies in artificial wooden boxes that can be readily translocated from one place to another for pollination services as they adjust to new surroundings quickly.
- During the blossoming season, rent stingless bees to various crops for better yields in terms of quality and quantity.
- Collaborate with diverse stakeholders, including agri-institutes, seed firms, farmers, farmer-producer organizations (FPOs), playhouses, etc.

Acknowledgments The authors thank the faculty and Department of Zoology at Dr. Babasaheb Ambedkar Marathwada University, Chhatrapati Sambhajinagar, for providing research facilities for beekeeping studies.

References

Abdulkadyrova MA, Dikinov AH, Tajmashanov H, Shidaev LA, Shidaeva EA (2016) Global food security problems in the modern world economy. Int J Environ Sci Educ 11(12):5320–5330

Ahvo A, Heino M, Sandström V, Chrisendo D, Jalava M, Kummu M (2023) Agricultural input shocks affect crop yields more in the high-yielding areas of the world. Nat Food 4(12):1037–1046. https://doi.org/10.1038/s43016-023-00873-z

Allen-Wardell G et al (1998) The potential consequences of pollinator declines on the conservation of biodiversity and stability of food crop yields. Conserv Biol 12(1):8–17. https://doi.org/10.1111/J.1523-1739.1998.97154.X

Anis MMS, Norfarizan-Hanoon NA (2022) Interrelated of food safety, food security and sustainable food production. Food Res 6(1):304–310. https://doi.org/10.26656/fr.2017.6(1).696

Atsharyan A, Baghdasaryan A, Melikyan A, Paytyan T, Movsesyan A (2023) The design project of a device to solve the problem of pollination connected with bee extinction. E3S Web Conf 460:01008. https://doi.org/10.1051/e3sconf/202346001008

Benites WA, dos Lima JCS, Gonçalves CR, De Oliveira FF, Júnior VVA (2023) Urban parks as a functional area for the maintenance and diversity of bee populations. Contrib a Las Ciencias Soc 16(10):20292–20310. https://doi.org/10.55905/revconv.16n.10-098

Bueno FGB et al (2023) Stingless bee floral visitation in the global tropics and subtropics. Glob Ecol Conserv 43:e02454. https://doi.org/10.1016/j.gecco.2023.e02454

Cheng Y-L et al (2016) We are IntechOpen, the world's leading publisher of Open Access books built by scientists, for scientists TOP 1%. INTECH 11(Tourism):13. [Online]. Available: https://www.intechopen.com/books/advanced-biometric-technologies/liveness-detection-in-biometrics

Chuttong B et al (2022) Foraging behavior and pollination efficiency of honey bees (Apis mellifera L.) and stingless bees (Tetragonula laeviceps species complex) on mango (Mangifera indica L., cv. Nam Dokmai) in Northern Thailand. J Ecol Environ 46(Tanda 2021). https://doi.org/10.5141/jee.22.012

Ciar RR et al (2013) Foraging behavior of stingless bees (Tetragonula biroi Friese): distance, direction, and height of preferred food source. 1:1–60. [Online]. Available: http://arxiv.org/abs/1310.3919

Cortopassi-Laurino M, Imperatriz-Fonseca VL, Velthuis HW, Nogueira-Neto P (2001, October). Stingless bee rearing as an activity for sustainable development. In: International Apicultural Congress, vol 37

Dawoud B (2020) Let's protect our pollinating insects to increase agricultural production

El Bilali H, Callenius C, Strassner C, Probst L (2019) Food and nutrition security and sustainability transitions in food systems. Food Energy Secur 8(2):1–20. https://doi.org/10.1002/fes3.154

Heard TA (1999) The role of stingless bees in crop pollination. Annu Rev Entomol 44(131):183–206

Inouye DW, Ogilvie JE (2017) Pollinators, role of ☆. Ref Modul Life Sci. https://doi.org/10.1016/B978-0-12-809633-8.02309-8

Kevan PG, Clark EA, Thomas VG (1990) Insect pollinators and sustainable agriculture. Am J Altern Agric 5(1):13–22. https://doi.org/10.1017/S0889189300003179

Khan SA (2023) Review of the role of bees as ecosystem engineers in nature. Sci Rev Biol:2–8. https://doi.org/10.57098/SciRevs.Biology.2.4.1

Komlatskiy G, Makarova T (2023) Pollination by bees in industrial crop production. BIO Web Conf 66:1–5. https://doi.org/10.1051/bioconf/20236612001

Kuhn-Neto B, Contrera FAL, Castro MS, Nieh JC (2009) Long distance foraging and recruitment by a stingless bee, Melipona mandacaia. Apidologie 40(4):472–480. https://doi.org/10.1051/apido/2009007

Lobell DB, Cassman KG, Field CB (2009) 'Crop yield gaps: their importance, magnitudes and causes' NCESR publications and research. Paper 3. Annu Rev Environ Resour 334:1–26. https://doi.org/10.1146/annurevfienviron.041008.093740

Mc Carthy U, Uysal I, Badia-Melis R, Mercier S, O'Donnell C, Ktenioudaki A (2018) Global food security – issues, challenges and technological solutions. Trends Food Sci Technol 77:11–20. https://doi.org/10.1016/j.tifs.2018.05.002

Narbona E et al (2020) A comparative study of food source selection in stingless bees and honeybees: Scent marks, location, or color. https://doi.org/10.3389/fpls.2020.00516

Noiset P, Cabirol N, Rojas-Oropeza M, Warrit N, Nkoba K, Vereecken NJ (2022) Honey compositional convergence and the parallel domestication of social bees. Sci Rep 12(1):1–6. https://doi.org/10.1038/s41598-022-23310-w

Omuse E, Niassy S, Union A, Nkoba K (2022) A fungal-based pesticide does not harm pollination service provided by the African stingless bee Meliponula ferruginea on cucumber (Cucumis sativus) original article A fungal - based pesticide does not harm pollination service provided by the African st. Apidologie 53:28. https://doi.org/10.1007/s13592-022-00938-1

Osterman J et al (2021) Global trends in the number and diversity of managed pollinator species. Agric Ecosyst Environ 322. Elsevier B.V. https://doi.org/10.1016/j.agee.2021.107653

Rajendran L (2019) Colony propagation in stingless bees, Tetragonula iridipennis Colony propagation in stingless bees, Tetragonula. J Entomol Zool Stud 7:754

Rani AL (2023) Latest trends in sciences agriculture. Integrated Publications

Reilly JR et al (2020) Crop production in the USA is frequently limited by a lack of pollinators. Proc R Soc B Biol Sci 287(1931):2–9. https://doi.org/10.1098/rspb.2020.0922

Reyes-González A, Camou-Guerrero A, Reyes-Salas O, Argueta A, Casas A (2014) Diversity, local knowledge and use of stingless bees (Apidae: Meliponini) in the municipality of Nocupétaro, Michoacan, Mexico. J Ethnobiol Ethnomed 10(1). https://doi.org/10.1186/1746-4269-10-47

Shanahan M (2022) Honey Bees and industrial agriculture: what researchers are missing and why it's a problem. J Insect Sci 22(1). https://doi.org/10.1093/jisesa/ieab090

Slaa EJ et al (2006) Stingless bees in applied pollination: practice and perspectives to cite this version: HAL Id: hal-00892203 review article Stingless bees in applied pollination: practice and perspectives of POLLINATION IN COMMERCIALLY GROWN. Apidologie 37:293–315

Thakur M (2012) Bees as pollinators-biodiversity and conservation. [Online]. Available: http://www.interesjournals.org/IRJAS

Vazhacharickal PJ, Sajan JK (2016) Nesting behavior and habitats of stingless bees (Trigona iridipennis Smith) in Kerala: current status of knowledge. Grin Publisher, p 32

Wakjira K et al (2021) Smart apiculture management services for developing countries—the case of SAMS project in Ethiopia and Indonesia. PeerJ Comput Sci 7:1–25. https://doi.org/10.7717/PEERJ-CS.484

Green Growth: India's Strategy for a Sustainable Future

Isita Mukherjee

Abstract

The economy and environment are suffering from continuous depletion around the globe. In light of India's anticipated growth trajectory, environmental sustainability has emerged as a critical issue. Thus, the nation's attempts to achieve economic development while concurrently preserving the environment and natural resources for future generations are the main focus. The main focus is promoting renewable energy, reducing pollution, improving waste disposal, and safeguarding wildlife. Green growth in India seeks to reconcile economic expansion with environmental preservation and advance sustainable development. It is anticipated that the wasteful present production system will be transformed into a regenerative one in keeping with the worldwide trend toward sustainability. These all-encompassing actions address India's proactive approach to combating climate change and its commitment to achieving a net-zero carbon footprint by 2070.

Keywords

Carbon foot print · Economy · Green growth · Development · Green energy · Mission

1 Introduction

At the Fifth Ministerial Conference on Environment and Development (MCED), which took place in Seoul in March 2005, the idea of "green growth" was discussed. 52 Asian and Pacific region governments and other stakeholders decided to pursue "green growth" to achieve a better future. To that end, a regional implementation plan for sustainable development and a ministerial statement (the Seoul Initiative Network on Green Growth) were approved.

By incorporating the synergies between the environment and economic development, MCED aimed to achieve sustained economic development through the green growth strategy. The green growth method gained support from several global events and initiatives aimed at putting this concept into practice.

The Global Green Growth Institute (GGGI), for instance, was founded in June 2010 to achieve green economic growth in conjunction with the reduction of poverty, the creation of jobs, social inclusion, environmental sustainability, the mitigation of climate change, the loss of biodiversity, and the security of access to clean energy and drinkable water. The Green Growth Knowledge Network (GGKN), a worldwide knowledge-sharing network, was founded by the World Bank, UNEP, OECD, and the GGGI in Mexico. In India, for the first time in 2013, the Thirteenth Finance Commission (TFC) proposed the idea of "green growth." There are as many as 13 distinct definitions of green growth, including ones provided by significant international organizations involved in this topic.

Definitions

UNESCAP: "Growth that emphasizes environmentally sustainable economic progress to foster low-carbon, socially inclusive development."

World Bank: "Growth that is efficient in its use of natural resources, clean in that it minimizes pollution and environmental impacts, and resilient in that it accounts for natural hazards and the role of environmental management and natural capital in preventing physical disasters."

OECD: "Fostering economic growth and development while ensuring that natural assets continue to provide the resources and environmental services on which our well-being relies."

I. Mukherjee (✉)
Department of Economics, Vijaygarh Jyotish Ray College, Kolkata, India

J. A. Parray et al. (eds.), *Environmental Landscape and Sustainable Biodiversity for Healthy Green Growth*, Sustainable Landscape Planning and Natural Resources Management, https://doi.org/10.1007/978-3-031-75405-0_17

GGGI: "Green growth is the new revolutionary development paradigm that sustains economic growth while at the same time ensuring climatic and environmental sustainability. It focuses on addressing the root causes of these challenges while ensuring the creation of the necessary channels for resource distribution and access to basic commodities for the impoverished."

Source: Green Growth Knowledge Platform. https://sdgs.un.org/

Since the Industrial Revolution, the fossil fuel-oriented economy and mass production system have allowed humans to live vibrant lives. However, due to industrialization, such a traditional method of economic development faced obstacles such as a shortage of natural resources and environmental contamination. The world now faces an overwhelming necessity to abandon the outdated paradigm for development and adopt new plans for sustainable growth.

Global warming is accelerated by greenhouse gases released during economic development activities. Furthermore, the entire planet is being impacted by climate change, which results from natural and human-caused factors. These factors include rising sea levels and variations in rainfall. A group of people cannot resolve such global concerns because of their externality. Instead, every nation should cooperate and make a concerted effort.

Until now, the leading industrialized nations have deliberated and devised strategies for combating climate change.

The percentage of the population living in poverty and the rate of population increase are comparatively higher in developing nations than in developed nations. In that sense, addressing the current poverty level in developing nations requires sustainable development. However, because they are more interested in fostering growth than addressing climate change, some emerging nations are not actively collaborating with industrialized nations to address the issue. Therefore, international collaboration is required to share ideas between developed and developing nations to sustain progress successfully.

Even with the international discussion, the phrase "green growth" is still poorly defined. Terms associated with green growth, like "green economy" or "eco-efficient economy," are difficult to understand even by the government and international organizations. Therefore, it is necessary to define those terms more precisely to reach sustainable development through appropriate policies.

Though necessary, more than green growth is needed to achieve sustainable development. Social development has yet to be integrated into green growth to the same extent as environmental issues or economic growth. Social development is neglected in favor of growth achieved through improved market systems and ecological concerns, which is known as "green growth." To attain sustainable development's ultimate aim, simultaneously pursuing eco-friendly initiatives, social development, and economic progress is imperative.

2 Concept and Importance of Green Growth

Green growth is significant because it combines innovative elements such as investments, green legislation, and innovation to produce sustainable economic growth. The green growth technique is typically used to achieve several goals related to sustainable development, such as reducing or eliminating greenhouse gas emissions, increasing resilience to extreme events or disasters, and delaying the beginning of long-term changes through more effective resource management. Vital benchmarks for green growth include raising living standards, increasing GDP sustainably and equitably, and appreciating the often invisible natural resources that have supported economic success for generations. While it is still emerging, India's ideal model of green growth will be shaped by the experiences of other established and developing nations (GGGI 2014).

A sustainable development strategy, "green growth" greatly emphasizes economic expansion while addressing poverty, social inclusion, and environmental and climatic sustainability. Green growth's primary goal is to pinpoint the underlying causes of ecological and socioeconomic problems and solve them through resource distribution, guaranteeing that people experiencing poverty have access to energy and necessities. Green growth creates new opportunities through sustainable technology and companies, but it also reduces the danger of depletion of natural resources and the problems caused by climate change. With the help of more intelligent, environmentally friendly, and inclusive methods and technology, the concepts of green growth revolutionize how economic values are created, how natural resources are priced, and how resources are distributed.

The two most urgent issues affecting most developing nations are poverty and extreme weather occurrences brought on by climate change. Green growth addresses these problems and can enhance livelihoods. By endorsing practices that naturally improve climate outcomes, various pathways of green growth can help impoverished communities become more resilient to external shocks and increase their economic viability by maximizing the value of their natural capital. Not only poverty reduction and increased climate resilience but also green growth raises the Happiness Index by enhancing people's general health.

As the availability of natural resources and environmental services is crucial for sustainable economic growth, the "theory of change" of green growth can be summed up as an integrated approach to sustainable economic prosperity that is characterized by:

- Enhancement of the quality and quantity of these resources and environmental services.
- Higher productivity from both natural and artificial resources.
- Increased funding for innovative technologies makes room for nature-based solutions.
- Reduced market volatility and increased economic growth through more effective resource allotment.

3 Approaches and Strategies of Green Growth

After analyzing the lessons learned from other nations that have already implemented it, a variety of approaches, strategies, and guidelines are developed to help developing countries (those on the path of development can initiate the process through green, low-carbon, and renewable methods) have a road map for green growth planning. The operational concepts already suggested or implemented offer host governments value propositions for stakeholder cooperation, capacity assistance, and strategic analysis.

The green growth process begins with diagnosing developmental issues for green growth. Next, a thorough assessment of opportunities for green growth at the macro and sectoral levels is conducted. A series of implementation actions are then prioritized and sequenced, institutional and human capacity for implementation is built, and progress is made.

The procedures are designed to:

- Comprehend and draw from prior green growth planning experiences
- Describe a planning framework that is empirically based and goes from strategy development and assessment to operationalization and implementation
- Determine precise solutions for issues unique to a given nation
- Recognize resources and instruments for conducting analysis and other techniques for planning for green growth

4 Principles of Green Growth Planning

- Compliance with national developmental plans, both active and in the works. The procedure is run by the government, both locally and nationally.
- Wide-ranging cooperation and involvement with various stakeholders, such as donors, entrepreneurs, and members of civil society.
- Place for experimentation and creativity.
- Validated analysis and empirical evidence only and political neutrality with an agenda.
- Integration of different sectors and levels with institutional cooperation across them.
- Adaptability to the unique requirements of a nation.
- Planners and legislators could select the best solution with these guidelines. These would also balance the national, environmental, social, and economic priorities. These principles will remove the riskiest routes to achieve green growth objectives while maintaining flexibility and providing solid options.

5 Implementation of Green Growth Planning

5.1 Step 1: Evaluation of the Situation

An essential first stage in the planning process for green growth is a situation analysis, which aids in determining the current state of a nation's economy. Through these procedures, all parties involved—including institutions enforcing regulations and policies—could identify the sectors and parties actively engaged in economic activities. An analysis of partner countries' strengths, weaknesses, opportunities, and dangers can aid in understanding the political environment, which is essential for preparing for green growth. Thus, this stage guarantees that fundamental issues were resolved before implementing green growing processes.

5.2 Step 2: Political Alignment and Stakeholder Mapping

Essential components of green growth planning include alignment with current and upcoming initiatives across multiple geographies and stakeholder involvement across sectors, including various government agencies, NGOs, and institutes. They cultivate a sense of ownership among the stakeholders in the planning process, improving collaboration and knowledge exchange. Even though stakeholder deliberation is ongoing, mapping is a crucial first step in green growth planning. This stage is only required to encourage stakeholder coordination in poor nations.

5.3 Step 3: Evaluation of Green Growth Opportunities

The process of assessing the opportunities for green growth involves examining the current state of the major industries, the needs of the country, and pertinent best practices that will propel green growth.

This step involves several analyses, such as:

Global Best Practice
Policy planning and programming benefit from the best practices and green growth initiatives already in place in industrialized nations. Comparing and adopting nations with comparable economic conditions, resource capacity, and governance structures is possible.

National Standards
Prioritizing sectors requires an awareness of national baselines through green growth indicators to reflect the current state of the nation's political, social, environmental, and economic situations.

Study of Economic Trends
To better comprehend the potential paths for future appropriate policy development, it is imperative to evaluate the nation's current economic status, including current development routes, the environment for economic growth, existing financial policies, etc.

Assessment of Environmental and Climate Risk and Vulnerability
Ecological risk and susceptibility profiles vary based on the local geographic and climatic scenarios. Therefore, understanding the possible ways environmental and climatic hazards can paralyze the economic or social system and preparing in advance are paramount.

Study of Social Development
This is critical in ensuring social inclusion, gender equality, and poverty alleviation.

Evaluation of Data Gaps
Genuine data availability is crucial for all other processes since only high-quality data can guarantee accurate problem identification and plan development.

5.4 Step 4: Planning for Green Growth Project Implementation

This is the most critical stage of all. It is only possible to achieve green growth by implementing concepts or methods. Implementing a plan correctly will also ensure that the goals are met. The project design and execution phase, governance structure development, and policy implementation are examples of the implementation stage.

6 Green Growth and Sustainable Development

6.1 Sustainable Development

People now have comfortable and affluent lifestyles as a result of economic progress. However, all of that came at a hefty cost. There are concerns about whether economic progress can continue in the coming years due to the emergence of environmental pollution and resource depletion. The Club of Rome produced *The Limits to Growth*, a book that addressed these issues, in 1972.

According to the report, nations had to opt for a zero-growth approach to save the ecosystem. Economic growth only results in pollution, so the world cannot sustain any more growth. Since developed nations have already experienced sufficient growth, this may seem logical to them. On the other hand, where economic development is a pressing necessity for emerging countries, more is needed.

The 1987 WCED publication *Our Common Future* addressed sustainable development more than any other in modern history and provided a definition most frequently employed by governmental bodies and other organizations. According to the paper, "Sustainable development is the development that satisfies current needs without jeopardizing future generations' ability to satiate their own needs." Because it gives no clear instructions, the definition needs to be clarified. It does, however, undoubtedly pave the way for other governments and international organizations to begin debating how to promote sustainable development.

Based on an extensive examination of several conversations held in various countries and international organizations, sustainable development involves three essential pillars. They are economic, social, and environmental or eco-friendly development. Only when these three elements are in a positive cycle and balanced with one another can the planet attain sustainable development.

6.2 Comparison

Figure 1 shows that green growth and sustainable development are not two different ideas (TERI and GGGI 2015a).

One instrument for sustainable development is "green growth." However, as was already said, social development is only considered a little in green growth. As a result, current green growth plans need to be revised. Concerns related to social development include income inequality, social inte-

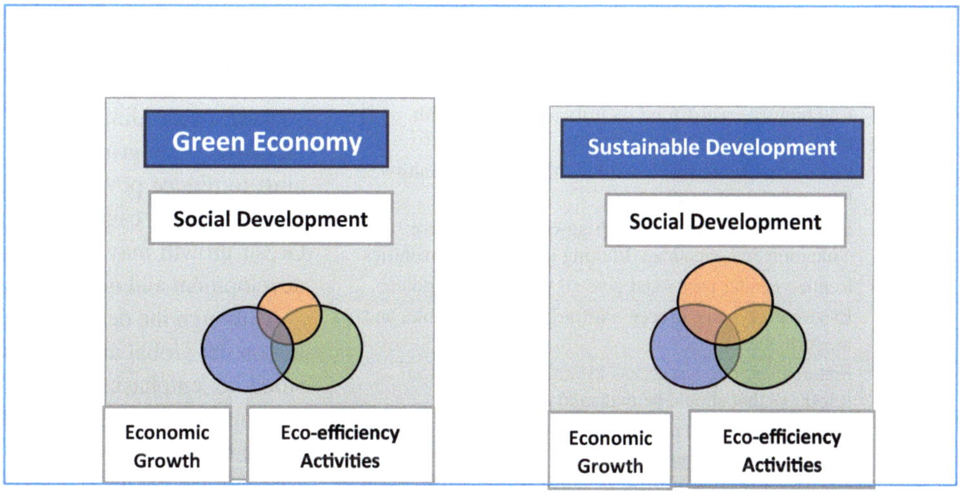

Fig. 1 Green growth and sustainable development

gration, the development of capabilities, absolute and relative poverty, and various social conflicts.

Job creation is one of many solutions to this. Green growth should soon evolve into a holistic development approach that can address many problems. It should also incorporate a detailed action plan. It will only be acknowledged as a notion consistent with sustainable development and fully prepare nations to enact policies.

Terms such as "green economy," "low-carbon green growth," and "ecological efficiency" are used all around the world and are similar to "green growth." Rather than arguing nomenclature, nations need to eliminate gaps and integrate to encourage green growth, or the "green economy," to emerge as an essential and effective instrument for sustainable development. All countries share the objective of sustainable development based on three pillars: social, environmental, and economic growth.

6.3 Determining the Routes and Locations of Entry for Green Growth

There are many routes and entry points for green growth, but choosing the right ones is essential based on the issues unique to a particular nation or region and the resources at hand. Stakeholder mapping and green growth potential evaluation will aid in determining the best entry points and appropriate paths for any given area.

A few instances of green growth pathways are as follows:

- The creation of green jobs and green investments
- Green economy governance in harmony with the environment
- Development of the blue economy with sustainability
- Natural resource management and the green environment
- Green education and the development of human resources
- Managing green technologies effectively
- Advancement of the social safety system through green growth
- Maintaining and preserving green cultural heritage and national identity, good governance, and green growth

After analyzing the current plans and strategies, the highest priority sectors for activities to achieve climate change adaptation and sustainable development should be determined. Policies for the nation's adaptation to climate change should include measures to lessen vulnerability in disaster risk management, infrastructure, livestock, agriculture, energy, forestry, gender equality, health, and water. A few key elements should be included in the national action plans, such as climate-wise agriculture, inclusivity, decision-making at different levels based on need, and access to resources and opportunities related to livelihoods and disasters.

7 Green Growth and Policy Imperatives in India

The Sustainable Development Goals are introduced in the New Agenda, the outcome document of the international talks of the Post-2015 Development Agenda. Concerning the new global agenda, the 169 targets and 17 Sustainable Development Goals established in September 2015 demonstrate the breadth and ambitions of the participating states. It is the first time in the worldwide development discourse that member states have placed such importance on environmental issues and have emphasized the three pillars of sustainable development. The 17 goals (TERI and GGGI 2015a) that find a place in the Outcome Document of the intergovernmental negotiations on the Post-2015 Development Agenda are listed in Table 1.

World leaders and other influential groups have also discussed the goal of concerted climate action by the worldwide

Table 1 Sustainable development goals

No.	Goal
Goal 1	Eliminate poverty worldwide
Goal 2	End hunger, ensure food security and better health, and advance sustainable agriculture
Goal 3	Ensure that everyone, regardless of age, has a healthy life and promote well-being
Goal 4	Ensure everyone can access high-quality, inclusive education and encourage lifelong learning opportunities
Goal 5	Realize gender parity and give all women more power
Goal 6	Ensure everyone can access affordable, sustainable, and modern energy
Goal 7	Ensure everyone has access to water and sanitation resources that should be managed sustainably
Goal 8	Encourage full and productive employment and equitable and sustainable economic growth
Goal 9	Create a robust infrastructure, encourage equitable and sustainable industrialization, and stimulate creativity
Goal 10	Diminish disparities both among and across nations
Goal 11	Make cities and human settlements safe, strong, and sustainable
Goal 12	Verify the sustainability of the production and consumption trends
Goal 13	Act rapidly to mitigate the effects of climate change
Goal 14	Preserve and responsibly utilize marine resources, the seas, and the oceans for sustainable development
Goal 15	Preserve, repair, and encourage the sustainable use of subastral ecosystems; oppose desertification; manage forests sustainably; stop and reverse land degradation; and stop biodiversity loss
Goal 16	Encourage open and peaceful societies for long-term growth, guarantee everyone's access to justice, and create inclusive, accountable, and prosperous institutions at all levels
Goal 17	Strengthen implementation strategies and reinvigorate the international alliance for sustainable development

Source: Transforming Our World: The 2030 Agenda for Sustainable Development

community to keep the global average temperature increase below two degrees Celsius—the universally acknowledged buffer against the worst effects of climate change. These changes focus on the significance of regional and global environmental concerns worldwide. The integration of ecological sustainability into national economic and development plans is crucial.

At the 2005 Ministerial Conference on Environment and Development in Asia and the Pacific (MCED), 52 nations and other stakeholders accepted the idea of "green growth" in the policy discourse.

The United Nations Environment Programme (UNEP) publicized the phrase "green economy" and the concept of "green stimulus packages" to identify areas where significant public investment may initiate a green economy on a global scale. UNEP's Green Economy Initiative (GEI) was introduced in October 2008 to offer policy support and analysis for investments in ecologically problematic industries and greening those sectors.

Green growth is acknowledged to have the ability to support the aim for inclusive development while also guaranteeing economic growth and environmental resilience of societies, according to the OECD in the year 2012. Having inclusive green growth makes sense. Inclusive green growth aims to reduce poverty, promote fairness, and create opportunity without causing irreparable environmental damage. Green growth may be seen as a tool to achieve sustainable development and not as a competing paradigm.

Although the definition of "green growth" varies depending on the global and national context, it is generally understood to emphasize economic growth and environmental sustainability strongly. The definition of green growth given by the United Nations Economic and Social Commission for Asia and the Pacific (UNESCAP) is "growth that emphasizes environmentally sustainable economic progress to foster low-carbon, socially inclusive development." Systemic change must be implemented to transform into a global green economy, building on national efforts to assemble regional and international collaboration.

The World Bank describes green growth as "Efficient in the use of natural resources that minimizes pollution and environmental impacts, and that is climate resilient." According to the Organization for Economic Cooperation and Development (OECD), green growth is "promoting economic growth while reducing pollution and greenhouse gas emissions, minimizing waste and inefficient use of natural resources, and maintaining biodiversity." The organization was permitted a pivotal role in this discourse in 2009 by the OECD Ministerial Council Meeting. Subsequently, the organization endorsed other worldwide initiatives aimed at promoting green growth.

The Global Green Growth Institute (GGGI) was founded in 2012 as an international organization to integrate environmental sustainability and economic development in response to the growing global conversation on green growth.

Several governments, like those in Korea and Japan, have illustrated political will and taken the initiative to move toward low-carbon, green growth. In response to the global financial crisis, the Republic of Korea (RoK) declared "low-carbon green growth" as the new national development target in 2008. 2009 saw the state implement the National Strategy for Green Growth and the Five-Year Plan for Green Growth. Accompanying this was adopting the Framework Act on Low Carbon, Green Growth.

Shifting to a green economy accelerates economic growth while lowering the crisis risk progressively built into the existing financial system. Nations like the United States of America and the majority of the EU28 have obtained high levels of human development, although at the sacrifice of a larger ecological footprint. The task facing developing countries is to reduce their environmental footprint per person while maintaining a high standard of living for their people when improving the environment. Governments' susceptibility to prioritize financial health may complicate the technical shift needed for green growth, as fiscal deficits and public debts are among the major obstacles to national governance. This makes it challenging to focus on creating a sustainable and fair economy, which would entail spending public money on social and environmental solutions. Since it is frequently discussed that as the world moves toward a capitalist system that cannot function without high growth and profits, some relevant systemic challenges are linked to green growth—adopting an utterly renewable energy pathway would seem difficult (Hoffmann 2011). Furthermore, opposing viewpoints highlight the inevitable trade-offs that come with green growth, which raises the social costs related to it. It should be ensured that those with low incomes cannot bear the cost of maintaining growth in a green economy since this could make it regressive for the impoverished population (Dercon 2012).

Additionally, green growth has come under fire and has been designated as a "reductionist approach," which may not be able to handle the complexity of climate change.

The effectiveness of sustainable development, or "green," measures are frequently undermined due to their mainstream and neoliberal orientation, which is preponderantly framed within a techno-bureaucratic policy framework, as noted by Sood and Banerjee (2012). The literature has used empirical data to emphasize that resource productivity and efficiency increases have limitations.

The international community may significantly contribute to developing nations by assisting partner nations in achieving specific green growth objectives. According to Sood and Banerjee (2012), political economy is necessary to comprehend ecology. Their research indicates that understanding the connections between growth methods, livelihoods, social exclusion, unemployment, displacement, and other related concerns is crucial for policymakers. Therefore, developing sector-specific policies and advancing green growth require political leadership. Political commitment also plays a significant role in determining how financial resources are allocated toward the greening of the economy (Ekins 2001).

Experts have also discussed green growth and how it may help developing nations with dynamic, emerging economies like India. In the framework of India's green growth process, two economic growth patterns—production and consumption—become fundamentally important. India's economy is green in many respects because it is a notable exporter of services and importer of manufactured goods; consumption still needs to become more environmentally friendly. The nation could move toward being greener more gradually if it were to leapfrog consumption.

The Indian Constitution has specific arrangements aimed at safeguarding and enhancing the condition of the environment. Article 48-A (Indian National Bar Association 2013) of the Constitution says, "The state shall endeavor to protect and improve the environment and safeguard the forests and wildlife of the country." Article 51-A (g) (Indian National Bar Association 2013) says, "It shall be the duty of every citizen of India to protect and improve the natural environment including forests, lakes, rivers, and wildlife and to have compassion for living creatures." These provisions highlight the national consensus on the importance of environmental protection.

Within the Central Government's administrative framework, the Ministry of Environment, Forest and Climate Change (MoEFCC) serves as India's nodal body for organizing, promoting, coordinating, and supervising the execution of the country's environmental, forestry, and climate change activities.

The Indian government introduced the National Action Plan on Climate Change (NAPCC) in 2008. NAPCC marks a salient turning point for integrating climate change into development. The NAPCC emphasizes the co-benefit strategy for low-carbon development, which uses distributed and decentralized energy systems to expand access to energy, provide energy security, and reduce local pollution. Detecting policies that successfully battle climate change and advance development goals is the cornerstone of the NAPCC technique. To cope with climate change, the NAPCC now has eight national missions that set objectives for adaptation and mitigation. Solar energy, energy efficiency, water, the Himalayan ecology, sustainable agriculture, sustainable habitat, Green India, and strategic knowledge are the eight missions of this plan. The National Water Mission and National Mission would be revised, and the government would establish new missions on wind energy, health, waste-to-energy, and coastal areas.

India disclosed its domestic mitigation action under the Copenhagen Accord, stating that it aims to lower its GDP's emissions intensity by 20–25% by 2020 compared to the level 1 emissions in 2005. More recently, India said in its Intended Nationally Determined Contributions (INDCs) that it would cut the GDP's emissions intensity from level 2 in 2005 by 33–35% by 2030.

A stable, sufficient, and secure fiscal foundation would be required with state funding contingent on environmental performance to provide local public goods for green growth; the third tier of Indian governance must be strengthened, composed of urban and rural local bodies (municipalities and panchayats).

When the money gets to the local level, things will happen rather quickly. Thus, regional resource availability primarily drives the shift toward sustainable development and green growth (Kelkar 2012).

The concept of green growth was first considered by the Government of India's Thirteenth Finance Commission within the framework of policy debate.

Rethinking growth methods in light of their effect(s) on environmental sustainability and the environmental resources accessible to the underprivileged and vulnerable populations is known as "green growth" (paragraph 3.15, Thirteenth Finance Commission). So, it is evident that "environment sustainability" and "poor and vulnerable groups" are fundamental to discussions of green growth in the Indian policy context.

According to the TFC Report, the Commission's overarching strategy is to support "inclusive and green growth" to advance fiscal federalism and more vital local institutions. This fundamental perspective has guided the Commission's suggestions on intergovernmental fiscal arrangements and recommendations (TFC 2009).

The Thirteenth Finance Commission also cites global experience, demonstrating that fiscal measures in numerous nations included environmentally friendly components. It suggested methods to encourage budgetary policies that reinforce initiatives for energy conservation, renewable energy use, soil conservation, afforestation, and more efficient and reasonably priced access to clean water at various governmental levels. Expanding the monetary basis at all governmental levels could spur innovation, mainly to enable the authorities to guarantee services like better access to clean water, better solid waste management, and green infrastructure.

Box 1: "Green Growth" as Recognized by the Ministry of Environment, Forest and Climate Change (MoEFCC), Government of India

The Ministry of Environment, Forest and Climate Change has recognized "green growth" in the Ministry's vision. The four aspects that cover the Ministry's vision include:

- Having sustainable development.
- We have growth and environment protection.
- Needed development without destruction.
- Our vision is green growth and the eradication of poverty.

Source: Based on MoEFCC (2015)

The Thirteenth Finance Commission and the Ministry of Environment, Forest and Climate Change have articulated the significance of inclusivity in driving green growth in India.

The report of the Fourteenth Finance Commission (FFC 2015) emphasized the need for a new institutional framework that might act as a forum for incorporating environmental and economic considerations in decision-making. In addition, the FFC advises governments to consider complete volumetric measurement of irrigation water usage, independent of the existence of water regulatory agencies. With the introduction of the Commission's forward-looking incentive-based grant, the States will be rewarded for having quality forest cover, as defined by Moderate and Very Dense forest cover. The FFC's devolution formula allows the states to consider forest conservation.

7.1 India Development Profile (Lawrbit 2023)

With 1.29 billion people living there as of 2014, India accounted for 17.84% of the global population (World Bank Development Indicators). Economic growth appears to have accelerated globally, and 2015–2016 should see additional improvements. According to the IMF in the year 2015, global growth in 2014 was a meek 3.4%. The global economy is expected to rise roughly 3.5% in 2015 and 3.8% in 2016. In its World Economic Outlook report, the IMF notes that advanced economies with historically wide production disparities face downside risks.

The Union Budget 2023s central theme is the seven "Saptarishi" priorities: inclusive development, achieving the last mile, infrastructure and investment, unleashing the potential, youth power, the financial industry, and green growth. The finance minister used "green" and "sustainable" in her budget speech. She said green growth could be revolutionary during "Amrit Kaal." Green growth initiatives aim to ensure that natural resources sustainably realize their full economic potential.

Furthermore, the Budget highlights the following four revolutionary opportunities to achieve India@100:

1. Women's economic empowerment via self-help groups
2. PM Vishwakarma Kaushal Samman (PM VIKAS)
3. Tourism promotion in mission mode
4. Green growth

"Green growth" is a term used to explain an economic growth strategy that prioritizes sustainable development while reducing adverse environmental effects. India ranked 169th out of 180 nations in the 2022 environment performance index. Air quality, biodiversity and habitat, fisheries, waste management, ecosystem services, and climate change were among the rankings' variables. Despite being unsatisfactory compared to many other minor economies on the list, India's economy is the fifth largest in the world.

8 India's Initiative to Promote Green Growth

The Prime Minister's ambition for "life," or lifestyle for the environment, is to stimulate a movement toward environmentally friendly living. India is actively seeking the "Panchamrit" and net-zero carbon emissions by 2070 to spearhead the global transition to a green industrial and economic system.

Furthermore, India is implementing multiple regulations and initiatives to optimize energy utilization throughout diverse economic domains, encompassing green buildings, equipment, farming, mobility, and fuels. These green growth initiatives help lessen the economy's carbon intensity and open up large-scale green job prospects.

8.1 Proposed and Ongoing Initiatives (Lawrbit 2023)

1. *National Green Hydrogen Mission*

    - To help the economy shift to low carbon intensity, lessen dependency on imports of fossil fuels, and position the country as a market and technology leader in this developing sector, a budget of Rs 19,700 crores has been set aside for the National Green Hydrogen Mission.
    - Target 2 is a yearly production of 5 MMT by 2030.

2. *GOBARdhan scheme*

    - As part of the GOBARdhan Scheme, 500 new "waste to wealth" facilities—200 compressed biogas (CBG) plants and 300 community-/cluster-based plants—will be constructed to foster the circular economy. A total of 10,000 crores were invested.
    - A 5% CBG requirement will eventually be implemented for all companies that market biogas and natural gas.

3. *Bhartiya Prakritik Kheti bio-input resource centers*

    Over the next 3 years, 10,000 bio-input resource centers will be created to assist one crore farmers in switching to natural farming. This would create a dispersed national network for manufacturing pesticides and micro-fertilizers.
4. *MISHTI*

    The Mangrove Initiative for Shoreline Habitats & Tangible Incomes (MISHTI) would transform mangrove protection by pooling resources from MGNREGS, CAMPA Fund, and other sources. MISHTI will support planting mangroves on saltpans and along India's coastline.
5. *PM PRANAM*

    The "Prime Minister Programme for Restoration, Awareness, Nourishment, and Amelioration of Mother Earth" encourages states and union territories to support the moderate application of chemical and nonchemical fertilizers.
6. *Amrit Dharohar*

    The Amrit Dharohar program, a central government program to protect wetlands, was announced in the Union Budget for 2023–2024. Over the next 3 years, Amrit Dharohar will be implemented to spur the best possible use of wetlands and enhance biodiversity, carbon stocks, ecotourism opportunities, and revenue production for surrounding communities.
7. *Other Investments in Green Energy*

    The Ministry of Petroleum and Natural Gas has set aside Rs 35,000 crores for priority capital investments to get net zero and for energy transition. These goals aim to decrease greenhouse gas emissions to as close to zero as possible, with any leftover emissions being reabsorbed from the atmosphere by places like forests and oceans, as well as energy security.

    - Durability gap funding for battery energy storage systems with a 4000 MWH capacity is intended to provide capital support to PPP (Public et al.) projects that would not otherwise be financially viable.
    - The interstate transmission infrastructure for the evacuation and grid integration of 13 GW of renewable energy from Ladakh is estimated to cost Rs 20,700 crores (with Rs 8300 crores in national funding).

8.2 India's Green Growth Strategy (The Hindu 2024)

In Budget 2024, India's Finance Minister announced a comprehensive "green growth" plan to achieve the nation's Net Zero goal by 2070. The plan allocates substantial funds for biomanufacturing, electric vehicles, coal gasification and biogas mixing, offshore wind energy, and biofuels. In Budget 2024, the Finance Minister unveiled a detailed plan to reach India's ambitious Net Zero target as a first step toward fostering a sustainable future for the country. A component of the Union Budget 2024 that places a strong prominence on "green growth" is this announcement. With a focus on investigating India's enormous offshore wind energy potential, the Finance Minister has confirmed significant funding to the green energy industry.

One of the top seven goals in the recent Budget is green growth, including green credits, energy, mobility, and agriculture.

Three pillars form the foundation of India's green growth and energy transfer:

1. Enlarge the production of renewable energy
2. Reducing the fossil fuels in the economy
3. Transition to a domestic gas-based economy

Recent budget measures highlighted the plan, including energy gasification, rooftop solar systems, PM KUSUM Yojana, ethanol unification, and array storage space. The path to achieving the same is being covered by additional programs, including green credits for industry, PM PRANAM Yojna for farmers, GOBARdhan Yojna for villages, city scrapping laws, green hydrogen, and wetland protection.

India has been the leading economy with the augmentation of renewable energy capacity since 2014. Nine years ahead of schedule, India has met its aim of having 40% of installed electricity capacity come from nonfossil fuels. India plans to mix 20% ethanol in gasoline from 2025 to 2026 rather than 2030, achieving its goal of 10% ethanol blend in gasoline 5 months ahead of schedule. It is anticipated to attain 500 GW of capacity by 2030. The launch of E20 fuel and the emphasis on biofuels have given investors new avenues to follow. Prioritizing national canal expansion reinforces environmentally responsible freight handling and water transportation. India needs to raise battery capacity to 125 GWh over the next 6–7 years.

- Funding for battery energy storage systems has also started.
- 4000 MWh of capacity is funded by viability gap funding (VGF).
- VGF support and further promotion of PSPP are essential for India to meet its energy storage commitment targets.
- To promote green mobility, imports of resources, merchandise, and apparatus requisite for producing lithium-ion cells for electric vehicle batteries are exempt from customs duty.

India is on pace to construct 5 MMT of green hydrogen as part of the National Green Hydrogen Mission. A sum of Rs 19,000 has been allocated to encourage the market. Additional options like producing electrolyzers, raw steel, and long-range fuel cubicles are also deemed significant. The world and the country are moving toward a "green lifestyle" according to the vision "LiFE" or "Lifestyle of Environment."

Viability gap funding for developing one gigawatt (GW) of offshore wind energy is one of the noteworthy projects. It is anticipated that this project will be needed to help India diversify its renewable energy sources and lessen its reliance on fossil fuels. The audacious plan to establish coal gasification and liquefaction operations with the capacity to process 100 metric tons by 2030 is another essential element of the strategy. This measure is expected to promote greener energy resources and reduce India's reliance on ammonia, methanol, and natural gas imports.

To drastically reduce carbon emissions, the government plans to make it mandatory to mix biogas with compressed natural gas (CNG) for transportation and piped natural gas for residential usage. This strategy is anticipated to strengthen the biogas sector and enhance air quality, supporting the circular economy. Funding will be available to acquire biomass aggregation necessary to produce bioenergy. Farmers will be encouraged to join the bioenergy supply chain by taking this action and developing a viable and sustainable agricultural paradigm. There will be a big push to the electric vehicle (EV) ecosystem since the government plans to improve and bolster its infrastructure for producing and charging EVs. This is anticipated to accelerate the country's EV adoption rate and lessen greenhouse gas emissions associated with transportation.

To reduce urban pollution and promote environmentally friendly public transportation, the government plans to encourage the usage of electric buses. There will be a new program that focuses on foundries and biomanufacturing.

9 Conclusions

The country's economy and environment are agonizing from continuous depletion. In light of India's anticipated growth trajectory, environmental sustainability has emerged as a critical issue. Thus, the nation's attempts to achieve economic development while concurrently preserving the environment and natural resources for future generations are the main focus of Budget 2023 and 2024. The main focus is promoting renewable energy, reducing pollution, improving waste disposal, and safeguarding wildlife. Green growth in India seeks to reconcile economic expansion with environmental preservation and advance sustainable development. The primary goal of Budget 2024 is to encourage green growth and innovation in the biotechnology sector. It is anticipated that the wasteful present production system will be transformed into a regenerative one in keeping with the worldwide trend toward sustainability. These all-encompassing actions address India's proactive approach to combating climate change and its commitment to achieving a net-zero carbon footprint by 2070.

References

Dercon S (2012) Is green growth good for the poor? World Bank: Policy Research Working Paper No. 6231, 1–22

Ekins P (2001) Economic growth and environmental sustainability: the prospects for green growth. Routledge, New York

Fourteenth Finance Commission (2015) Available: https://fincomindia.nic.in/asset/pdf/commissionreports/14thFCReport.pdf

GGGI (2014) GGGI 2014 annual report. Global Green Growth Institute, Seoul

GGGI (2019) GGGI strategic outcomes guideline frameworks and methodologies for development impact estimation. Global Green Growth Institute, Seoul

Government of India (2001) Census of India. Office of Registrar General and Census Commissioner, Government of India, New Delhi

Government of India (2006) Population projection for India and states 2001–2026. Government of India, New Delhi

Government of India (2015) Economic survey 2014–15. Oxford University Press, New Delhi

Hoffmann U (2011) Some reflections on climate change, green growth illusions, and development space. United Nations Conference on Trade and Development

Indian National Bar Association (2013) Constitutional provisions for the protection of the environment with relevant case laws. Indian National Bar Association

Lawrbit (2023) Green growth: one of the key priorities leading India in its "Amrit Kaal". https://www.lawrbit.com/article/green-growth-one-of-the-key-priorities-leading-india-in-its-amrit-kaal/

Ministry of Environment and Forests (MoEF) (2006a) Report of the National Forest Commission. Ministry of Environment and Forests, Government of India, New Delhi

Ministry of Environment and Forests (MoEF) (2006b) National Environment Policy 2006. Ministry of Environment and Forests, Government of India, New Delhi

Ministry of Environment and Forests (MoEF) (2010a) India: greenhouse gas emissions 2007. Ministry of Environment and Forests, Government of India, New Delhi

Ministry of Environment and Forests (MoEF) (2010b) National Mission for a Green India (Under the National Action Plan on Climate Change). Ministry of Environment and Forests, Government of India, New Delhi

Ministry of Environment and Forests (MoEF) (2012) India's Second National Communication to the United Nations Framework Convention on Climate Change. Ministry of Environment and Forests, Government of India, New Delhi. Available online at http://unfccc.int/resource/docs/natc/indnc2.pdf

Ministry of Environment and Forests (MoEF) (2014a) Implementation guidelines for National Mission for a Green India (GIM). Ministry of Environment and Forests, Government of India, New Delhi. Available online at http://www.moef.gov.in/sites/default/files/GIM_Implementation_Guidelines.pdf

Ministry of Environment and Forests (MoEF) (2014b) India's fifth national report to the convention on biological diversity. Ministry of Environment and Forests, Government of India, New Delhi

Ministry of Environment, Forest and Climate Change (MoEFCC) (2015) Development without destruction: a saga of turnaround

Ministry of Law and Justice (MoLJ) (2007) The Scheduled Tribes and Other Traditional Forest Dwellers (Recognition of Forest Rights) Act, 2006. Gazette of India, Ministry of Law and Justice. Available online at http://tribal.nic.in/writereaddata/mainlinkFile/File1033.pdf

Ministry of Power (MoP) (2013) Smart grid vision and roadmap for India. Ministry of Power, Government of India, New Delhi

Ministry of Tribal Affairs (MoTA) (2015) Status report on implementing the Scheduled Tribes and Other Traditional Forest Dwellers (Recognition of Forest Rights) Act, 2006 [for the period ending 31.01.2015]. Ministry of Tribal Affairs, Government of India

NATCOM (2012) India Second National Communication to the United Nations Framework Convention on Climate Change. Ministry of Environment & Forests, Government of India

National Action Plan on Climate Change (NAPCC) (2008) National Action Plan on Climate Change. Prime Minister's Office, Government of India, New Delhi

National Transport Development Policy Committee (NTDPC) (2012) Working group on urban transport. National Transport Development Policy Committee, New Delhi

Kelkar Committee Report (2012) Available: https://dea.gov.in/sites/default/files/Kelkar_Committee_Report.pdf

Sood A, Banerjee P (2012) The political economy of green growth in India. Social Policies for Inclusive and Sustainable Development by United Nations Research Institute for Social Development

TERI and GGGI (2015a) Green Growth and Sustainable Development in India. Towards the 2030 Development Agenda: Summary for Policymaker. Implemented by The Energy & Resources Institute (TERI) and Global Green Growth Institute (GGGI)

TERI and GGGI (2015b) Green Growth and Sustainable Development in India. Implemented by The Energy & Resources Institute (TERI) and Global Green Growth Institute (GGGI)

The Energy and Resources Institute (TERI) (2009) Looking back to change track. TERI Press, New Delhi

The Energy and Resources Institute (TERI) (2014a) The energy & environment data directory and yearbook (TEDDY) 2013/14. The Energy and Resources Institute (TERI), New Delhi

The Energy and Resources Institute (TERI) (2014b) India-California Air Pollution Mitigation Program: options to reduce road transport pollution in India. University of San Diego, California Air Resources Board

The Energy and Resources Institute (TERI) (2015a) TERI energy and environment data diary and yearbook (TEDDY) 2014/15. The Energy and Resources Institute (TERI), New Delhi

The Energy and Resources Institute (TERI) (2015b) Understanding energy risks in India: towards a safe and informed energy program (for the Atomic Energy Commission). The Energy and Resources Institute (TERI), New Delhi

The Hindu (2024) The Finance Minister announced green measures in the interim budget. https://www.thehindu.com/business/budget/green-measures-announced-by-finance-minister-in-interim-budget/article67799660.ece

Thirteenth Finance Commission (2009) Available: https://fincomindia.nic.in/commission-reports-thirteenth

UNESCAP (2012) Green growth, resources and resilience-environmental sustainability in Asia and the Pacific. United Nations and Asian Development Bank

United Nations (n.d.) Green growth knowledge platform. https://sustainabledevelopment.un.org/index.php?menu=1447

United Nations Economic and Social Commission for Asia and the Pacific (UNESCAP) (2007) Application of economic instruments for green growth. Background note for Second Policy Forum of the Seoul Initiative Network on Green Growth. UNESCAP

Vishnu S, Victor MA, Manida M (2023) Green growth: India's strategy for green economy. Int J Emerg Knowl Stud 2(8):99–203

FSC
www.fsc.org
MIX
Papier aus verantwortungsvollen Quellen
Paper from responsible sources
FSC® C105338

If you have any concerns about our products,
you can contact us on
ProductSafety@springernature.com

In case Publisher is established outside the EU,
the EU authorized representative is:
Springer Nature Customer Service Center GmbH
Europaplatz 3, 69115 Heidelberg, Germany

Printed by Libri Plureos GmbH
in Hamburg, Germany